T0319593

Open Channel Design

Open Channel Design

Fundamentals and Applications

Ernest W. Tollner

University of Georgia
Athens
GA, USA

This edition first published 2022
© 2022 John Wiley & Sons Ltd

Registered Offices
John Wiley & Sons, Inc., 111 River Street, Hoboken, NJ 07030, USA
John Wiley & Sons Ltd, The Atrium, Southern Gate, Chichester, West Sussex, PO19 8SQ, UK

Editorial Office
9600 Garsington Road, Oxford, OX4 2DQ, UK

For details of our global editorial offices, customer services, and more information about Wiley products visit us at www.wiley.com.

Wiley also publishes its books in a variety of electronic formats and by print-on-demand. Some content that appears in standard print versions of this book may not be available in other formats.

Library of Congress Cataloging-in-Publication Data applied for

Hardback ISBN: 9781119664246

Cover Design: Wiley
Cover Image: © E W Tollner

Set in 9.5/12.5pt STIXTwoText by Straive, Pondicherry, India
Printed and bound by CPI Group (UK) Ltd, Croydon, CR0 4YY

C9781119664246_010921

Contents

Preface

With many excellent texts on Open Channel Hydraulics and Sediment Transport, why is another book needed? Available texts represent excellent tools for graduate instruction. We stand on the shoulders of giants in the field such as V.T. Chow, H.A. Einstein, Jr., to mention only a few.

Undergraduates generally find the available books to be somewhat intimidating. Available texts tend not to have accessible supporting software. In a knowledge domain where most problems require iterative solutions, a need exists for software to fill a void, especially for undergraduates. The presentation of concepts in Open Channel hydraulics in available texts is more oriented to graduate students who have a solid knowledge of basic concepts.

This text supports a split-level class that is mostly undergraduate in composition. Planning for the book began just before the COVID-19 pandemic. The move to online courses in response to the COVID-19 pandemic caused many educators to rethink course delivery. In-class and online education, in our experience, is most effective when content is delivered to undergraduates in modules sequentially build on the previously presented material. In our view, an online presentation stresses the need to be as sequential as possible because student interaction becomes more challenging than face-to-face delivery.

Another guiding factor in the book organization was to present many design approaches for uniform flow as earlier as possible. Chapter 4 mostly completes the coverage of uniform flow. Early uniform flow presentation enables students to have a toolbox for solving many practical design problems early in the semester. The front-loading of uniform flow allows students to begin working on design projects early in the term. We then present nonuniform flow and unsteady flow topics, enabling their addition to design projects as needed. Graduate students start to work on topics in Chapter 10, which flows from Chapter 4. Graduate students also do extra work on topics related to nonuniform and nonsteady flows as the course continues.

A variety of Excel spreadsheets supports the concepts presented in each chapter. Public domain software examples (HEC-RAS and HY-8) support advanced analyses. Students may expect to use these analyses in subsequent classes, such as capstone design, when applicable. Mathematica notebooks support selected theoretical analyses. The Mathematica notebooks are useful for graduate student analysis of more advanced concepts.

Chapter 10, covering alluvial transport processes, is written following a different approach compared to other chapters. Undergraduates could readily understand the incipient motion concepts and the design of channels using Shields type analysis. They could readily grasp selected regression approaches for sediment transport computation. On the other hand, graduate students could spend considerable time looking at the more advanced sediment transport and river mechanics aspects.

This text attempts to present a highly sequential course with affordable and challenging supporting software for undergraduate and graduate students. We leave advanced topics such as density currents, scour, and convection–diffusion of pollutant constituents to other texts. Emphasis on 3-D computational modeling is left to other books as well.

The author takes responsibility for the material presented. Please call my attention to any errors discovered. I look forward to learning about your experiences using the text and eagerly desire to hear suggestions for future improvements.

Ernest W. Tollner

Acknowledgments

I am grateful for the can-do attitude of my parental family, who immigrated from northeastern Germany late in the 1800s. They were dedicated to the proposition that the virtues in the US Declaration of Independence enabled improving one's life compared to living on what bore resemblances to a feudal estate. The general farm and dairy background made possible by my parents, Ernest and Ruby Tollner, both now deceased, was of incalculable value to this undertaking. This view has sharpened as time passed.

Likewise, the broad-based agricultural and biological engineering and civil engineering experiences gained while studying under the guidance of B.J. Barfield, Tom Haan (Ag. Engineering, University of Kentucky), and Dr. David Gao (Civil Engineering, Univ of Kentucky), Drs. Charlie Busch and Dave Hill (Agric. Engineering, Auburn University) and Drs. Joseph Judkins and Fred Molz (Civil Engineering, Auburn) were inspiring and formative. Colleagues Brahm Verma and Dale Threadgill, who mentored me for 35 years at the University of Georgia, have been helpful in uncountable ways. Dr. Steve McCutcheon (Civil Engineer, USEPA, retired) continues to share many valuable insights.

Many thanks to my CVLE 4210/6210 class for serving as guineas for the text's trial run. They made valuable suggestions for improving the flow of the book. In particular, graduate students Will Mattison, Whitney Phelps, Shep Medlin, and Matthew Terrell come to mind. They also suggested improved figures and tables, for which I am grateful.

I appreciate the patience and encouragement of my wife, Caren. To God be the Glory! I put this work forward as an offering to all. May all on His journey leave behind a more sustainable environment! May we dress and keep His magnificent creation in a sustainable way for all to better serve!

About the Companion Website

This book is accompanied by a companion website.

www.wiley.com/go/tollner/openchanneldesign

This website includes:

- A variety of Excel spreadsheets to support the concepts presented in each chapter.
- Public domain software examples (HEC-RAS and HY-8) to support advanced analyses.
- Mathematica notebooks to support selected theoretical analyses.

1

Basic Principles and Flow Classifications

Hydraulic Engineering has served humanity all through the ages by providing drinking water and protective measures against floods and storms. In the course of history, it has made the water resource available for human uses of many kinds. Biswas (1970) chronicles contributions since Hammurabi (c. 1700 BCE) hydraulic engineering over the centuries. In a survey of the University of Georgia Libraries' holdings under "land-use change," some 8000 articles discuss facets of the hydrologic cycle and associated runoff. Simons and Senturk (1992) provide a synopsis of contributions to sediment transport science in streams that date back to work in China to date back to 4000 BCE. Students who seriously pursue open channel hydraulics and sediment transport should explore works such as those mentioned. Management of the world's water is a complex task, and both its scope and importance continue to grow as we strive for sustainable stewardship of our abode.

Over time humanity has not only diverted and used the waters of the world for its purposes but, by engaging nature into its service, has turned deserts into fertile land. Natural habitat is threatened in more and more parts of the world by an ever-growing human population. Thus, long-term needs are food, water, shelter, and an aesthetically pleasing, healthy, nurturing environment.

Open channel flow is, in brief, a flow where the fluid has a free surface, where the free surface of the flow is subject to atmospheric pressure. Problems covered include flow in a conduit when the conduit is not full, such as in a storm sewer. The primary fluid of interest is water, although any fluid could, in principle, be addressed. This chapter introduces concepts that are developed as we progress through the text.

Why do we consider open channels when one can simply take an earthmover and create a conveyance? The hydraulic engineer meets engineering, economic, and social objectives in the client's and society's best interests. Sound engineering should result in sustainable design. Figure 1.1 shows a channel under construction with a channel design that conveys water while meeting aesthetic and sustainability goals.

Here are some definitions:

Ditch – an excavated water conveyance often installed without extensive advanced engineering design.

Natural channel – watercourses that exist naturally, such as gullies, brooks, rivers, streams, or estuaries. These are often analyzed to determine flows associated with high-water marks.

Open Channel Design: Fundamentals and Applications, First Edition. Ernest W. Tollner.
© 2022 John Wiley & Sons Ltd. Published 2022 by John Wiley & Sons Ltd.
Companion website: www.wiley.com/go/tollner/openchanneldesign

Figure 1.1 A trapezoidal waterway under construction (*Source:* Photo courtesy of Mr. Greg Jennings).

Ephemeral channel – a natural channel that does not continuously flow.

Perennial channel – the natural channel that continually flows.

Artificial channel – watercourse developed by humankind such as navigation, irrigation, drainage, or closed conduit (e.g., flow does not fill the entire channel such as a sewer or culvert) channels. Hydraulically designed conveyances with attributes strategically chosen to meet the client's engineering, economic, and social objectives (Graf and Altinakar 1998). Designed channels not continuously flowing are *intermittent*.

We address the hydraulic engineering needed to satisfy social, economic, and engineering needs in this text. We focus mainly on one-dimensional solutions in this introductory text. Once solved using tabular solutions, the one-dimensional problems (along the channel) are easily solved with spreadsheets and other equation processing software. A commonly used public domain software, HEC-RAS, can, in some cases, facilitate two-dimensional solutions.

Fluid Mechanics Foundations

<u>Fluid statics</u> is a crucial limiting case. It describes fluid pressures and forces when the fluid velocity is zero. The pressure at a point is the product of fluid unit weight, γ, and depth, symbolized as d, b, or y. The unit weight in SI and imperial units is $9810\,\text{N/m}^3$ or $62.4\,\text{lb/ft}^3$. Figure 1.2 shows a hydrostatic distribution on a channel wall and bottom. Figure 1.3 shows the hydrostatic forces on a sloping wall where one partitions the effects between vertical and horizontal vector sums. Acceleration causes deviations from the purely hydrostatic pressure distribution. Sluice gates and transitions due to structures and slope changes result in acceleration. Thus, the pressure is not hydrostatic near sluice gates and other facilities.

One can easily show that the effective center of force is 1/3 up from the channel bottom. The 1/3 rule applies when the wall or gate object is rectangular. Symmetric gates of other shapes have force centers at the center of pressure, which deviates from the 1/3 point. In general, the center of pressure is computed using the following equation:

$$y_\text{p} = \frac{I_\text{G}}{\bar{y}A} + \bar{y} \tag{1.1}$$

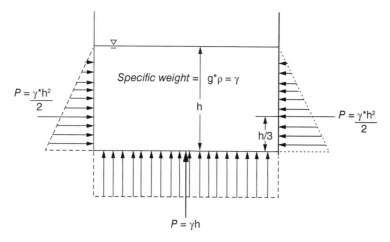

Figure 1.2 Free body diagram of a static fluid.

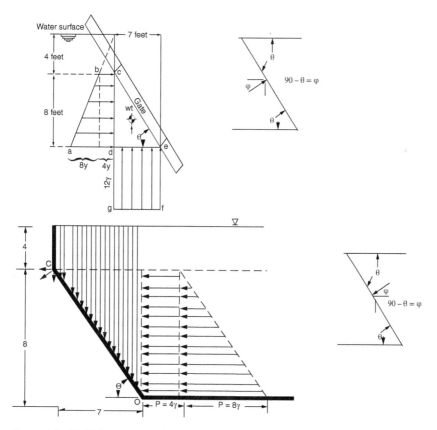

Figure 1.3 Fluid forces exerted on a submerged gate.

where I_G is the moment of inertia ($b * y^3/12$ for a rectangular gate where b is length and y = depth), $\bar{y}$ is the depth to the centroid of the object (half the distance to the bottom for a rectangle), A is the area of the port or gat object (length $b \times$ depth y for a rectangular gate), and y_p is the distance below the surface to the centroid where the total hydrostatic force is concentrated.

It is easily shown that y_p is 2/3 the depth from the top or 1/3 of the bottom for the rectangular gate. Equation 1.1 gives the distance along the slope for non-vertical gates. One can separate the forces into horizontal and vertical vector components and apply the 1/3 rule to the vertical component, giving a similar value to Equation 1.1. The value of Equation 1.1 comes to the fore when the gate is nonrectangular. One may then consult a statics text for the moment of inertia for the shape in question. Nonsymmetric gates (along the length b-axis) have a product of inertia, which shifts the center of pressure a distance from the centroid along the width b of the gate. We are not concerned with nonsymmetric gates. One may apply fundamental statics analysis to compute forces to secure closed gates in a channel with ponded, static conditions. A spreadsheet is provided which analyzes the forces required to fasten a symmetric gate on sloping walls.

A partially opened gate obviously does not represent a static condition. We evaluate forces associated with moving water in our chapter on rapidly varied flows. As in closed conduits flowing full, flows in open channels may be laminar or turbulent. The flow is laminar when viscous forces dominate inertial forces in determining flow behavior. Flows are turbulent when inertial flows dominate viscous forces. The Reynolds number, is expressed as follows:

$$R = \frac{\rho V L}{\mu} = \frac{V L}{\nu} \tag{1.2}$$

where R is the Reynolds number $(-)$, ν is the kinematic viscosity $((L/T^2)$, typically $1.93E-06\,m^2/s$ or $1.93E-05\,ft^2/s)$, μ is the dynamic viscosity $(FT/L^2$, typically $3.75E-05\,lbf\ s/ft^2$ or $1.79E-03\,N\ s/m^2)$, ρ is the fluid density (M/L^3), typically $1.94\,slugs/ft^3$ or $1000\,kg/m^3$, L is the characteristic length, typically depth (L), and V is the velocity (L/T).

Units for viscosity are quite varied, depending on the usage of force or mass units. Dynamic viscosity may be expressed in mass units as $M/(LT)$. One may perform an internet search to find these expressions in desired units. We include a spreadsheet showing viscosity and density as a function of temperature for SI and Imperial units. We generally ignore temperature effects in most open channel applications.

As in closed conduit flow, flows with $Re < 2000$ are generally laminar, and flows with $Re > 10000$ are generally turbulent. The region $2000 > Re > 10000$ is a transition zone. Consider the Darcy–Weisbach formula given as follows:

$$h_f = f\frac{LV^2}{d_o 2g} \tag{1.3}$$

where h_f is the head loss (L), f is the friction factor $(-)$, d_o is the diameter of the pipe (L), g is gravity (L/T^2), and L is the length over which the head loss occurs (L).

Defining the slope as h_f/L, d_o as $4R$ (where hydraulic radius R is more fully defined later), one may write Equation 1.3 as follows:

$$f = \frac{8gRS}{V^2} \tag{1.4}$$

Simons and Senturk (1992) compare friction factor vs. Re pipes and channels with varying roughness heights. Their figure is reproduced in Figure 1.4. The Darcy–Weisbach friction factor may be related directly to the Manning state equation friction term in the turbulent zone. The concept of channel roughness is more fully developed in the next chapter.

Figure 1.4 shows notable similarities with the Darcy–Weisbach–Moody diagram for closed conduit pipe flow. Most flows generally occur in the regime where Re > 4000, which enables one to relate the friction factor to roughness height. Most flows of interest involve turbulent flows, except for shallow sheet flows and flow found in natural treatment systems. The above results generally apply to steady and unsteady flows.

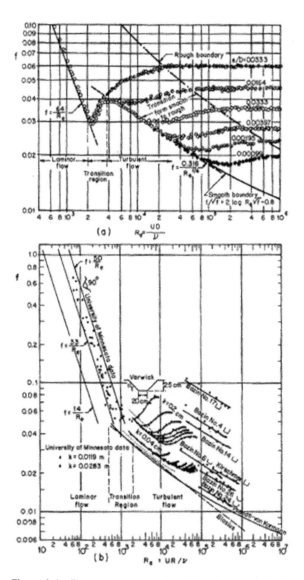

Figure 1.4 Flow resistance in pipes (a) and channels (b) (*Source:* From Simons and Senturk (1992); used with permission of Water Resources Publishers).

In addition to inertial flows and viscous flows, a third flow force, gravity forces, is vital in open channel flows. Another dimensionless number, the Froude number, is essential for further classifying turbulent flows. The Froude number appears as follows:

$$F = \frac{V}{\sqrt{gL}} \tag{1.5}$$

Parameter L is a characteristic length, which is frequently taken as hydraulic depth. The hydraulic depth is the channel depth for a rectangular channel and is more fully defined later for nonrectangular channels. Graf (1971) provides Reynolds and Froude similitude relations, shown in Table 1.1.

Similitude is a basis for studying scale models of various structures in a lab. Flowrate, velocity, time, and force may be measured and then scaled back to the lab's prototype structure. An accurate model requires the satisfaction of multiple dimensionless numbers. However, without varying fluids properties and gravity, the satisfaction of true similarity is not practical, so we focus on key numbers in ranges that give reasonably good results. Most applications involve nondistorted models with predominant Reynolds or Froude similarity. For example, given a prototype structure with basic dimensions of 100 and 50 m, handling a turbulent $10\,m^3/s$ flow. If one chooses to scale it to 10 and 5 m, one can use Table 1.1 to see scale relations between the model and prototype. The required flow to operate the Froude similitude model is then $Q_m = Q_p\,(L_m/L_p)^{5/2}$ or $10*(1/10)^{5/2} = 0.03\,m^3/s$. Forces, velocities, and times can be measured in the model and scaled back to the prototype. For best results,

Table 1.1 Modeling ratios for undistorted fixed bed models with Reynolds or Froude similarity.

	Model parameters	
Parameter of interest	Reynolds number	Froude number
Velocity, $\dfrac{V^M}{V^P}$	$\dfrac{L^P}{L^M}\dfrac{\rho^P}{\rho^M}\dfrac{\mu^M}{\mu^P}$	$\left(\dfrac{L^M}{L^P}\right)^{1/2}$
Flowrate, $\dfrac{Q^M}{Q^P}$	$\left(\dfrac{L^M}{L^P}\right)^2\dfrac{\rho^P}{\rho^M}\dfrac{\mu^M}{\mu^P}$	$\left(\dfrac{L^M}{L^P}\right)^{5/2}$
Force, $\dfrac{F^M}{F^P}$	$\left(\dfrac{L^M}{L^P}\right)^3\left(\dfrac{\mu^M}{\mu^P}\right)^2\dfrac{\rho^P}{\rho^M}$	$\dfrac{L^M}{L^P}\dfrac{\rho^M}{\rho^P}$
Time, $\dfrac{t^M}{t^P}$	$\left(\dfrac{L^M}{L^P}\right)^2\dfrac{\rho^M}{\rho^P}\dfrac{\mu^P}{\mu^M}$	$\left(\dfrac{L^M}{L^P}\right)^{1/2}$
Roughness n, $\dfrac{n^P}{n^M}$	–	$\left(\dfrac{L^P}{L^M}\right)^{1/6}$

Nomenclature: μ is dynamic viscosity, ρ is density, L is length, t is time, Q is flowrate, V is velocity, and n is the Manning roughness, discussed in a subsequent chapter. Superscript P refers to the prototype, and M refers to the scaled model. Be sure to check that the Manning equation applies in the case of Froude similarity.
Source: Based on Graf (1971) with modifications.

flows should be turbulent in the model, assuming the prototype is in the turbulent regime. If the prototype flow is laminar, then the model flow should also be laminar. One would use Reynolds's similarity in laminar flow cases.

If the flow in the prototype is laminar instead of turbulent, one can consider Reynolds's similarity instead of Froude similarity. Henderson (1966) devotes an entire chapter to similitude studies. Currently, there is one remaining hydraulics laboratory in the United States[1] that offers design services based on similitude.

Flows with a Froude number greater than 1.1 are supercritical. A supercritical flow is typically a shooting flow. Placing one's hand in a strongly shooting flow results in flow tending to go up one's arm. Flows less than 0.9 are subcritical or tranquil. Moving one's hand back and forth in a tranquil flow produces waves that move upstream. Flows with $F \approx 1$ are denoted to be critical flows, which results in a standing wave. At the critical flow condition, $V = \mathrm{Sqrt}(g * D)$.

The critical flow state is useful for the design of flow measurement devices. The critical flow condition gives a unique relationship between flow and depth. Flow measurement structures are further developed in a later chapter.

It will be shown that $\mathrm{Sqrt}(g * D)$ is the velocity of a gravity wave. Gravity waves can lead to pulsations in channels, which can pose structural stability issues for lined channels and erosion hazards, particularly for unlined channels. To avoid the possibility of flow pulsations, we try to avoid designs leading to $0.9 < F < 1.1$ in channels. Design implications are discussed in due course.

<u>Flow regimes:</u> Given the laminar-turbulent and subcritical–supercritical flows, one may encounter the following flow regimes:

- Subcritical-laminar – may occur with low slopes and low flows; the constructed wetland is a possibility.
- Supercritical-laminar – may occur with steep slopes and thin sheet flows.
- Subcritical-turbulent – the typical flow regime of most engineered channels.
- Supercritical-turbulent – may occur with steep slopes.
- Transitional – supercritical or subcritical – may occur near flow measurement devices.

Figure 1.5 shows the four main regimes (excluding the transitional case). Note the subcritical-turbulent and supercritical-turbulent flow regimes highlighted in Figure 1.5. Turbulent subcritical flow is the primary regime considered in this course. Other flow regimes, such as subcritical laminar flow, occur in constructed wetlands. Supercritical laminar flow is rare.

Hydrologic Foundations

Perhaps the most frequent channel design application is to determine the depth of a specified flow in given channel geometry. The client chooses the channel flowrates. For drainage purposes, peak flows are determined using runoff models such as the Rational equation, NRCS Curve number flow equations, or other more advanced hydrology models. Details on

---

1 Hydraulic Engineering Research Unit, United States Department of Agriculture, Agricultural Research Service, 1301 N. Western Road, Stillwater, OK 74075. The web address is https://www.ars.usda.gov/plains-area/stillwater-ok/hydraulic-engineering-research/.

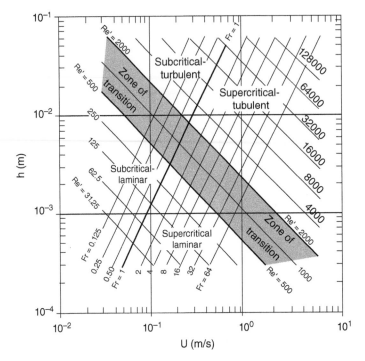

Figure 1.5 The four regimes of open-channel flow plotted as a function of velocity and depth, with Reynolds and Froude numbers denoted on the plot. Some vary the transition zone to include Reynolds numbers as much as 10000. A Froude number range of 0.9–1.1 is suggested as a practical transition zone for critical flow (*Source:* From Graf and Altinakar (1998); used with permission of John Wiley & Sons).

runoff estimation are contained in works such as Tollner (2016) or Huffman et al. (2013) and not discussed here.

Another application is to determine a flowrate resulting from rainfall required to cause flow at a given depth in a watershed. For example, one may assess flowrate based on a high watermark in a natural channel having a given geometry, slope, and roughness.

Presentation Organization

Research in all facets of Open Channel Hydraulics has steadily progressed through the nineteenth and twentieth centuries and continues through the present. Works such as Chow (1959), Henderson (1966), Simons and Senturk (1992), French (2007), and Sturm (2010) are but a few of the works providing interesting recent historical insights into how Open Channel Hydraulics has matured over time. From a practical viewpoint, the NRCS (2007) provides an excellent overview of the design of natural channels and stream restoration. Concepts developed in the past but not widely used are likely useful in certain situations; thus, much of this history warrants preservation. A basic bookshelf in open channel hydraulics should contain works such as those above, guiding one while looking for additional insights via the internet.

This introductory text focuses on providing that which is needed to facilitate access to more advanced discussions. Jobson and Froehlich (1988) present an applied treatment of open channel hydraulics. Still, they do not bring computational tools available today to bear. This text aims to provide a resource that uses modern computation resources to avoid excessive coding but facilitates learning and applying the basic principles of continuity, energy, and momentum transfer. This chapter aims to introduce fundamental concepts in outline form, which the author expands through the remainder of the text.

Flows may be broadly classified as steady (dQ/dt and dy/dx both constant), gradually varying (dQ/dt constant but dy/dx may vary), and unsteady (dQ/dt varies, and dy/dx usually varies). Q represents flowrate, y represents depth, and t represents time. Steady flow provides the basis for many basic channel designs. Energy and continuity generally suffice

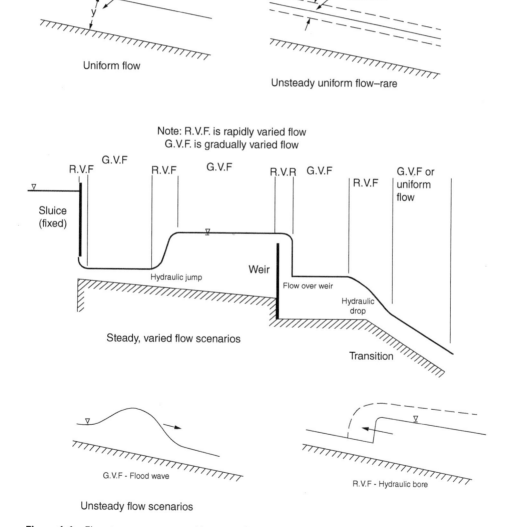

Figure 1.6 Flow types encountered in open-channel flow situations.

for analyzing these problems. Also, flows may be rapidly varying, which typically requires the momentum equation. Spatially varying flows (dQ/dx vary as does dy/dx) require energy and momentum tools introduced but not substantially developed.

Figure 1.6 provides a schematic view of the above flow conditions. Turbulent flows are the primary focus, with little mention of laminar flows due to limited practice applications. Continuity, energy, and momentum transfer, coupled with the Manning state equation for uniform flow, provide the basis for walking through the above classifications as well as the classifications that follow. The text follows the above flow regime outline.

This text does not claim to provide an exhaustive review of any topic related to open channel hydraulics. References cited herein are of value to those choosing to delve more deeply into any of the issues. There are some contemporary topics not addressed in this text. They include the diffusion of substrates and density currents, which is well discussed by French (2007). Wave motion, significant in coastal engineering, is not addressed; however, Henderson (1966) provides excellent coverage. Unsteady flow analysis, limited by the need for numerical solutions, has been greatly augmented by finite difference and finite element solution approaches. Jain (2001) and Graf and Altinakar (1998) provide a detailed analysis of the dam-break problem, which this text covers superficially. Simons and Senturk (1992) and Sturm (2010) provide much more detailed coverage of scour and aggradation topics. Chaudhry (1993) gives an advanced treatment of unsteady flow analysis, including the dam-break problem. Computer simulation approaches such as HEC-RAS (Brunner 2016) embody many numerical advances for unsteady flow analysis.

Problems and Questions

1 Define the difference between a ditch and a channel.

2 Give examples of situations that might lead to (i) steady flow and (ii) unsteady flow.

3 Give examples of conditions that might lead to (i) laminar flow and (ii) turbulent flow.

4 Show that the Reynolds and Froude numbers are both dimensionless. Be prepared to explore other units of viscosity to be consistent with force and mass units.

5 Using a fluid mechanics or other text available in your school library, prepare a paper elaborating on the development of the Reynolds number in fluid mechanics.

6 Using fluid mechanics or other text available in your school library, prepare a paper showing the dimensional analysis leading to the Reynolds and Froude numbers.

7 Given a sediment detention pond that is $50 \times 20 \times 1\,\text{m}$ and handling a flow of $0.3\,\text{m}^3/\text{s}$. Develop a 1/5th scale model of this pond. What are the dimensions of the model? What flow is needed to operate the model, assuming Froude similarity? Estimate the same for Reynold's similarity.

8 Analyze the pressure on a rectangular gate slanted at 30° from the vertical. The top of the gate at 6 ft below the surface, and the gate's bottom is 12 ft below the surface. The wall is slanted inward, such that the topwidth is greater than the bottom width. The gate is 8 ft wide.

9 Use Equation 1.1 to estimate the center of pressure of a 2 ft diameter circular gate on a vertical wall with the gate center being 10 ft below the water surface. Redo the problem given the wall being 60° from the horizontal (inward at the bottom) and the gate center being 10 ft below the surface. HINT: The projected port becomes elliptical. The width would remain the same, but the vertical dimension would shrink. Figure the major and minor axes and develop the center of pressure from the projected surface.

References

Biswas, A.K. (1970). *History of Hydrology*. Amsterdam: North-Holland.

Brunner, G.W. (2016). *HEC-RAS River Analysis System User's Manual*. Davis, CA: US Army Corps of Engineers, Hydrologic Engineering Center.

Chaudhry, M.H. (1993). *Open-Channel Flow*. Englewood Cliffs, NJ: Prentice-Hall.

Chow, V.T. (1959). *Open Channel Hydraulics*. New York, NY: McGraw-Hill.

French, R.H. (2007). *Open Channel Hydraulics*. Highlands Ranch, CO: Water Resources Publications, LLC.

Graf, W.H. (1971). *Hydraulics of Sediment Transport*. Highlands Ranch, CO: Water Resources Publications, LLC.

Graf, W.H. and Altinakar, M.S. (1998). *Fluvial Hydraulics: Flow and Transport Processes in Channels of Simple Geometry*. Chichester: Wiley.

Henderson, F.M. (1966). *Open Channel Flow*. New York, NY: Macmillan.

Huffman, R.L., Fangmier, D.D., Elliot, W.J., and Workman, S.R. (2013). *Soil and Water Conservation Engineering*, 7e. St. Joseph, MI: American Society of Agricultural and Biological Engineers.

Jain, S.C. (2001). *Open-Channel Flow*. New York: Wiley.

Jobson, H.E. and Froehlich, D.C. (1988). *Basic Hydraulic Principles of Open Channel Flow*. US Geological Survey Open File Report 88-707. Reston, VA.

NRCS. (2007). *Stream Restoration Design of the National Engineering Handbook*. Part 654 of the USDA National Engineering Handbook, Washington, DC. https://www.nrcs.usda.gov/wps/portal/nrcs/detail/national/water/manage/restoration/?cid=stelprdb1044707 (accessed April 2016).

Simons, D.B. and Senturk, F. (1992). *Sediment Transport Technology: Water and Sediment Dynamics*. Littleton, CO: Water Resources Publications.

Sturm, T.W. (2010). *Open Channel Hydraulics*, 2e. New York, NY: McGraw-Hill.

Tollner, E.W. (2016). *Engineering Hydrology for Natural Resources Engineers*, 2e. Chichester: Wiley.

2

Channel Fundamentals*

This chapter presents classic design approaches for lined and unlined (earthen) channels. Methods offered herein provide the basis for common design approaches and provide a foundation for more advanced procedures covered in later chapters.

Goals

- To design simple lined channels given the lining, slope, channel cross section, and flow rate.
- To design earthen channels under the same conditions listed above, where the velocity represents a constraint, which is Earthen Channels I.
- To design channels that convey a stated flow rate in conditions that minimize cross-sectional area and wetted perimeter.
- To understand the assessment of flows in natural channels.
- To understand some of the limitations of relationships used in channel design.

Channel Elements and Nomenclature

Natural channels are *ephemeral* if they do not continuously flow. They are *perennial* if they always flow or are below the water table. Designed channels not continuously flowing are *intermittent*. Otherwise, they are designated *continuous*.

One can describe flow channels with the following elements shown in Figure 2.1:

Flow depth (y) is the actual depth in the channel (L);

Area (A) is the portion of the channel cross section wetted by the fluid (L^2);

Wetted perimeter (P) is the wetted segment of the channel perimeter (L);

Hydraulic radius(R) is the flow area divided by the wetted perimeter (L);

Top width (t) is the length of the cross section exposed to air; Figure 2.2 shows that top width is equal to the area differential with respect to depth ($t = dA/dY$). This equality is useful in some manipulations of the energy and other equations (L).

---

*Draws heavily from Tollner (2016), chapter 8.

Open Channel Design: Fundamentals and Applications, First Edition. Ernest W. Tollner.
© 2022 John Wiley & Sons Ltd. Published 2022 by John Wiley & Sons Ltd.
Companion website: www.wiley.com/go/tollner/openchanneldesign

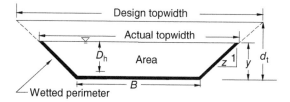

Figure 2.1 Channel elements of a trapezoidal channel. The hydraulic depth is a calculated quantity. The slope (S_o) is measured perpendicular to the cross-sectional area shown.

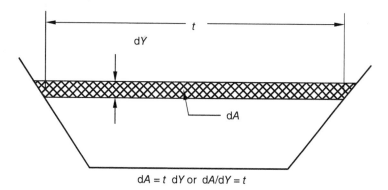

$$dA = t \ dY \ or \ dA/dY = t$$

Figure 2.2 A definition sketch showing how top width relates to differentials of depth and area.

Side slope (z) is the slope of the side of a trapezoidal, triangular, or rectangular channel. Designers express side slope as z:1, run to rise, where z is the horizontal run, and the vertical rise is unity. The rectangular cross section has a $z = 0$. Side slope z is dimensionless.

Channel gradient or slope (S_o) is the channel's longitudinal slope, normal to the cross-sectional area, along the flow direction. The slope is dimensionless.

Flow rate (q) is the channel flow rate (L^3/T). One may approximate flow rate by multiplying velocity times the cross-sectional area of the channel.

Hydraulic depth (D_h) is defined as the cross-sectional area (A) divided by the top width (T), used in Froude number calculations (L). Refer to Figure 2.2. Hydraulic depth is useful for computing the Froude number of a non-rectangular cross section.

Freeboard is the extra depth increment added to the final design to account for settling, wave action, possible sedimentation, capillary action in the soil, and the design safety factor. Freeboard may be a constant amount, fraction of design depth, fraction of the velocity-depth product, or other agency policy-determined amount.

Section factor for critical flow Z – useful for computing the flow of a given cross section occurring at a Froude number of one. Critical flow $= Zg^{0.5}$, where g is gravity. One can easily show that $q/A = V = D^{0.5}g^{0.5}$ or $V = $ Sqrt($D*g$), which only occurs at a Froude number of one.

Velocity (v) is the channel mean velocity (L/T). We may approximate velocity by recording the distance moved by a float per unit time.

Trapezoidal, triangular, circular, and parabolic cross sections are the standard cross sections. Table 2.1 gives relationships for computing hydraulic elements for typical

Table 2.1 Hydraulic elements for selected prismatic channel cross sections.

Cross section	Cross-sectional area A	Wetted perimeter P	Hydraulic radius $R = A/P$	Top width T	Hydraulic depth $D_h = A/T$	Section factor for critical flow $Z = AD^{0.5}$
Trapezoid	$by + zy^2$	$b + 2y\sqrt{z^2 + 1}$	$\dfrac{by + zy^2}{b + 2y\sqrt{z^2 + 1}}$	$b + 2zy$	$\dfrac{by + zy^2}{b + 2zy}$	$\dfrac{\left[\{(b+zy)y\}\right]^{1.5}}{b + 2zy}$
Triangle	zy^2	$2y\sqrt{z^2 + 1}$	$\dfrac{zy^2}{2y\sqrt{z^2 + 1}}$ $\approx y/2$ if $z > 3$	$T = 2yz$	$\dfrac{y}{2}$	$\dfrac{\sqrt{2}}{2}zy^{2.5}$
Rectangle	by	$b + 2y$	$\dfrac{by}{b + 2y}$ $\approx y$ when $(b/y \approx 25)$	b	y	$by^{1.5}$

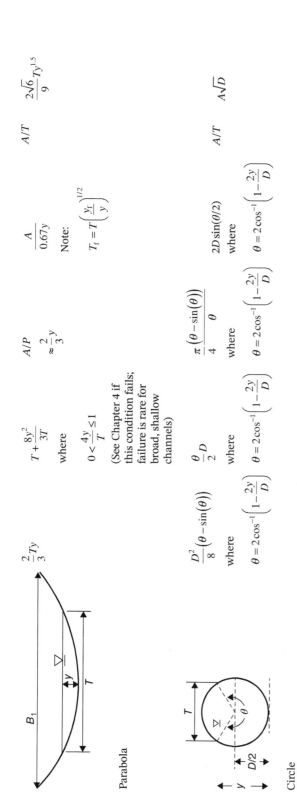

Parabola

$$\frac{2}{3}Ty$$

$$T + \frac{8y^2}{3T}$$

where

$$0 < \frac{4y}{T} \le 1$$

(See Chapter 4 if this condition fails; failure is rare for broad, shallow channels)

A/P

$$\approx \frac{2}{3}y$$

$$\frac{A}{0.67y}$$

Note:

$$T_f = T\left(\frac{y_f}{y}\right)^{1/2}$$

A/T

$$\frac{2\sqrt{6}}{9}Ty^{1.5}$$

Circle

$$\frac{D^2}{8}\big(\theta - \sin(\theta)\big)$$

where

$$\theta = 2\cos^{-1}\left(1 - \frac{2y}{D}\right)$$

$$\frac{\theta}{2}D$$

where

$$\theta = 2\cos^{-1}\left(1 - \frac{2y}{D}\right)$$

$$\frac{\pi}{4}\frac{\big(\theta - \sin(\theta)\big)}{\theta}$$

where

$$\theta = 2\cos^{-1}\left(1 - \frac{2y}{D}\right)$$

A/P

$$2D\sin(\theta/2)$$

where

$$\theta = 2\cos^{-1}\left(1 - \frac{2y}{D}\right)$$

A/T

$$A\sqrt{D}$$

Note: Except for the circle, freeboard is necessary. The above relations can be used to compute the freeboard once a depth adjustment is determined. The parabolic cross-section requires special handling to maintain the parabolic geometry with freeboard. The generalized parabolic relationship is shown above for the parabolic top width using the subscript "f".

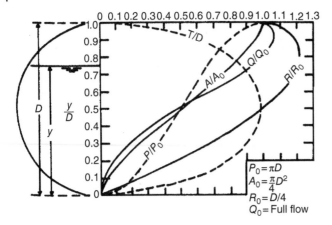

Figure 2.3 Hydraulic elements for a circular cross section as a fraction of hydraulic elements at full pipe flow computed as shown.

channel cross sections. Some parameters in Table 2.1 are defined as needed in later chapters. Figure 2.3 shows hydraulic elements of a circle (constant Manning's n) in a graphical form. An approximate relationship between hydraulic radius and depth for the triangular and parabolic geometries often suffices when channels are wide and shallow. When the trapezoidal channel is wide, depth well approximates the hydraulic radius. Graphical solution aids show the various component sensitivities. Works such as Sturm (2010) provides visual aids for several standard channel cross sections. This text offers a rich set of Excel® and Mathematica® programs for solving common channel design problems, which are referenced in upcoming sections. Also included is a first look at HEC-RAS (Brunner 2016) to model a channel and quickly analyze a steady uniform flow case.

A channel section increasingly used is the natural or sustainable channel, as shown in Figure 2.4. Figure 1.1 shows a sustainable channel under construction. The sustainable channel shown in Figure 2.4 has a cross section that is a combination of cross sections in Table 2.1, in that there is a primary channel embedded within a larger channel. The main channel with the stream slope generally conveys the one-year storm, while the larger channel, with a slope of the valley, conveys the more extreme events. Both channel cross sections tend to be parabolic. Reasons for this parabolic cross section will be apparent in our discussion of tractive force in a following chapter. The sustainable channel has the advantages of aesthetics and less erosion, reducing sediments and nutrients in downstream flow. Sustainable channels are usually designed with empirical relationships based on a design approach known as "Regime Theory," dating back to Blench (1952). Regime theory is further discussed in Chapter 10, where an approach summarized by Huffman et al. (2013) is presented. Analyzing an existing channel is discussed in this chapter.

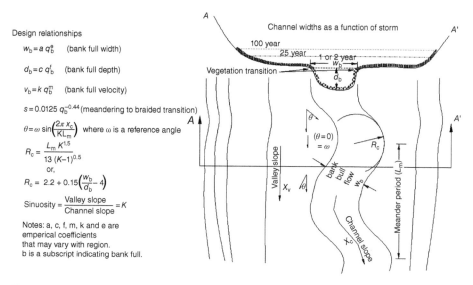

Design relationships

$w_b = a\, q_b^e$ (bank full width)

$d_b = c\, q_b^f$ (bank full depth)

$v_b = k\, q_b^m$ (bank full velocity)

$s = 0.0125\, q_b^{-0.44}$ (meandering to braided transition)

$\theta = \omega \sin\!\left(\dfrac{2\pi\, x_c}{KL_m}\right)$ where ω is a reference angle

$R_c = \dfrac{L_m\, K^{1.5}}{13\,(K-1)^{0.5}}$

or,

$R_c = 2.2 + 0.15\!\left(\dfrac{w_b}{d_b} - 4\right)$

$\text{Sinuosity} = \dfrac{\text{Valley slope}}{\text{Channel slope}} = K$

Notes: a, c, f, m, k and e are emperical coefficients that may vary with region.
b is a subscript indicating bank full.

Figure 2.4   Hypothetical natural channel cross section and a plan view showing nomenclature and design relationship as summarized by Huffman et al. (2013). Used with Permission.

General Flow Relationships

Continuity equation (CE): One can write the continuity equation in macroscopic and microscopic terms. We limit the discussion to the macroscopic form below. Equation 2.1 shows the macroscopic version of the CE[1] when the fluid is incompressible:

$$A_1 v_1 = A_2 v_2 = q \tag{2.1}$$

where A_i is the cross-sectional area (L^2) at point i,
   v_i is the velocity (L/T) at point i, and
   q is the flow rate (L^3/T).

Uniform Flow Relationships

Chezy uniform flow equation: Some engineers outside the United States employ the Chezy relationship for channel design. The Chezy equation has theoretical roots and finds usage in some conceptual work. Consider the free body diagram in Figure 2.5 (exaggerated to enhance the forces). Evaluating forces along the direction of flow, one can write Equations 2.2 and 2.3, assuming uniform flow:

$$F_{\text{drag}} = 0.5\rho C_d A_d v^2; \quad A_d = P_{\text{wtd}}L \tag{2.2}$$

---

[1]The steady-state CE given in Equation 8.1 was apparently first used by Hero of Alexandria prior to 150 BCE (Biswas, 1970).

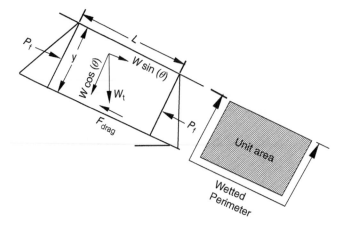

Figure 2.5 Simplified forces bearing upon uniform flow in an open channel with the slope exaggerated.

$$F_{drag} = WSin(\theta) + P_1 - P_2 \tag{2.3}$$

where F_{drag} is a resisting force,

P_{wtd} = wetted perimeter,

W_t = weight of water in the control volume between points 1 and 2, and

P_1, P_2 is the pressure force at points 1 and 2.

The pressure forces cancel and, thus, do not contribute. The component of the weight parallel with the channel bottom counters the resisting fluid drag force. Writing F_{drag} as $C_D PL\rho\, v^2$ and expressing W_t as $\rho g A L$ and substituting into Equation 2.3 gives the first two terms of Equation 2.4:

$$\rho g A L \sin(\theta) = C_d P L \rho v^2 \quad \left[\approx \gamma R S = \tau = C_d \rho v^2 \right] \tag{2.4}$$

where L is an arbitrary length,

C_d is a drag coefficient,

$\sin(\theta) \approx \tan(\theta)$ = slope,

P is the water density,

A is the cross-sectional area, and

g is gravity.

From the first two terms of Equation 2.4, recall the definition of the hydraulic radius R as A/P. Assume $\sin(\theta) = \tan(\theta)$ = slope S. Next, solve for v and consolidate constants. We then have Equation 2.5:

$$v = C_{Chezy}\sqrt{RS_o} \tag{2.5}$$

where R is the hydraulic radius (L),

S_o is the slope (−),

v is the velocity (L/T), and

C_{Chezy} is the roughness coefficient, sometimes called the Chezy coefficient ($L^{0.5}/T$).

The two terms on the right-hand side of Equation 2.4 show the forces expressed as stress on the channel side.

Manning uniform flow: The Manning uniform flow equation is the standard uniform flow state equation in the United States. Some additional equations relate roughness and channel geometry in the discussion of alluvial channels. According to Henderson (1966), an Irish engineer named Manning found the Chezy equation, multiplied by $R^{1/6}$, to provide consistently better agreement with experimental data. The relationship thus bears the Manning name. Equation 2.6 shows the Manning equation:

$$v = \frac{\varphi_{\dim}}{n} R^{2/3} \sqrt{S_o} \qquad (2.6)$$

where $\varphi_{\dim} = 1.486$ (this is $3.28\,\text{ft/m})^{0.333}$ for English (ft/s) units and 1 for SI (m/s) units.
The constant "n" is the channel roughness.
V is the mean velocity of the channel flow.

The roughness factor n is a straightforward modification of the Chezy roughness C, where $1/C$ was experimentally observed to depend on $R^{1/6}$. The unit for n is $\text{s/ft}^{0.33}$. In the flow form, by multiplying both sides of the equation by area and applying the CE to the left side, we have Equation 2.7.

$$q = \frac{\varphi_{\dim}}{n} A R^{2/3} \sqrt{S_o} \qquad (2.7)$$

Frequently one sees the term "Conveyance." King and Brater (1963) define conveyance (K_{conv}) as (Equation 2.8a and b):

$$K = \frac{\varphi_{\dim}}{n} A R^{2/3} \qquad (2.8a)$$

$$q = K_{conv} \sqrt{S_o} \qquad (2.8b)$$

where q is the flow rate (L^3/T),
K is the conveyance (L^3/T), and
A is the cross-sectional area (L^2).

Manually solving for velocity or flow with known depth and channel parameters is straightforward with the Manning equation.

As an aside, there are many equations for describing flow in open channels. Simons and Senturk (1992) summarize some 38 equations relating velocity to slope, roughness, and cross-section attributes. Most of the equations relate to some aspect of the earthen channel where the flow can influence roughness. Chapter 10 revisits roughness in channels with moveable beds.

Manning equation computations: Solving for depth requires a rearrangement of the equation to have terms containing depth on the left and all other known parameters on the right, as shown in Equation 2.9:

$$AR^{0.67} = \frac{nq}{\varphi_{\dim} \sqrt{S_o}} \qquad (2.9)$$

Table 2.2 Typical Manning equation roughness *n* values for a variety of artificially lined channels or conduits.

Lining material	Minimum *n*	Maximum *n*[a]
Artificially lined channels		
Asphaltic concrete	0.012	0.018
Concrete	0.015	0.03
Smooth metal	0.011	0.015
Corrugated metal	0.021	0.026
Plastic	0.012	0.014
Shotcrete	0.016	0.017
Planed wood	0.01	0.015
Unplaned wood	0.011	0.015
Pipe materials		
Cast iron pipe	0.011	0.015
Clay or concrete drain	0.011	0.02
Clay or vitrified sewer	0.01	0.017
Corrugated plastic tube	0.015	0.02
Corrugated metal ring	0.021	0.026
Corrugated metal helical	0.013	0.015

[a] Use the maximum *n* in cases of intermittent flows; use an average of the maximum and minimum *n* values for continuous flows.
Source: Adapted from French (1985) and Sturm (2010).

Substitute appropriate channel elements into the above relationship's left-hand side and put the known values on the right-hand side. Then, solve for the *y* value satisfying the equation.

Tables 2.2 and 2.3 summarize values for *n* with artificially and naturally lined channels. Table 2.2 gives *n* values for artificial lining materials. Lane (1955) tabulated value for *n* for natural surfaces, as shown in Table 2.3. One may also consult works such as Barnes (1967), French (2007), or Sturm (2010) for pictorial guidance. The Barnes (1967) photograph series is available in color on the web at http://wwwrcamnl.wr.usgs.gov/sws/fieldmethods/Indirects/nvalues/. A comprehensive stream morphology approach such as Rosgen (1996) presents perhaps the most systematic approach to the Manning *n* determination in natural watercourses.

In channels with a sand or gravel bed, Jobson and Froehlich (1988) summarize some relationships between particle size and Manning's *n*, with selected relationships given in Equation 2.10.

$$n = 0.034 d_{50}^{1/6} \text{ for gravel bed river} \tag{2.10a}$$

$$n = 0.032 d_{d90}^{1/6} \text{ for sand mixture in a flume} \tag{2.10b}$$

$$n = 0.039 d_{d75}^{1/6} \text{ for canal lined with cobbles} \tag{2.10c}$$

Table 2.3 Permissible velocity and Manning roughness values for natural after aging with flow depths less than 0.9 m (3 ft).

Material	Approximate roughness n values for stability design. See below for capacity recommendation[a]	Clearwater permissible velocity v_p values (m/s)[b]	Water-transporting colloidal silts permissible velocity v_p values (m/s)[c]	Water transporting noncolloidal silts, sands, gravels, or rock fragments
Fine sand, noncolloidal	0.02	0.46	0.76	0.46
Sandy loam, noncolloidal	0.02	0.53	0.76	0.61
Alluvial silts, noncolloidal	0.02	0.61	0.92	0.61
Ordinary firm loam	0.02	0.61	1.07	0.69
Silt loam	0.02	0.61	0.91	0.61
Volcanic ash	0.02	0.76	1.07	0.61
Shales and hardpans	0.025	0.76	1.83	1.52
Stiff clay, very colloidal	0.025	1.14	1.53	0.91
Alluvial silts, colloidal	0.025	1.14	1.53	0.91
Fine gravel	0.02	1.83	1.83	1.14
Graded loam to cobbles when noncolloidal	0.03	0.76	1.53	1.52
Graded silts to cobbles when colloidal	0.03	1.14	1.68	1.52
Coarse gravel, noncolloidal[d]	0.025	1.22	1.83	1.98
Cobbles and shingles[d]	0.035	1.22	1.68	1.98

[a] Suggested n value for the capacity design of intermittent channels is 1.25 times the table value not to exceed 0.06. For soils not listed, use judgment. If soil is known to be easily erodible, use sandy loam values. If soil is erosion-resistant, treat as shale/hardpan.
[b] Convert velocities to ft/s by multiplying values by 3.28.
[c] Use velocities with silt unless runoff comes from a lined surface such as a parking lot or specialized situations such as aquaculture dictate the flow velocity; convert velocities to ft/s by multiplying values by 3.28.
[d] See references such as Simons and Senturk (1992) or Goldman et al. (1986) for design procedures for linings composed of larger aggregate materials.
Source: Adapted from Lane (1955), supplemented with Table 8.3 of NRCS (2007).

Diameters d_{50}, d_{75}, and d_{90} represent the sediment distribution sizes passing 50, 75, and 90%, respectively. The grain diameter is expressed in feet for these empirical relationships.[2] Jobson and Froehlich (1988) show additional relationships. Equations 2.10 have a theoretical foundation shown in Equation 2.10d, provided by Chow (1959) and Sturm (2010).

$$n = \frac{(R/k)^{1/6} k^{1/6}}{21.9\log\left(\dfrac{12.2R}{k}\right)} \tag{2.10d}$$

$$n = \frac{d_{50}^{1/6} \dfrac{K_n}{\sqrt{8g}} \left(\dfrac{R}{d_{50}}\right)^{1/6}}{0.794 + 1.85\log\left(\dfrac{R}{d_{50}}\right)} \tag{2.10e}$$

where R is the hydraulic radius and k is the roughness height.

K_r is $\left(1 - \dfrac{\sin^2\theta}{\sin^2\phi}\right)^2$ where θ is the side slope angle, and ϕ is the angle of repose for the granular media. The parameter K_r is revisited in the tractive force chapter.

When one separates $k^{1/6}$ from 2.10d, taking roughness k as d_{50}, the constant in the Strickler equation (2.10a) approximates that in 2.10d. The same is true for the Sturm equation.

Another method for determining a roughness method, attributed to Cowan (1956), is given in Equation 2.11.

$$n = \left(n_b + n_1 + n_2 + n_3 + n_4\right)m \tag{2.11}$$

where n_b = base n value,
  n_1 = adjustment for irregularity,
  n_2 = adjustment for cross-section variation,
  n_3 = adjustment for effect of obstructions,
  n_4 = adjustment for vegetation, and
  m = adjustment for degree of meandering.

Table 2.4 provides suggested values for the n_i and m.

Especially for novices, one should use resources such as Barnes (1967), various tables, relationships such as those in Equation 2.11, and the Cowan method. A recommended way to arrive at a roughness distribution is to have several experienced engineers collaborate to estimate the n value based on a study of photographs, tables, and field experience.

---

[2]To convert to meters, one may multiply the constant in Equations 10 by 3.28 ft/m to the 1/6 power. For example, Equation 2.10a is $0.034*3.28^{1/6}$ or $0.041d^{1/6}$ where d is in meters.

Table 2.4 The Cowan (1956) adjustment factors for estimating the Manning *n*.

Channel conditions		Values	
Material involved	Earth	n_b	0.02
	Rock cut		0.025
	Fine gravel		0.024
	Coarse gravel		0.028
Degree of irregularity	Smooth	n_1	0.000
	Minor		0.005
	Moderate		0.010
	Severe		0.020
Channel cross-section variation	Gradual	n_2	0.000
	Occasional alternations		0.005
	Frequent alternations		0.010–0.015
Obstruction effects	Negligible	n_3	0.000
	Minor		0.010–0.015
	Appreciable		0.020–0.030
	Severe		0.040–0.060
Vegetation degree	None	n_4	0.000
	Low		0.005–0.010
	Medium		0.010–0.025
	High		0.025–0.050
	Very high		0.050–0.100
Degree of Meandering	Minor	m	1.0
	Appreciable		1.15
	Severe		1.30

Source: Adapted from Chow (1959).

Theoretical Considerations

The Manning equation roughness value is generally variable. The Manning equation is empirical, and the addition of the $R^{1/6}$ factor is only approximate. Particularly with alluvial where the bed moves, the practical *n* value can vary in response to bedforms present. The alluvial flow chapter discusses Bedform effects.

  <u>Manning n and friction factor f:</u> Flows behave (see Figure 1.5) like flows in closed conduits described by the Darcy–Weisbach–Moody diagram. Recall that the Darcy–Weisbach chart shows a friction factor that is approximately constant with the Reynolds number on the graph's upper right area. The upper right segment of the open channel variant of the friction factor diagram is the complete turbulence zone. It is the region where Manning is applicable. Most engineering applications are turbulent. By starting with the Darcey–Weisbach equation and substituting slope for h/L and rearranging to isolate velocity, one can easily show that the following correspondence exists between the Manning *n*, friction factor *f*, and the Chezy coefficient *C*:

$$\sqrt{\frac{8}{f}} = \frac{NR^{1/6}}{n\sqrt{g}} = \frac{c}{\sqrt{g}} \tag{2.12}$$

Further research by Kazemipour and Apelt (1979) has refined the friction factor for pipe friction (f_m) to channel friction (f_c) ratio as a function of the wetted perimeter, bottom width, and depth. Equation 2.13 gives the Kazemipour and Apelt (1982) findings for a rectangular channel.

$$\frac{f_c}{f_m} = \frac{\left(\dfrac{P}{B}\right)^{1/2}}{\dfrac{1}{\left(\dfrac{B}{D} + 0.3\right)} + 0.9} \tag{2.13}$$

Sturm (2010) notes that the ratio changes from 1.04 to 1.10, as B/D goes from 1 to 40. Equation 2.12 is usually sufficient for situations requiring a Moody friction factor.

Manning equation validity: The Manning equation, or any equation that describes frictional losses with roughness height alone, presumes a fully developed turbulence. In this zone, the loss depends only on roughness and not Reynolds number. One may check the Re of flow to ensure that the flow is turbulent. The product of $\dfrac{\sqrt{gRS} * n}{\upsilon}$ should exceed 100 (Jain 2001), where υ is the kinematic viscosity ($1.7E{-}6\,\mathrm{m^2/s}$ or $1.8E{-}5\,\mathrm{ft^2/s}$ for water). When the product is less than 100, the flow is in the transition zone or laminar zone. Flow is laminar if the product is less than four. Notable applications where flows may not be turbulent are the constructed wetland (prolonged flows). Powell (1950) presented an approach for handling flows in the transition and laminar flow zones. The formula is an implicit function of the Chezy C value, given in Equation 2.14.

$$C = -\log\left(\frac{C}{4R} + \frac{\varepsilon}{R}\right) \tag{2.14}$$

where C is the Chezy coefficient $\left(= \sqrt{\dfrac{8g}{f}} \text{ or } \dfrac{\phi}{n} R^{1/6}\right)$

f is the Darcy–Weisbach friction factor,
n is the Manning n,
g is gravity,
N_{dim} is a dimensions factor (1 for SI; 1.486 for imperial),
R is the hydraulic radius, and
ε is the roughness height.

A spreadsheet is provided in the downloadables that solves the Manning equation in the transition region, where we have used the above relationships between C and n.

Boundary layer effects: The Manning equation was derived based on an assumption of uniform flow. Does uniform flow exist in practice? Suppose a discharge in a channel encounters a step change in the bottom height. The boundary development following the step is given by Equation 2.15 (Chow 1959).

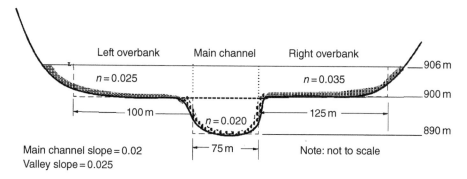

Figure 2.6 Section AA′ of the compound sustainable channel cross section shown in Figure 2.4.

$$\delta = \frac{0.024\,x}{\left(\dfrac{x}{k}\right)^{0.13}} \tag{2.15}$$

where δ is the height of the boundary layer to the point of free stream flow. The boundary is fully developed when δ is the flow depth.

x is the distance from the step and

k is the roughness height (may be taken as d_{50}).

Equation 2.15 indicates that the distance x downstream from the step change where δ becomes equal to the channel flow depth can be a considerable distance from the step. Uniform fully developed flow is a practically useful concept but, in theory, rarely occurs. Soon after the step, a laminar sublayer forms, which is revisited in Chapter 5.

Channels may have wetted perimeters with varying roughness conditions. Although many approaches with varying n values have been proposed, a straightforward method for estimating roughness is to do a weighted average of n with the wetted perimeter. Alternatively, one may weigh the n value with flow hydraulic radii or area of flow sections (see Figure 2.6 and surrounding discussion). French (1985) gives additional discussion. He also elaborates on estimating the Manning n of ice-covered channels.

Parameter selection: Side slope z is selected depending on contractor equipment, agency preference, and, in the case of the unlined channel, the soil condition. Table 2.5 provides guidance for various soil conditions in the case of the general trapezoidal channel. Topography onsite usually determines the slope. The flow rate is often determined by the watershed's hydrology, demands such as irrigation needs (western US), urban water supply needs, and other factors. Hydrology texts such as Tollner (2016) and other references cited elsewhere discuss flow rate estimation.

Natural, Compound, or Sustainable Channels

Sustainable channels are usually composed of several distinct cross sections, sometimes referred to as compound channels. Figure 2.6 is a cross section from the compound channel shown in the plan view in Figure 2.4. Dashed lines are shown, which subdivide

Table 2.5 Constraining z values for general trapezoidal channels, natural linings as specified, or optimum y/b values in the case of other indicated geometries.

	Side slopes – horizontal: vertical is Z:1	
Soil or lining type	In trapezoidal channels no deeper than 1.2 m, z should not be less than the following:	In trapezoidal channels 1.2 m and deeper, z should not be less than the following:
Rock	0	0
Peat and muck	0^a	0.25^a
Heavy clay	0.5^a	1
Earthen with stone lining or earth for large channels	1	1
Clay or silt loam	1	1.5
Sandy loam	1.5	2
Loose sandy soil or porous clay	2	3
Channel lined with gravel, cobbles	$4-6^b$	$4-6^b$
Any channel to be routinely crossed	4 or greater	4 or greater
Optimum (stabilized sides) trapezoid	$z = 0.5774; b/y = 1.15$	
Optimum (stabilized sides) triangle	$z = 1; b = 0$	
Optimum (stabilized sides) rectangle	$b/y = 2; z = 0$	
Lined (stable) circular channel	$y = D/2$	

[a] In the case of an optimized geometry, use the optimal z if it equals or exceeds the tabulated amount.
[b] Side slopes suggested for gravel and cobble lining are approximate; more precise stability designs require more detailed design approaches such as those of Simons and Senturk (1992).
Source: Adapted from Chow (1959) and Sturm (2010).

Figure 2.6 into rectangular sections, left and right overbanks, and the main channel. One has to evaluate the appropriateness of subdividing the channel or treating the cross section as a single channel. Jobson and Froehlich (1988) present Equations 2.16, using the nomenclature defined in Figure 2.7.

$$\frac{D}{d} \geq 2 \tag{2.16a}$$

$$\frac{W_o}{d} \geq 5 \tag{2.16b}$$

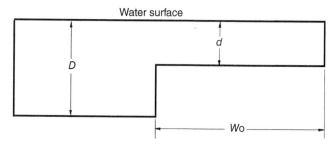

Figure 2.7 Schematic of a compound sustainable channel.

If both Equation 2.16a and b are satisfied for one of the cross sections, that cross section should be treated separately. Example 2.1 demonstrates the procedure.

Example 2.1 Analyze the flow shown in Figure 2.6. We first evaluate Equation 2.16a and b. For the left overbank, $D/d = 16/6 > 2$ and $W_o/d = 100/6 > 5$ and for the right overbank, $D/d = 16/6 > 2$ and $W_o/d = 125/6 > 5$. Thus, each overbank is treated separately. Note that had one overbank not satisfied Equation 2.16, it would be combined with the main channel and treated as one unit.

We first assume the slope is the same for each overbank. The overbanks are treated as wide rectangular channels. We develop a computational table to compute the flow with a single slope of the channel and with the complex slopes:

Channel section	Slope	Hydraulic radius[a] (m) $\approx d$ in over banks	Area (m²)	Conveyance (m³/s; see Equation 2.8)	V (m/s)	Q (m³/s)
Single slope of the main channel						
Left	0.02	6	600	79 246	18.67	11 207
Right	0.02	6	750	70 755	13.34	10 006
Main	0.02	12.6	1200	324 886	38.3	45 945
Totals	–		–	474 887		67 159
Stream and valley slope						
Left	.015	6	600	79 246	16	9705
Right	.015	6	750	70 756	12	8666
Main	0.02	12.6	1200	324 886	38.7	45 946
Totals	–	–	–	–		64 317

[a] The hydraulic radius of the riverbanks here is determined to be wide channels, with $R \approx d$. The main channel's hydraulic radius is the section ($75 * 16 \text{ m} = 1200$) divided by the main channel's wetted perimeter below the overbank elevation ($75 + 10 + 10 \text{ m}$). The rationale for not including the main channel's entire depth is that there is negligible flow resistance in the main channel above the overbank bottoms.

Conditions in Example 2.1 represent an extreme storm (e.g. a 100-year return period) in a mountain valley. The velocity is highly variable in the channel sections. One can add the conveyances and multiply by the square root of the slope to obtain the total flow.

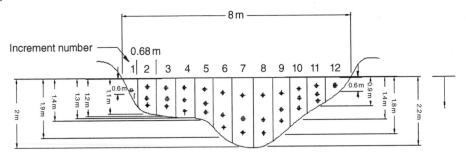

Legend
+ 0.2 depth and 0.8 depth
+ 0.6 depth (from surface)

Figure 2.8 Channel bathymetric survey and flow rate measurement schematic.

Now, suppose we admit the notion that the valley slope is measured straight down the valley. In contrast, the main channel slope is measured along the pathway of the main section. One can then use the Manning equation to calculate the flow of each section and add the flows. The total flow when using the valley slope for the over banks and the main channel slope for the main area results in a flow of 70 272 m³/s.

The desire to estimate flows in a natural channel leads to measuring flows in natural channels. A detailed approach to measuring flow velocity in a stream is shown in Figure 2.8. The stream is segmented, and velocities are measured at the 0.2 and 0.8 depth in each segment using a current meter or other instrument and averaged. One may replace the average with a single measurement at 0.6 D to represent the average segment velocity. A GPS and Sonar for making rapid bathymetry measurements are shown in Figure 2.9. The GPS and Sonar approach to bathymetry represents a reasonably agile approach for developing the cross-section database needed for modeling natural streams using methods such as HEC-RAS.[3] We take a more in-depth look at velocity profiles in our discussion of the energy equation.

Lined Channels, Optimum Channels, and Velocity Constraints

The flow, the downhill gradient, and the applications influence the design process. The overarching goal is to have the physical and economic feasibility of water conveyance solutions.

This text evaluates three flow solution types:

Lined channels (type A), where stability is not an issue due to the presence of an artificial or natural rock lining, and one desires to compute depth with other geometric elements known.

---

[3] Hydrologic Engineering Center – River Analysis System, Developed by the US Army Corps of Engineers, Hydrologic Engineering Center, 609 Second Street, Davis CA 95616. The web address is www.hec.usace.army.mil

Lowrence sonar

Ag114
GPS

Figure 2.9 Sonar system coupled with precision GPS unit for bathymetric (depth) measurements in a stream. *Source:* Courtesy of Mr. Ken Swinson.

Constrained velocity (type B), where natural lining stability is critical, and one simultaneously solves for depth and a geometric parameter.

Optimal channel (Type C), where the optimal cross section is critical.

Figure 2.10 shows the interrelationship between these problem categories. The Type A solution approach is foundational to Type B and C solutions. A given channel design may or may not satisfy Type B and Type C solution criteria simultaneously. One must be able to apply each design approach in creative ways to determine the best engineering solution.

Channel design requires the channel to carry the design flow without eroding. One may stabilize an earthen channel by lining or by satisfying velocity constraints in erodible linings. Channels having erodible linings require evaluation under multiple criteria, especially when the flows are occasional.

Type A – Solutions involving the Manning equation and CE without additional constraints: Solution Type A applies where channels are lined with rock or artificially lined with a nonerodible material. All geometric elements except one are fixed with known flow or velocity. When velocity or flow rate is unknown, and all geometry is known, the solution is straightforward. One may use the Type A solution procedure to solve problems involving:

- channels where all elements except depth are known (e.g. the channel construction machinery determines the bottom width and side slopes; see ASABE 1986, 1998);
- trapezoidal drainage ditches with entering tile lines requiring specified depth. Solve for required bottom width given a needed depth and side slope along with roughness and flow rate;

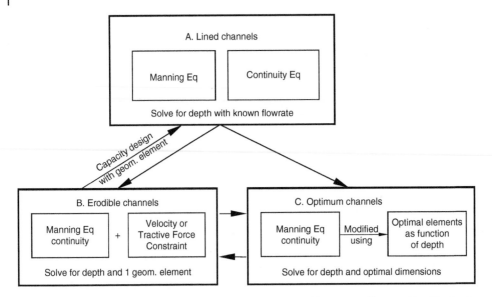

Figure 2.10 Interrelationships among the channel design problem types discussed in this chapter. Satisfaction of Type B may or may not lead to satisfaction of Type C, and the Optimum C may be erodible (B). After satisfying Type B, one may revisit Type A, particularly when flow is occasional and a capacity evaluation is appropriate.

- determination of existing channel performance when the flow rate changes or lining (e.g. roughness) increases due to age or other conditions;
- the determination of channel performance in the capacity condition (discussed below) where other criteria (e.g. Type B) dictate the bottom width (trapezoid) or side slopes (triangular).

External circumstances often dictate channel, slope, roughness, flow rate, side slope, and bottom width. One can solve for depth iteratively. One may then readily compute other hydraulic elements. The solution steps are:

1) Writing the Manning equation as $v = (\varphi_{dim}/n)R^{2/3}S^{1/2}$ or $q = (\varphi_{dim}/n)AR^{2/3}S^{1/2}$. Select n for the material and increment upward if the flow is occasional.
2) If all geometry is known, slope, roughness v, or q unknown, solve one equation in step 1 to get the unknown directly.
3) If one knows the roughness, slope, and flow, and one other geometric entity is unknown, use the Manning equation's flow form and isolate the known quantities. There is an Excel spreadsheet in the downloadables for solving these problems. Add a freeboard to the depth and recompute the major channel elements.
4) Check the Froude number. In the case of near-critical flow (e.g. $0.9 > Fr > 1.1$), modify roughness by changing material, change the slope by rerouting or inserting periodic drop structures. One may also change the bottom width or use some combination of methods to reduce the velocity and move the Froude number away from the critical region. Velocities should not exceed 2.1 m/s or the stated permissible velocity (unlined, discussed below) or 5.5 m/s in lined channels. Froude numbers should not exceed 0.9 in unlined channels (see Sturm 2010; Chin 2013). Designs involving critical or supercritical flow are beyond the scope of this text.

The downloadables contain spreadsheets for the solution of Type A lined channels. The spreadsheets compute freeboard and costs, provided one has estimates of the lining cost per unit area and the excavation and spoil management cost per unit volume. These spreadsheets use the Solver package, which comes with Excel. There is also a test for the applicability of the Manning equation (fully developed turbulent flow) taken from Sturm (2010) and Chin (2013). The downloadables contain selected Mathematica notebooks for solving the Manning equation for a lined channel.

Example 2.2 Given a trapezoid channel, $n = 0.025$, slope $= 0.1\%$, steady flow $= 28.3\,\text{m}^3/\text{s}$, bottom width $= 6.1\,\text{m}$, $z = 2$, freeboard factor is 20% of design depth. Find the normal depth. Substituting into $nq/\varphi\text{Sqrt}(s) = AR^{2/3}$ or,

$$0.025(28.3\,\text{m}^3/\text{s})/(1)(\text{Sqrt}(0.001))) = 22.37 = (6.1\,\text{m}\ y+2y^2)((6.1\,\text{m}\ y+2y^2)/(6.1\,\text{m}+2y\,(\text{Sqrt}(1+2^2))))^{0.67}$$

Estimate depth for starting the iteration by making the wide channel assumption where hydraulic radius R is equal to the depth and $z = 0$ in Equation 2.10. Then, Equation 2.9 appears as:

$$22.37 = (6.1\,\text{m}\ y)\ y^{0.67}$$ and $y = 2.18\,\text{m}$. Start the iteration with this value for depth or use the spreadsheet.

An iterative solution for y yields $y = 1.9\,\text{m}$ and $D = 2.286\,\text{m}$. The example is adaptable to other channel cross sections by using appropriate hydraulic elements.

Check the Froude number: $Fr = v/\text{Sqrt}(gD_h)$; compute A to be $18.85\,\text{m}^2$; $v = 1.5\,\text{m/s}$; $t = 13.7\,\text{m}$; $D_h = 18.85\,\text{m}^2/13.7\,\text{m} = 1.38\,\text{m}$; $Fr = (1.5\,\text{m/s})/\text{Sqrt}(9.81\,\text{m/s}^2 * 1.38\,\text{m}) = 0.41$, therefore the flow is subcritical. For occasional flows, increment the n by 25% before calculating y. Spreadsheets are available in the downloads for these computations.

Example 2.3 Given a trapezoid channel, $n = 0.025$, slope $= 0.1\%$, steady flow $= 28.3\,\text{m}^3/\text{s}$, $z = 2$, freeboard factor is 20% of design depth. Tile lines enter the channel at a depth of 2 m requiring the design depth of 2.5 m (without freeboard). What is the required bottom width?

Starting with Equation 2.9 and substituting known quantities gives

$$0.025(28.3\,\text{m}^3/\text{s})/(1)(\text{Sqrt}(0.001))) = (b(2.5\,\text{m})+2(2.5\,\text{m})^2)((b(2.5\,\text{m})+2(2.5\,\text{m})^2)/(b+2\,(2.5\,\text{m})(\text{Sqrt}(1+2^2))))^{0.67}$$

An iterative solution gives a bottom width of approximately 2.3 m. The Froude number is 0.40 and the depth with 20% freeboard is 3 m.

Example 2.4 Analyze a uniform flow in a 4:1 trapezoidal channel with $n = 0.015$, slope $= 0.01$, bottom width of 10 ft, with 20 cfs.

Open HEC-RAS and consider the "DesignFeatures" project. A cross section of a 4:1 trapezoid with a 10 ft bottom width has been prepared at station 101. The cross section was copied to a distance of 100 ft downstream to appear at station 1. Elevations were all adjusted (see the options under the geometric editor) by a negative 1 ft, resulting in a slope of 0.01. Under "Tools," an interpolation editor was used to create the intermediate cross sections. Now go to the Run menu and select "Hydraulic Design Functions," apply geometry, and enter the slope and discharge data. Figure 2.11a shows a set of HEC-RAS design tool menus and outputs.

(a)

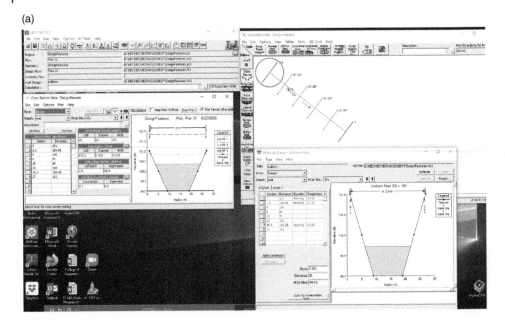

(b)

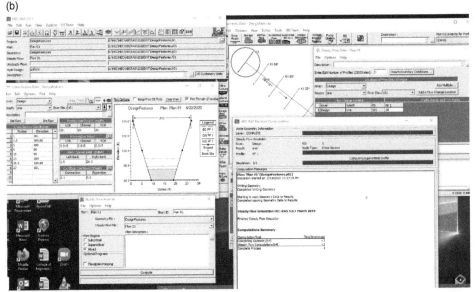

Figure 2.11 (a) Collection of HEC-RAS menus and outputs resulting from the uniform flow design tool. (b) Collection of HEC-RAS menus and outputs resulting from the uniform flow analysis. *Source:* HEC-RAS Computer program.

One can also set up the "Steady Flow" analysis in HEC-RAS. Go to "Edit Steady Flow Data" and enter the flow under the "PF1" field. Open "Reach Boundary Conditions" and set the normal depth at both boundaries since we are analyzing uniform flow. Look for the slope prompt, and enter $S = 0.01$ at the upper and lower boundaries. Go to the "Run" and select "Steady Flow Analysis." Select the "mixed" flow regime (unsure at this point about the nature of the flow) and press "Compute." Figure 2.11b shows a set of menus and outputs for the HEC-RAS steady-flow analysis. Thus far, we have a quick and superficial view of HEC-RAS. We encourage you to dig into the HEC-RAS documentation to become more comfortable with HEC-RAS channel analysis. A step-by-step procedure for using the HEC-RAS hydraulic design function to measure depth in a simple channel is given in Table 2.6.

Table 2.6 Using HEC-RAS for a simple uniform flow assessment.

1) Load HEC-RAS by clicking on the following icon:

2) Click on file – new project

3) Select directory and folder (or create a folder)

4) Name the project

5) Under Edit on the main menu, choose Geometric Data to build the channel.

6) On the Geometric Data pallet, click on River Reach to define the stream's pathway.

7) Draw the reach as a straight line by holding the left mouse as you drag to define the stream. Double click when done.

8) Name the river and reach

9) Click on the cross-section editor

10) Under options on the cross-section editor, add a new cross section

11) Enter the river station. Since the program moves downstream, give a number such as 2000 for the upstream station.

12) Now, we build a trapezoidal cross section point by point. For convenience, start at station 0 and call the elevation 100 ft. To get a $z = 4$, go 20 ft and drop the elevation by 5 ft. To have a bottom width of 15 ft, add 15–20 and hold the elevation at 95 ft. We complete the trapezoid by adding 20 ft and raising the elevation to 100 ft.

13) We want to define the end of the reach to be 2000 ft downstream. Enter 2000 in the LOB, Channel, and ROB fields (LOB is left overbank and ROB is right overbank).

14) Select an appropriate n value, say 0.02, for a lined channel, and enter the fields' values.

15) Define the point of the LOB and ROB. Since we have no flood plain here, enter the beginning station (0) and ending station (55).

16) Apply the data and close the editor. Click on the geometric data pallet and see the cross section's location on the upper end of the reach.

17) One can click the red circle and reopen the cross-section data screen.

18) On the cross-section data screen, click on options, copy current cross-section data. In the River Station field, type 0. Say OK and close the cross-section editor.

(Continued)

Table 2.6 (Continued)

19) We want to impart a slope to the channel. For a 0.1% slope on a 2000 ft channel, we need to drop the elevation by 2 ft. Open the Station zero cross section, click options, and select adjust elevations. Enter-2, then apply data.
20) We can use the interpolation tool to add cross sections to the reach. On the Geometric Data screen, select tools, XS interpolation.
21) Pick the within a reach option. Set the minimum distance between cross sections as 20 ft. Click on Interpolate XS's, close, and see the interpolated cross sections.
22) Pick one of the cross sections by clicking on one of them.
23) On the main menu, click on run and look for Hydraulic Design Functions. Click on type and select uniform Flow. We can set the slope to any value for this analysis. Choose 0.2% or 0.002. Enter a Discharge, say 500 cfs. Apply Data, then Compute.
24) Snip the Hydraulic Design screen, which shows the depth. Click on and snip the Report Screen.
25) On the main menu, click on File, Save Project. Be sure you create a directory for the project as several files must stay together.

Waterways with parabolic and triangular cross sections have straightforward iterative solutions. The parabolic cross section is not generally used for conventionally lined channels. It is often used with grass-lined waterways, which are discussed in detail in the grassed waterway chapter. For a triangular channel, one may solve for depth with known side slope z. For a circular conduit, where diameter determination is the goal, one must insert functions of hydraulic radius, wetted perimeter, and cross-sectional area vs. depth (see Table 2.1 or Figure 2.4) into Equation 2.10 to iterate for depth at a known pipe diameter. A spreadsheet or another package is helpful for these solutions in practice. Spreadsheets and Mathematica notebooks are available for each of these cases. In the software folder, under Excel, look at the LinedChannels folder. Under Mathematica, look at the folder for each cross section and select the notebook solving for y (or d). The notebooks enable users to see solution flow. Additional details are contained in the Mathematica notebooks.

For these and other channel spreadsheets, an iterative guess depth is required. The solver converges to a reasonable answer (usually small depth and broad top with), an unreasonable depth (considerable depth and small top width), or fail to converge. Lack of convergence usually associates with an unreachable permissible velocity (see Type B), often due to a low slope. The depth and other parameters are constrained in the solver to positive values. Starting with an initially small depth usually results in reasonable solutions.

Type B Manning equation and CE subject to velocity constraints – erodible waterway stability design: Type B solution approaches apply when lining stability is at issue. We call this Earthen Channel I design.

Examples include:

- Grassed waterways (manipulate hydraulic radius usually; discussed in Chapter 3);
- Naturally lined surface drainage channels;
- Channels are usually short (e.g. field-scale), and spoil material handling is generally not an issue;
- Channels flow intermittently; one designs for stability and capacity. Changing the Manning n to a higher value or using vegetated waterways, assuming unmowed vegetation, usually comprises the capacity design.

With the Manning equation, one may manipulate hydraulic radius, slope, or surface to force a design to conform to a permissible velocity or tractive force requirement. As an aside, one may ask how the permissible velocities were established. Permissible velocity values were determined using surveys to practitioners (Fortier and Scobey 1926). Table 2.3 provides permissible velocity values for various channel types. The NRCS (2007) recommends lowering the permissible velocity in curved sections. They provide reduction factors depending on the soil type and curve radius. For example, one may reduce the hydraulic radius by forcing a channel to flow broad and shallow. One may vary the slope sometimes. Restrictions on roughness (e.g. using channel linings of various sizes to change the Manning *n*) are possible; we leave the discussion of varying roughness for more advanced treatments.

This chapter focuses mainly on velocity control with the management of hydraulic radius. We leave the discussion of slope for velocity control to the vegetative waterways and terracing chapters. Set up the permissible velocity method-hydraulic radius control with the following steps:

1) Solve the Manning equation for $R = [(nv_p)/(\varphi_{dim}\ S^{1/2})]^{1.5}$. Dimensions are ft-s or m-s.
2) Solve the CE for *A* from $A = q/v_p$.
3a) For *Trapezoid*: Write *A* and wetted perimeter equations in terms of bottom width, *z*, and depth. One usually knows *z*. See Equation 2.16:

$$A = by + zy^2 \qquad (2.16c)$$

Then, solve for *d* and *b* after substituting and rearranging the relationships into a form solvable with the quadratic formula as follows (Equation 2.17):

$$R = \frac{A}{\left(b + 2y\left(z^2 + 1\right)^{0.5}\right)} \qquad (2.17)$$

Solve the relationship for *b* as a function of *y* (Equation 2.18);

$$b = \frac{\left(A - zy^2\right)}{y} \qquad (2.18)$$

Substitute back into relationship for *R* (Equation 2.19):

$$A = R\left\{\left[\left(A - zy^2\right)/y\right] + 2y\left(z^2 + 1\right)\right\} \qquad (2.19)$$

Rearranging Equation 8.21 as a quadratic equation in *y* results in Equation 2.20:

$$y^2\left[z - 2\left(z^2 + 1\right)^{0.5}\right] + y\left(A/R\right) - A = 0 \qquad (2.20)$$

Using the quadratic formula (note sign change in Equation 2.19) results in Equation 2.21:

$$y = \frac{-b_1 \pm \left(b_1^2 - 4a_1c_1\right)^{0.5}}{2a_1} \qquad (2.21)$$

where $a_1 = [2\ (Sqrt(z^2 + 1)) - z]$; $b_1 = -A/R$; $c_1 = A$. \qquad (2.22a,b,c)

Put the sensible value of y (usually smallest positive value) unless other factors such as the need to accommodate deeply buried tile drain. Use Equation 2.18 to solve for bottom width b. Appendix B contains a closed form of the solution shown in Equation 2.21.

Note: Sometimes, the trapezoid may degenerate to the case where $b = 0$. A negative under the radical in the above quadratic formula suggests a degenerating trapezoid. If so, one may fix the bottom width to zero or setting the bottom width at a bulldozer blade width (e.g. 1.2 m). *The negative radical implies that the stream only flows at a less-than-critical velocity.* Design for capacity by increasing n by 25%, then use Type A approaches. Chapter 3 (Grassed Waterways and Bioswales) continues this discussion. The available spreadsheet fails to converge to a solution in this instance. One may consider lowering the permissible velocity to some value above 0.6 m/s (1.4 ft/s). It is desirable not to go below this lower velocity threshold to avoid excessive sediment deposition (if sediment is present).

3b. For a *parabola*: One can use the approximate relation between hydraulic radius and depth (see Table 2.1) to solve for depth and then use the top width relationship. Solve for T, then add freeboard. The lower velocity threshold discussed with the trapezoid applies here as well.

3c. For a *triangle*: Using the approximate relation (see Table 2.1) between hydraulic radius and depth, solve for depth. Check the solution by solving the relationship for depth, then solve for top width, and add the freeboard. Complete the capacity design, as described in 3a. As with the trapezoid and the triangle cross sections, the lower velocity threshold applies to the triangle also.

Spreadsheets are available in the downloadables (See the "PermissibleVelocityEarthen" folder under Excel in the software) for each of the above cases. Mathematica notebooks are available and are located in each cross-section folder. Notebooks provide depth and slope (fixed bottom width) and depth and bottom width (fixed slope). Note that the Mathematica solutions are generally quite long and too complicated for routine use outside of the Mathematica notebook.

Example 2.5   Given an occasional flow of 100 cfs (2.83 m^3/s) over a 2% slope with a natural trapezoidal channel with a $z = 4$. The loamy soil permissible velocity is 2 ft/s (0.61 m/s) and $n = 0.03$. Determine the flow depth.

$$A = (100\,\text{ft}^3/\text{s})/\,(2\,\text{ft/s}) = 50\,\text{ft}^2; R = \{0.03(2\,\text{ft/s})]/[1.49(0.02)^{0.5}]\}^{1.5} = 0.15\,\text{ft}$$

Compute the depth using the quadratic equation (Equation 2.21):
$$a_1 = 4.24;\ b_1 = -333.33\,\text{ft};\ c_1 = 50\,\text{ft}^2$$

This gives $y = 0.1538\,\text{ft}, 78.46\,\text{ft}$. Reject the second solution as impractical.[4] Solving for b with the first solution using Equation 2.19 gives

$$50\,\text{ft}^2 = by + zy^2;\ 0.15\,\text{ft} = R = A/P = 50\,\text{ft}^2/(b + 2y\text{Sqrt}(z^2 + 1))$$

---

[4] Computations with open channel design can be greatly facilitated using equation processing software; however, beware of the pitfalls. Symbolic math packages produce elegant solutions of the equations in some cases (see Appendix B). Spreadsheets referenced in Appendix B can converge to extraneous values. For example, the quadratic equation can lead to two real solutions of the trapezoid yielding a permissible velocity. One will be broad and shallow, the other narrow and deep. Both have the same hydraulic radius and area. One would choose the shallow and broad channel for construction ease. In the case of the triangular channel, we get four roots, some of which are imaginary, when using the exact definition of hydraulic radius. The designer must beware!

One may solve for b, using Equation 2.18, which yields:

$b = [50\,\text{ft}^2 - 4(0.15\,\text{ft})^2]/0.15\,\text{ft} = 333\,\text{ft}\ (101.5\,\text{m})$.

The bottom width is quite large (unless this is a constructed overland flow waste treatment system!) Consider lining the channel with coarse gravel, noncolloidal, having a $v_p = 1.83\,\text{m/s}$ (6 ft/s) and an $n = 0.025$. Proceeding as above, $A = 16.66\,\text{ft}^2$ and $R = 0.6\,\text{ft}$, $a_1 = 4.24$, $b_1 = -27.77\,\text{ft}$, and $c_1 = 16.66\,\text{ft2}$ This leads to $y = 0.67\,\text{ft}$ and $b = 22.2\,\text{ft}$ before freeboard, which is more tenable.

If the expression under the radical above ($b^2 - 4ac$) were negative, the trapezoid would have been impossible. One could then set the bottom width and move onto the capacity design phase. One would design for capacity by increasing n by, say, 25% and using the bottom width determined in the stability design. Capacity design requires a Type A solution.

The capacity design $AR^{0.67} = 1.25n\ q/\varphi$ Sqrt $S = 1.25(0.03)(100\,\text{cfs})/1.486$ Sqrt$(0.02) = 17.844 = (by + zy^2)[(by + zy^2)/(b + 2y\text{Sqrt}(z^2 + 1))]^{0.67}$. Substituting $b = 22.2\,\text{ft}$ and iterating gives $y = 0.90$. Adding freeboard of 20% gives $y_{fb} = 1.08\,\text{ft}$.

Note: The selection of the appropriate permissible velocity is critical. *In cases when the radical is not negative, the computed bottom width for stability is sensitive to the chosen permissible velocity. Use a well-chosen permissible velocity.*

Example 2.6 Suppose Example 2.5 (Loamy soil lining) had specified a triangular cross section.

$R = y/2$ from Table 2.1, therefore $y = 2(0.15) = 0.3\,\text{ft}$.

$A = zy^2$, therefore $Z = 50/0.3^2 = 555$; $T = 2(0.3)\ (555) = 333.3\,\text{ft}$. Acceptable. We leave calculations for a channel lined with coarse gravel to the student. Increase n by 25% and carry the stability z value over to the capacity design.

Example 2.7 Suppose Example 2.5 (Loamy soil lining) had specified a parabolic shape.

Using relationships for a parabolic cross section results in a depth of $y = 1.5\,R$ (see Table 2.1) $= 0.225\,\text{ft}$. $T = 50\,\text{ft}^2/(0.67*0.225\,\text{ft}) = 332\,\text{ft}$. Stability calculations for a channel lined with coarse gravel are routine, as are the capacity design and freeboard calculations. We discuss the capacity design for parabolic channels in Chapter 3. The parabolic spreadsheet provides similar results.

The example problem required a wide, shallow channel to convey the flow at the specified velocity. In practice, we would look for an alternative route having a lower slope or an alternative lining.

Occasional flow channels usually accumulate debris between runoff events, increasing the resistance to flow. Bedforms, discussed in Chapter 10, also may develop and contribute to effective roughness. To compensate, one may increase the Manning n value. Increasing the Manning by 25% not to exceed $n = 0.06$ for typical field-scale problems as a first step is suggested. This suggestion applies to solution Types A and C as well. For trapezoidal channels, the capacity design would follow the stability design. The purpose of the stability design is to set the bottom width. Use the determined bottom width along with the adjusted Manning n when determining the final depth and freeboard. With triangular channels, one may find the z in the stability calculation and determine the final depth using an adjusted Manning n. Spreadsheets referenced in Appendix B solve the fundamental problem on the first of two pages. Then, a depth reflecting the increased n

value is found on the second sheet. Comments regarding initial depth selection apply here as well. Some Mathematica notebooks are included as well. Stable earthen channel design is revisited in Chapters 4 and 10.

Type C – Solve the Manning equation and continuity channel subject to the minimal wetted perimeter with or without side slope z constraint; the economic channel: The Type C solution procedure is useful when one wants to minimize excavation and lining costs. Examples include:

- lined channels requiring minimal excavation, such as long irrigation canals or other water delivery channels; and,
- long, unlined channels such as drainage channels when the slope is low (e.g. <0.003) and permissible velocity (or tractive force, discussed) not exceeded.

Often the economic channel is the one carrying the flow with a minimum cross section. For naturally lined channels, the slope is frequently small (e.g. 0.003 or less) to be stable. One can show the best choice for the trapezoidal channel bottom width to depth (b/y). Equation 2.23 gives the ratio.

$$\frac{b}{y} = 2\left(\sqrt{1+z^2} - z\right) \tag{2.23}$$

One can show that the optimum z is 0.5774 (giving a $b/y = 1.15$), which is half an equilateral hexagon. The downloadables (Appendix B) provide a Mathematica workbook, which shows a derivation of Equation 2.23. Other Mathematica workbooks develop the optimum cross-section parameters for the triangle, the curb triangle (z of one side is zero and z of the other side is about 20), rectangle, parabola, and circle. Table 2.7 summarizes the optimum parameters for standard cross sections. An Excel spreadsheet comparing all the standard cross sections with a similar slope, flow rate, and roughness demonstrates the most economic cross section based on the wetted perimeter and cross-sectional area semicircle followed closely by the optimum trapezoid. The ease of construction of trapezoidal sections compared to semicircular sections explains in part why the trapezoidal channel is ubiquitous. The analysis assumes the lining cost per unit area in the bottom section is the same as unit costs for the sides. Chapter 4 further discusses channel costs. More advanced treatments account for variation in materials cost with the position on sides or bottom (e.g. see Chapter 4), and a Mathematica notebook (OptimumChannels folder) considers variable wetted perimeter and excavation costs. One can usually achieve a stable optimum unlined channel if the slope is less than 0.002 or 0.003 in cases where $n \approx 0.02$–0.025.

One usually knows the flow, roughness, and slope in cases where velocity is not limiting. The goal here is to minimize the removed area with a constraint on z. *Again, the optimum b/y ratio with the optimum z is b/y = 1.15.* Table 2.5 provides typical z values. To solve for the design depth:

1) Get b as a function of y from Equation 2.23 using z values suggested in Table 2.5. Substitute into the hydraulic elements relations R and A in Table 2.1.
2) Write the Manning equation, substituting in the above relations. Solve iteratively for depth and back solve for the bottom width. Add freeboard.

Table 2.7 Optimum hydraulic sections.

Cross section	Notes	Area A	Wetted perimeter P	Hydraulic radius R	Top width T
Trapezoid (half of hexagon)	$z = 0.5774$ $b/y = 1.115$ most common	$\sqrt{3}y^2$	$2\sqrt{3}y$	$\dfrac{y}{2}$	$\dfrac{4\sqrt{3}}{3}y$
Suboptimum trapezoid	z is selected based on soil or user requirement $\dfrac{b}{d} = 2\left(\sqrt{1+z^2} - z\right)$	See Table 2.1	See Table 2.1	See Table 2.1	See Table 2.1
Rectangle (half of square)	$z = 0$ $b/y = 2$	$2y^2$	$4y$	$\dfrac{y}{2}$	$2y$
Triangle (half of square)	$z = 1$	y^2	$2\sqrt{2}y$	$\dfrac{\sqrt{2}}{4}y$	$2y$
Semicircle	Most economical but expensive to install	$\dfrac{\pi}{2}y^2$	πy	$\dfrac{y}{2}$	$2y$
Parabola	Infrequently used	$\dfrac{4\sqrt{2}}{3}y^2$	$\dfrac{8\sqrt{2}}{3}y$	$\dfrac{y}{2}$	$2\sqrt{2}y$

Source: Data from Chow (1959). Most are derived in Mathematica notebooks. For the rectangular sections, using the specified z and/or b/y ratios will lead to values of A, P, and T that are consistent with the general formulae for the respective cross sections.

3) If naturally lined, be sure the permissible velocity does not exceed the allowed value for the wetted surface. If artificially bound, check the permissible velocity if applicable, and check the Froude number.

Example 2.8 Given a flow rate of 40 cfs (1.14 m³/s), $n = 0.025$, trapezoid cross section, slope = 0.25%, silty clay ($z = 1$ if depth less than 1.2 m or $z = 1.5$ otherwise); assume d is less than 1.2 m, so $z = 1$ initially.

$b = 2y$ [Sqrt(1 + 1²)–1] = 0.84y (Equation 2.19)

$R = (0.84y^2 + (1)y^2)/(0.84y + 2y\text{Sqrt}(1 + 1^2)) = 0.501y$ (Tables 2.1 and 2.6 and Equation 2.23)

$A = ((0.84y + (1)y)\,y = 1.84y^2$ (Tables 2.1 and 2.6 and Equation 2.18)

$q = (1/0.025)(0.501y)^{0.67}(1.84y^2)(0.0025)^{0.5} = 1.14$ m³/s (Manning with R and A)

Solving for y gives $y = 0.77$ m. Note that when using the optimum z value and the b/y optimum ratio, the area, wetted perimeter, and R values give values consistent with Table 2.7. Checking z; $z = 1$ OK since y is less than 1.2 m. If y were higher than 1.2 m, one would select the new z and resolve for y.

Velocity check: $v = (1/0.025)(0.501*0.77\,\text{m})^{0.67}\text{Sqrt}(0.0025) = 1.1$ m/s, less than permissible velocity for bare soil. If the velocity had been above, say 2 m/s, then one would (i) consider stabilizing lining or (ii) redesign the channel based on permissible velocity. Economic factors would determine the final design.

For the lined optimum triangle, use $z = 1$ in the hydraulic elements. Do not use the b/y relationship when the cross section is the triangle. For the optimum lined rectangle, $b = 2*y$. The optimum circular cross section is defined by $y = D/2$. In cases where the optimum channel carries occasional flows, increase the Manning n by 25% before starting the calculation as with the Type A solution.

How does one determine what solution type is applicable? Channels with slopes greater than 0.003 usually require the Type B approach. One can generally design channels with slopes less than 0.002 can using Type C solutions because they will not usually exceed the critical velocity. One should check the economics of channel cross sections designed with lining (Type A) and without lining (Type B and Type C) when the slope lies between 0.002 and 0.003. For example, the optimum channel serving as a long-term drainage ditch may be wet enough to justify the inclusion of biological habitat aspects into the mix of client desires and esthetic factors. The spreadsheets provided facilitate such comparisons. Mathematica notebooks may be found under the Optimums folder that derives optimum parameters for standard cross sections.

Flow losses: Earthen or vegetation lined channels often lose flow due to infiltration. Losses can be substantial in long channels. Earthen lined channels to convey drinking water or irrigation water are often found in arid areas. In humid areas, losses are frequently ignored with ephemeral drainage channels. Table 2.8 summarizes seepage loss for channels not affected by underlying groundwater tables. Soils more course than gravelly clay loam generally requires a lining material unless the primary objective is drainage.

Riprap (revetment) design (Earthen I continued): Earthen channels often require some type of lining material to keep velocities limited without requiring excessive width. One

Table 2.8 Seepage losses for channels not affected by groundwater tables.

Perimeter material	Seepage loss (L^3 water lost/L^2 perimeter/day)	
	(ft/day)	(m/day)
Impervious clay loam	0.25–0.35	0.076–0.107
Medium clay loam underlain with hardpan at depth within 2–3 ft below channel	0.35–0.5	0.107–0.152
Ordinary clay loam, sandy clay loam, or lava ash loam	0.5–0.75	0.152–0.228
Gravelly clay loam or sandy clay loam, cemented gravel, sand, and clay	0.75–1.0	0.228–0.305
Sandy loam	1.0–1.5	0.305–0.457
Loose sandy soils	1.5–1.75	0.457–0.533
Gravelly sandy soils	2.0–2.5	0.609–0.762
Porous gravelly soils	2.5–3.0	0.762–0.915
Very gravelly soils	3.0–6.0	0.915–1.83

Source: Based on French (1985), citing work by Davis and Sorenson (1969).

Figure 2.12   Riprap channel under construction. *Source:* Courtesy of the USDA.

may choose a rock lining or a vegetative lining. We focus here on rock linings and defer the discussion of vegetative linings to a later chapter. Figure 2.12 shows a riprap channel under construction, showing riprap installed over a layer of geosynthetic liner. One strategy is to select a trial rock diameter, then use a relationship such as Equation 2.10a–d to estimate the roughness, and then check the velocity. If the resulting channel velocity lies between 0.6 m/s (1.4 ft/s) and a maximum velocity set by the rock diameter given by an NRCS empirical relationship (see Haan et al. 1994), shown in Equation 2.24.

$$V_{\lim} = 12.84 * d_{50}^{0.51}$$

(2.24)

where d_{50} is rock diameter in ft (divide by 305 to convert mm to ft).

Spreadsheets are available in the downloads for the trapezoidal, parabolic, and triangular cross-section analyses. Having a velocity higher than 0.6 m/s limits the deposition of sediments. One can evaluate the capacity if the roughness increases by 25% due to debris accumulation for a final design if desired. Riprap should be placed on the channel bottom only unless the channel's z exceeds four to keep gravitational forces from causing slumping to the channel bottom. Riprap should be very angular as opposed to round for stability.

The Federal Highway Administration (FHA) and the NRCS have published empirical approaches for designing riprap-lined channels. These techniques, explained by Haan et al. (1994), are contained on a spreadsheet available in the downloads. Riprap channels are further discussed in Chapters 4 and 10.

Example 2.9   Given a requirement to convey a flow of 55 cfs down a slope of 7%, a pro-posed aggregate size $d_{50} = 50$ mm, and a desire to cross with equipment occasionally (set z to 4), find a bottom width and depth that conducts the flow safely using the Riprap design spreadsheet. Entering the data into RIPRAPTrapezoidalWaterwayDesignEarthenPermVel spreadsheet and solving, we find bottom width $b = 55.9$ ft and depth (stability) $= 0.19$ ft. If we increase n by 25% to account for debris accumulation, the depth increases to 0.21 ft. Compare these results using the FHA procedure on the FHA_NRCSRipRap sheet. Here, one must enter the bottom width and solve for d_{50}. Using the FHA solver, we find the pre-dicted d_{50} under the above bottom width results in a $d_{50} = 0.188$ ft or 57 mm. Using the NRCS procedure, we find that d_{50} is 0.12 ft or 37 mm. The more conservative larger size is recommended. NRCS (2007) suggests that the permissible velocity design method dis-cussed below be used when the channel boundary material is sand or smaller. Table 2.3 gives permissible velocity recommendations for a variety of materials. The method is useful for design checks. The NRCS recommends that the tractive force method, discussed in an upcoming chapter, be used when the boundary materials are larger than sand.

Channel curvature-subcritical flows: Much goes into installing a channel. Channels usu-ally have maximum curvature to contain the flow adequately. Table 2.9 provides guidance. Flows around curves create centripetal forces that result in a differential elevation in the bend. Jain (2001) recommends Equation 2.25 for predicting the elevation difference. The differential elevation should be added to the freeboard depth in the curve. Note that Equation 2.25 presumes subcritical flow. Supercritical flow requires more advanced treatment.

$$\Delta\text{depth} = \frac{V^2 T}{gR} \qquad\qquad (2.25)$$

In Equation 2.25, V is the mean velocity, T is the top width, g is gravity, and R is the radius of curvature. A stream flowing at 4 ft/s and having $T = 20$ ft and $R = 100$ ft experiences a differential elevation of about 0.1 ft. Figure 2.13 shows secondary currents induced by flow

Table 2.9   Maximum curvatures for curves as a function of slope and channel width.

Channel top width (m)	Slope (%)	Approx. maximum curvature (°)[a]	
		Deg (English)	Deg (Metric)
4–6	<0.06	19	62
4–6	0.06–0.12	14	46
5–11	<0.06	11	36
5–11	0.06–0.12	10	33
11	<0.06	10	33
11	0.06–0.12	7	23

[a] Degree of curvature in SI units is 3.28 times the curvature degree in English units.
Source: From USDA-NRCS (1973).

around a curve. The secondary currents affect the velocity profile in the curve. In situations where the freeboard factor is 1.2, the differential elevation is often neglected. In unlined channels, the curve induces currents that can change the channel geometry, further discussed in Chapter 10. NRCS (2007) provides design information related to curves in unlined channels.

Channel Installation

The NRCS (2007) provides excellent practical information related to earthen channel installation. Harris (1994) offers excellent advice on machine selection for channel construction. Figure 2.13 shows how channels can be laid out on the ground. Other surveying details, such as setting the slope stakes, are not shown but may be found in any surveying text. Sustainable channels would not exceed these maximums. Equipment for constructing channels varies widely, ranging from farm tillage machines to commercial equipment. Table 2.10 provides additional information (Figure 2.14).

When the engineer accepts a client's request to design a channel, one needs to understand the mindset. Prompts in Table 2.11 assist in teasing out the physical perspective. The discussion often helps the client better understand the problem they are posing. Complimentary to Table 2.11 is Table 2.12, which is a tool that facilitates the mapping of the client mindset to engineering factors that the design must accommodate.

Summary

This chapter has covered the design of simple channels under conditions of synthetic linings and earthen linings where velocities are limited. We have learned the design with optimal cross sections. Optimal cross sections are frequently possible in earthen channels

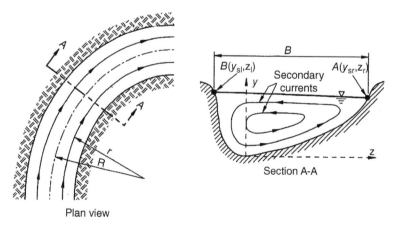

Figure 2.13 Portrayal of secondary currents induced in subcritical flow around a curve. *Source:* From Jain (2001); used with permission of John Wiley & Sons.

Table 2.10 Construction machinery for channels.

Function	Action		
	Continuous – excavation or transport; used on long channels.		Intermittent – alternate excavation and transport steps; used on short drainage channels or diversions.
Excavation			
	Wheel excavator – deep, narrow; used for subsurface drainage installation		Dragline and Clamshell
	Plow-type ditcher – small shallow such as terraces		These operate from a crane-type machine
	Excavating conveyors – defined cross section; they convey spoil material to waiting for transport or to the channel side; used for large irrigation or drainage canals		Backhoe – a general-purpose machine; a wide variety of hoe attachments; pivoting hoes excellent for finish work
	General-purpose motor blade grader – small, shallow channels		Crawler shovel – a wide variety of shovel attachments; pivoting hoes also available.
	Elevating grader – general; tillage for loosening soil followed by elevating conveyor belt		Motor scraper – general use; limited excavation except for shallow channels.
	Hydraulic dredge – removal of submerged sediments		Bulldozer – general-purpose; wide variety of blade attachments and rear spades available
	Rotary ditcher – deep, narrow channels		Dozer w/pullback blade – good for finish
			Motor scraper – reasonably good for excavation when fitted with spade attachment
			Crawler/wheel loaders – excavation and transport
Spoil spreading			
	Blade grader – general transport		Bulldozer – reasonably good for transport
	Pans – excavation and transport		Hauler – transport
	Tillage machines – finishing		Motor scraper – reasonably good for transport
	Terracing machines – specialized for terrace channels		Pull back blade – short distance transport
	Sheepfoot roller/wheel compactor – used for compaction of clayey/sandy soils.		

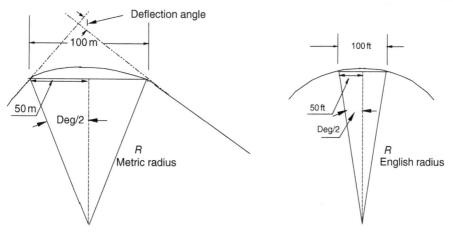

Figure 2.14 Schematic showing how the degree of curvature is defined for English and Metric units, where *I* is the deflection angle and Deg/2 is half the degree of curvature.

Table 2.11 Customer requirements prompt sheet for application to open channel design problems.

Customer requirements
Protect area
Water quality objectives
Incoming water quality
Drain upland area
Minimal hydraulic impact on lower areas?
Clearwater need or silts allowable?
Water conveyance efficiency requirement?
Land use in the watershed
Aesthetics
Livestock or other wildlife in the area?
Ecological impacts?
Safety requirements
Sensitivity to blasting and other noise
Is water to be sold to customers along the channel route?
Water value
Available $$ for excavation and spoil storage/transport
Flood plain impacts
Cost of land in the area
Relation to other watershed BMP's
The larger hydraulic picture of the region
State/Federal legislation/agency rules
Is it essential to obtain an optimal cross section?
Availability of maps, soil surveys, and aerial photos of the watershed and potential channel routes.
Channel capacity or multiple channel capacities
Cross-section requirements

Table 2.12 Engineering factors prompt sheet for application to open channel design problems.

Engineering factors
Grade direction
Life?
Curves allowed?
Ecological design criteria to be satisfied?
Roughness
Leakage allowed?
Water treatment within?
Sediment load?
Channel capacity at relevant design storms
Soil stability
Liner strength/endurance?
Materials specifications
Special velocity requirements?
Energy extraction or dissipation methods
Side slope limits?
Bottom width limits?
Soil conditions
Spoil storage/placement
Surrounding land use
Baseflow vs. storm of other flow
Temporary measures needed during construction
Flow measurements needed?
Accuracy?
Interfaces to other channels/structures/drainage (surface or underground)

if the slope is less than 0.003 or so. We have learned how to assess flows in natural channels. As stated at the outset, these methods provide the foundation for approaches that are generally more sustainable. The following chapters explore vegetated waterways, which represent a situation where the Manning n is variable. A design alternative for earthen channels and vegetated channels, the tractive force method, is covered in a later chapter. Additional discussion of the parabolic channel cross section is given in subsequent chapters. A further chapter covers alluvial channels, which transport much sediment. The NRCS (2007) provides a succinct design summary for various earthen channels that guided this discussion.

Maintenance is an essential aspect of channels functioning over time, be the channel lined or unlined. Lined channels must be kept free of debris. Permissible velocity was initially thought to be a rigid no-erode-erode threshold. The rigid threshold is, in reality, not the case. Tractive force and contemporary alluvial channel design methods discussed in subsequent chapters revisit the reduction of maintenance costs to maintain unlined channels. As with lined channels, unlined channels must be kept free of debris. Damage caused by excessive rainfalls or improper equipment necessitates maintenance.

Problems and Questions

Note: Draw sketches of all completed designs and provide supporting reasons for discussion questions. Show all work (or excel sheet excerpts) and offer a rationale for assumptions made. Remember to use appropriate significant digits. Feel free to use the resources of Appendices A, B, and C. Use a 20% freeboard unless otherwise indicated.

1. Use Cowan's method to estimate an n value for a stream reach selected by your instructor. Include values for all the Cowan factors in your submission. Form groups of two, with each one submitting a solution. Also, by referring to the USGS pictures, estimate an n for the stream.

2. What is the depth of flow in a rough ($n = 0.02$) concrete lined trapezoidal channel, $z = 3$, slope $= 0.002$, bottom width $= 1$ m while the flow is $2\,m^3/s$? Check the Froude No. What would be the depth and Froude No. if the lining were resurfaced with a smooth concrete surface ($n = 0.011$)? What surface would you recommend based on the Froude No.? Show your work leading to this recommendation.

3. Design a trapezoidal channel with $z = 4$ with a noncolloidal alluvial silt lining to move a 25 cfs flow down a 2% slope. HINTS: Type B for stability, then increment n by 20% (this simulates effects of aging and debris accumulation) for capacity using Type A approach; add freeboard.

4. Compute the most efficient bottom width for an open, effective trapezoidal channel with a flow rate of $2.5\,m^3/s$ in a silt loam soil. The slope is 0.09%. HINT: earth lining, low slope, Type C solution. Compute the depth if n increases to 0.035 due to age or debris accumulation. Assume slope, flow, and channel width remain the same. HINT: Now use the Type A approach.

5. (i) What is the design capacity of a steady flow irrigation canal with a total depth of 1.5 m, including a freeboard? The channel has a bottom width of 1.2 m, 2:1 side slopes, and a hydraulic gradient of 0.09%. Assume the freeboard is 20% of the depth and $n = 0.04$. (ii) Repeat for the case where the roughness degrades to $n = 0.045$ over time.

6. How deep is the $15\,m^3/s$, steady flow in a trapezoidal channel, $z = 4$? The canal is lined with smooth plastic, has a gradient of 1%, a freeboard factor of 20%, and a bottom width of 2 m. Is the resulting flow subcritical, critical, or supercritical? What if the roughness factor increases by 25% (This usually happens with age)?

7. You are given a trapezoidal, concrete-lined irrigation ditch on a 0.5% slope. The ditch capacity is to be a steady $6\,m^3/s$. Minimize the cross-sectional area to minimize the

concrete required as well as the excavation costs. What is the resulting depth of the trapezoidal channel, including a 20% freeboard? Compute the Froude number and interpret the result.

8 There is a need to move a 55 cfs occasional flow down a 7% slope in a silty clay loam soil. Since land is available for a potentially wide channel, it was decided not to pay the lining cost and use a natural waterway. Design a trapezoidal channel, which must be crossed, handling this flow rate for stability. HINT: This is a Type B problem. Next, since the flow is occasional, design the channel for capacity by increasing the n value used in the stability analysis by 25%. HINT: Capacity is a Type A problem, using the bottom width determined in the stability analysis.

9 A high watermark in a natural unlined parabolic channel in a silt loam soil with a fine sandy bottom suggests a 1 m deep peak flow for a recent runoff event. The top width at the high water mark was 2.75 m. The channel slope was found by surveying to be 2.5%. Estimate the flow. Was scouring (erosion) likely under the peak flow condition?

10 A concrete-lined irrigation ditch is being built on a 0.5% slope. The ditch capacity is to be 6 m^3/s. Use a freeboard of 10%. Minimize the cross-sectional area to minimize the concrete required as well as the excavation costs. Use (i) triangular channel, (ii) rectangular channel, (iii) trapezoidal channel, and (iv) circular channel cross sections. Assume concrete lining in all cases. Which of the optimal cross sections has the smallest wetted perimeter in the freeboard condition (except for the circular cross section)? Which of the above has the smallest cross-sectional area?

11 Design the riprap d_{50} needed for a drainage channel at a 10% slope that conveys 115 ft^3/s. The channel is rectangular and 18 ft wide. Use the FHA and the NRCS methods. Using the Trapezoidal spreadsheet for riprap design and the respective d_{50} values above, find the depth and bottom width.

12 Compute the most efficient bottom width for an open trapezoidal channel with a flow 2.5 m deep in a silt loam soil. What are the velocity and channel flow rate if the gradient is 0.09% and $n = 0.035$? How does the computed velocity compare with the permissible velocity? Make a recommendation regarding acceptance of the optimum or next steps.

13 Given the natural channel shown in Figure 2.15, determine the total flow under the indicated conditions.

14 A rectangular box culvert, 8 ft wide and 6 ft high, is designed for aquatic organism passage. The design flow is 200 ft^3/s. The structure is to be filled to a depth of 6″ with a 1-in. diameter aggregate. Assume the roughness of the aggregate controls the roughness. The 100 ft long structure lies on a 0.1% slope. As with the circular cross section discussed in class, assume the depth of flow plus the aggregate layer's thickness should not exceed 0.8 times the structure height. Will the structure handle the flow without exceeding 0.8 times the total height? If not, what recommendation would you suggest?

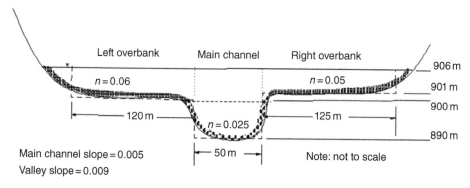

Left overbank   Main channel   Right overbank

$n=0.06$ $n=0.05$

906 m
901 m
900 m

120 m $n=0.025$ 125 m

890 m

Main channel slope = 0.005   50 m   Note: not to scale
Valley slope = 0.009

Figure 2.15 Natural channel cross section.

15 Project: A $10\,\text{m}^3/\text{s}$ flow is conveyed down a 1000 m direct distance with a fall of 1 m. Flow events are occasional. An alternate path of 12 000 m is available (e.g. a longer way around). The soil is shale. If the channel runs along the longer route, assume it must be crossed and built with a small dozer. Further, if riprap is used, one uses a side slope no steeper than z = 4. Crossing is not necessary along the steeper route. A dozer is available for constructing the short channel also. Excavation costs are $200 per m^3 soil and an additional $10.50 per linear m per unit wetted perimeter of concrete lining or $1.25 per linear foot per unit wetted perimeter for a rock/cobble lining. A grassed lining of Fescue has a cost of $1.75 per square meter. Identify a feasible and most economical channel design.

16 The area (A) and wetted perimeter (P) of a circular section are as follows:

$$A = 0.125\big(\beta - \sin(\beta)\big)D^2 \tag{P8.15a}$$

$$P = 0.5\beta D \tag{P8.15b}$$

where angle β is in radians and D is the diameter.
Using trigonometry, relate depth Y to hydraulic radius R. See Figure 2.16.

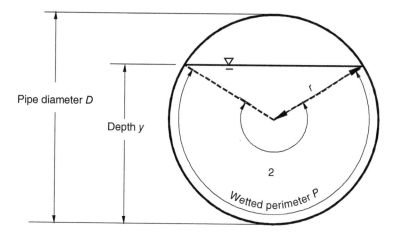

Pipe diameter D

Depth y

r

2

Wetted perimeter P

Figure 2.16 A definition sketch showing the key nomenclature for a circular channel.

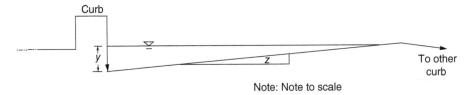

Figure 2.17 A schematic of a roadside curb.

17 Derive the hydraulic elements (A, WP, R) for a roadside curb, shown in Figure 2.17. Express R, A, and WP in terms of y and z.

18 Determine the minimum clean pipe diameter needed to convey a flow of 100 cfs over a slope of 0.01, a roughness of 0.024.

19 Determine the size of a circular conduit that contains a layer of sediment over the bottom of the pipe equal to a depth of 6 in. The flow is 100 cfs. Use a Manning $n = 0.03$ and a slope of .01. HINT: There is a Mathematica notebook in the CircularChannels folder named PartiallySubMCircCHelements that solves this problem, where one inserts a trial diameter. The depth of sediment plus water should not exceed 0.8 D. The program computes the Ased, WPsed, and Tsed, Atot, WPtot, and Ttot of the sediment first, then deducts As from At, deducts Wps from Wptot and adds Tsed to Wp to obtain the elements of the flow section. The actual hydraulic radius is then put into the Manning equation and solved for depth. Start with a large diameter, and gradually reduce diameter until the sediment plus flow depth is near the 0.8 D limit.

20 Project: A 100 m³/s flow is conveyed down a 1000 m direct distance with a fall of 100 m. Flow events are occasional. An alternate path of 10000 m is available (e.g. a longer way around). The soil is shale. If the channel runs along the longer route, assume it must be occasionally crossed, and built with a small dozer. Further, assume that if the riprap, if used, the side slope no steeper than $z = 6$ is used. Crossing is not necessary along the steeper route. A dozer is available for constructing the short channel also. Excavation costs are $200 per m³ soil and an additional $10.50 per linear m per unit wetted perimeter of concrete lining or $1.25 per linear foot per unit wetted perimeter for a rock/cobble lining. Identify a feasible and most economical channel design.

21 Use the HEC-RAS uniform flow design tool to compute the depth of flow in problem 5.

22 Compute the superelevation in a channel with a top width of 15 ft, a velocity of 4.5 ft/s, and a radius of curvature of 30 ft. Assume the flow is subcritical.

23 What additional factors would you add to the customer requirements or engineering requirements?

References

ASABE. (1986). Installation of flexible membrane linings. EP340.2 MAR86. American Society of Agricultural and Biological Engineers, St. Joseph, MI.

ASABE. (1998). Concrete slip-form canal linings. S289.2 FEB98. American Society of Agricultural and Biological Engineers, St. Joseph, MI.

Barnes, H.A. (1967). Roughness characteristics of natural channels. USGS Water Supply Paper No. 1849. United States Geological Survey, Department of Interior, Washington, DC.

Biswas, A.K. (1970). *History of Hydrology*. Amsterdam: North-Holland Publishing Company.

Blench, T. (1952). Regime theory for self-formed sediment bearing channels. *Transactions of the American Society of Civil Engineers* 117: 383–400.

Brunner, G.W. (2016). *HEC-RAS River Analysis System: User's Manual*. Davis, CA: US Army Corps of Engineers, Hydrologic Engineering Center.

Chin, D.A. (2013). *Water Resources Engineering*, 3e. Upper Saddle River, NJ: Prentice-Hall Publishing Co.

Chow, V.T. (1959). *Open Channel Hydraulics*. New York: McGraw-Hill.

Cowan, W.L. (1956). Estimating hydraulic roughness coefficients. *Agricultural Engineering* 37 (7): 473–475.

Davis, C. and Sorenson, K.E. (1969). *Handbook of Applied Hydraulics*, 3e. New York, NY: McGraw-Hill Book Company.

Fortier, S. and Scobey, F.C. (1926). Permissible canal velocities. *Transactions of the American Society of Civil Engineers* 89: 940–956.

French, R.H. (1985). *Open Channel Hydraulics*. New York, NY: McGraw-Hill Book Company.

French, R.H. (2007). *Open Channel Hydraulics*, 2nde. Highlands Ranch, CO: Water Resources Publications.

Goldman, S.J., Jackson, K., and Bursztynsky, T.A. (1986). *Erosion and Sediment Control Handbook*. New York, NY: McGraw-Hill Book Company.

Haan, C.T., Barfield, B.J., and Hayes, J.C. (1994). *Hydrology and Sedimentology of Small Catchments*. New York, NY: Academic Press.

Harris, F. (1994). *Modern Construction & Ground Engineering Equipment & Methods*, 2e. White Plains, NY: Longman Publishing Group.

Henderson, F.M. (1966). *Open Channel Flow*. New York, NY: Macmillan Publishing Co.

Huffman, R.L., Fangmier, D.D., Elliot, W.J., and Workman, S.R. (2013). *Soil and Water Conservation Engineering*, 7e. St. Joseph, MI: American Society of Agricultural and Biological Engineers.

Jain, S.C. (2001). *Open-Channel Flow*. New York, NY: Wiley.

Jobson, H.E. and Froehlich, D.C. (1988). Basic hydraulic principles of open-channel hydraulics. USGS Open-File Report 88-707, Reston, VA.

Kazemipour, A.K. and Apelt, C.J. (1979). Shape effects on resistance to uniform flow in open channels. *J. Hyd. Res.* 17 (2): 129–147.

Kazemipour, A.K. and Apelt, C.J. (1982). New data on shape effects in smooth rectangular channels. *J. Hydr. Res* 20 (3): 225–233.

King, H.W. and Brater, E.F. (1963). *Handbook of Hydraulics*. New York, NY: McGraw-Hill Book Company.

Lane, E.W. (1955). Design of stable channels. *Transactions of ASCE* 120: 1234–1260.

NRCS. (2007). Threshold channel design, Chapter 8 of Part 654. In: *Stream Restoration Design of the National Engineering Handbook*. Washington, DC: USDA National Engineering Handbook. The Stream Restoration Handbook. https://www.nrcs.usda.gov/wps/portal/nrcs/detail/national/water/manage/restoration/?cid=stelprdb1044707 (accessed March 2019).

Powell, R.W. (1950). Resistance to flow in rough channels. *Transactions of the American Geophysical Union* 31 (4): 575–582.

Rosgen, D. (1996). *Applied River Morphology*, 2e. Pagosa Springs, CO: Wildland Hydrology Press.

Simons, D.B. and Senturk, F. (1992). *Sediment Transport Technology: Water and Sediment Dynamics*. Littleton, CO: Water Resources Publications.

Sturm, T.W. (2010). *Open Channel Hydraulics*, 2e. New York, NY: McGraw-Hill Book Company.

Tollner, E.W. (2016). *Engineering Hydrology for Natural Resources Engineering*, 2e. Cambridge, UK: Wiley.

USDA-NRCS (1973). *Drainage of Agricultural Lands*. Port Washington, NY: Water Information Center.

3

Vegetated Waterways and Bioswales*

Vegetated waterways, erosion, and diversion terraces and, lately, bioswales are open channels with intermittent flows. The physics involved is similar. Between 1996 and 1997, the US land area protected by grassed waterways increased 20 700 acres for a total of 1 665 500 acres (NRCS 1998). Grassed waterways and terraces find extensive applications worldwide, although design methods are somewhat different (Hudson 1971). The bioswale and vegetated filter strip are becoming attractive grassed waterway applications in urban settings. NPDES permits mandate water quality controls in urban environments, and citizens expect esthetic low impact development (LID). Bioswale approaches represent extensions of vegetated waterways' success in rural areas. The design of the bioswales cross-section follows that of the channel, with some simplifications and additions. As with other open channels (Chapter 2), we focus on design methods commonly used in the United States. Vegetated channels represent a channel with a natural lining. Vegetation addition is a classical approach for increasing the sustainability economic desirability of earthen channels.

Goals

To identify situations where grassed waterways and bioswales may be most appropriate;
  To design vegetated waterways with trapezoidal, triangular, and parabolic cross-sections for specified peak runoff;
  To create bioswales for use in urban and suburban areas; and,
  To discuss channel maintenance and installation strategies.

Background

Vegetated waterways find primary usage in rural drainage situations. Applications include the renovation of gullies, conveying water from field terrace systems, carrying water from urban areas such as parks, and the like. Vegetated waterways are "ecologically friendly" (see Goldman et al. 1986). Figure 3.1 shows an application of vegetated waterways in agricultural

---

*Draws substantially from Tollner (2016), chapter 9.

Open Channel Design: Fundamentals and Applications, First Edition. Ernest W. Tollner.
© 2022 John Wiley & Sons Ltd. Published 2022 by John Wiley & Sons Ltd.
Companion website: www.wiley.com/go/tollner/openchanneldesign

Figure 3.1 A vegetated waterway in an agricultural field. Vegetated waterways protect naturally occurring draws and serve as discharges for surface outlet terraces. *Source:* Courtesy of NRCS.

settings. Grassed waterways with steeper side slopes, usually called diversions, divert surface runoff away from areas below the diversion. Vegetated waterways are attractive choices where land is costly or where flows are ephemeral. One might specify a small lined channel bottom if a permanent base flow prohibited the vegetation of choice. For a bioswale, one might use water-loving plants. A grassed waterway is a natural or constructed channel lined with erosion-resistant grasses. Grassed waterways are usually broad and shallow. A grassed waterway is typically 10–35 m wide and flows at depths less than 1 m. The usual storm recurrence interval is 10 years. Climate, topography and flow frequency influence vegetation.

Steep slopes require impractically wide channels, taking much land from agricultural or urban uses. The advantages of grassed waterways include potential forage production in the waterway. They serve as an esthetic enhancement in urban landscape design. Though more expensive than a bare earth channel, grassed waterways are less costly than artificially lined channels per unit length.

Channel Planning

In many respects, vegetative waterway design is like channel planning for any channel plus vegetation selection. Specific planning steps include the following.

Runoff determination: The design storm depends on the upstream and downstream issues. Traditionally, the 10-year return period with storm duration equal to the catchment time of concentration defines the design storm. The design storm may vary with the site (e.g., urban or rural) and the agency involved.

Channel slope: Determine the channel slope from the local topography. Select the gradient, which is based on topography and desired channel routing. Slopes exceeding 5% frequently lead to large widths and shallow depths. Grass vegetation selection is critical when the slope exceeds 5%. Sites may afford the possibility of taking a circuitous route with a more economical lining rather than a steeper, direct route, possibly requiring a more expensive lining.

Channel cross section: The cross section of the grassed waterway is usually trapezoidal, triangular, or parabolic. The chosen cross section often depends on available construction equipment and client's desires. Trapezoidal and triangular channels are standard. One must specify a 4–1 slope or flatter for farm or small construction equipment crossing. Trapezoid bottom width should equal or exceed the blade width (e.g., 1.2 m) of available bulldozers. Triangular channels are often constructed using graders. The parabolic shape applies to a natural channel or a designed channel with more natural esthetics. Triangular and trapezoidal channels tend to morph into the parabolic shape over time when maintenance is lacking. Parabolic channels can be difficult to cross with farm and small construction equipment.

Vegetation selection: The selection of sustainable vegetation depends on climate and soil conditions and tolerance to wet soil. Tables 3.1 and 3.2 tabulate the characteristics of commonly used plants. Vegetation may be either unmowed or mowed during a runoff event. Retardance classes A, (very tall grass) B, C, D, or E (close-mowed or burned grass/stubble) represent vegetation roughness states. We usually work with B (unmowed condition) or D (mowed condition). Local experience may dictate other retardance class selections. One must design for stability (mowed condition) and capacity (unmowed state). One cannot anticipate storm timing. NRCS (1966) provides graphical solutions to the Manning equation for the various retardance classes. The vegetated waterway spreadsheets (permissible velocity) represent the graphical solutions via formulas shown in the spreadsheets.

Design velocity for stability: The design velocity depends on soil erosivity, vegetation type, vegetation condition, and channel slope. Table 3.3 tabulates permissible velocity values. These data came from experimental studies. Gwinn and Ree (1980) and references cited within give additional details concerning design velocity determination.

Trapezoidal and triangular side slopes: Design for crossing with equipment requires a $z = 4$ or more. If a crossing is unnecessary, a parabolic channel or lower z values in the trapezoid or triangle are acceptable. The minimum z for the soil (refer to Table 2.5) must be less than or equal to the selected z. Many agencies have in-house guidelines for the choice of z.

Table 3.1 Vegetation selection guide.

Geographical region of the United States	Suggested vegetation
Northeastern	Ky bluegrass, redtop, tall fescue, white clover, birdsfoot trefoil
Southeastern	Ky bluegrass, tall fescue, Bermuda, reed canary, centipede
Western gulf	Brome, bluestem, native grass mixture, tall fescue
Southwestern	Wheatgrass, western and tall wheatgrass, smooth brome
Northern plains	Smooth brome, wheatgrass, redtop, switchgrass, native bluestem mixture

Source: Adopted following NRCS (1966).

Table 3.2 Vegetal cover classification according to retardance.

Retardance	Cover	Condition	Height (m)
A	Reed canary	Excellent stand, tall	0.9
A	Yellow bluestem		
B	Smooth brome	Good stand, not mowed	0.3–0.4
B	Bermuda	Good stand, tall	0.3
B	Native grass mix	Good stand, unmowed	
B	Tall fescue	Good stand, unmowed	0.5
B	*Lespedeza serica*	Good stand, not woody, tall	0.5
B	grass-legume mixture	Good stand, unmowed	0.5
B	Reed canary	Good stand mowed	0.3–0.4
B	Tall fescue/birdsfoot trefoil or Ladino clover	Good stand, unmowed	0.5
B	Blue gramma	Good stand, unmowed	0.3
B	Other class B mixtures	Good stand, unmowed	0.3
C	Bahia	Good stand, unmowed	0.2
C	Bermuda	Good stand, mowed	0.15
C	Redtop	Good stand, headed	0.4–0.5
C	Grass-summer legume mix	Good stand, unmowed	0.2
C	Centipede grass	Very dense cover	0.15
C	Ky Bluegrass	Good stand, headed	0.3–0.5
D	Smooth brome	Good stand, mowed	0.1–0.2
D	Bermuda	Good stand, mowed	0.1–0.2
D	Red fescue	Good stand, headed	0.3–0.5
D	Buffalo grass	Good stand, unmowed	0.1–0.2
D	Grass-legume fall mixture	Good stand, unmowed	0.2
D	*Lespedeza sericea*	After mowing	0.1
D	Most Class B grasses after mowing	After mowing	0.1
E	Bermuda grass	Good stand, closely mowed	0.05
E	Burned stubble	Scorched earth	–
E	Poor stand of vegetation	Spotty – much bare soil	–
E	Temporary covers for channels used in erosion control during construction	May often be spotty with much bare soil	–

Source: US Soil Conservation Service (1979).

Basic Design Procedures

Equations 2.15 through 2.20 guide the design process for trapezoidal vegetated waterways. The substitution of appropriate hydraulic elements for the triangle or parabola is to be understood. The Manning n is variable, with vegetation height, as shown in Figure 3.2. Vegetated channels

Table 3.3 Permissible velocities v_p for vegetated channels.

		Permissible velocity by slope and soil (m/s)[a]	
Grouped cover grasses	Slope range[b]	Erosion resistant soils[c]	Easily eroded soils[c]
	0–5	2.4	1.8
Bermuda grass	5–10	2.1	1.5
	>10	1.8	1.2
Bahia			
Buffalo grass			
Ky bluegrass	0–5	2.1	1.5
Smooth brome	5–10	1.8	1.2
Blue gramma	>10	1.5	0.9
Tall fescue			
Grass mixtures	0–5	1.5	1.2
Reed canary	5–10	1.2	0.9
Lespedeza sericea			
Weeping lovegrass			
Yellow bluestem			
Redtop	0–5[d]	1	0.8
Alfalfa			
Red fescue or mowed fescue			
Common lespedeza	0–5[e]	1	0.8
Sudangrass	0–5[e]	1	0.8

[a] Use velocities exceeding 1.5 m/s only where good covers and proper maintenance is assured.
[b] Do not use on slopes steeper than 10% except for vegetated side slopes in combination with a stone, concrete, or highly resistant vegetative center section.
[c] Easily eroded soils are those satisfying one or more of the following: (i) USLE K values exceeding 0.2 (English) or 0.0258 (Metric). Soils with very low organic matter or (ii) severely disturbed soils should be considered as easily erodible. Warner (1998) makes these recommendations for use in SEDCAD®.
[d] See footnote a, limiting the slope to 5%.
[e] Use on mild slopes or as temporary protection until permanent covers are established.
Source: US Soil Conservation Service (1979).

must be designed both for stability and capacity. One may solve the Manning equation for each retardance class in graphs given in NRCS (1966). Appendix B contains the NRCS (1966) charts for retardance classes A, B, C, D, and E. The charts were derived from the n vs. V–R relationship (Figure 3.2). Although the charts allow for visualization of relationships among the key variables, computer solutions now replace the charts. Appendix A software employs relations shown in Figure 3.2 in the vegetated waterway permissible velocity spreadsheets. The examples below show the calculation details and discuss the design spreadsheet aids.

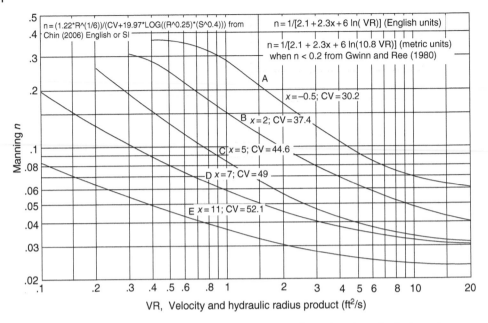

Figure 3.2 The relationship between the Manning roughness "*n*" and product of velocity and hydraulic radius for each of five vegetation roughness classes ranging from very high (A) to very low (E). Also shown are parameters useful for representing the curves in computer modeling software.

The design of vegetated waterways requires knowledge of local conditions to select the grass. One must know the expected maintenance level to anticipate possible mowing of the channel. As the channel's roughness depends on the chosen vegetation (and maintenance), researchers have developed special design procedures. Ree (1949) proposed the first methodology, later improved by Larson and Manbeck (1960), Gwinn and Ree (1980), and Chen and Cotton (1988). One designs channels for *stability* and *capacity*, manipulating hydraulic radius. Tollner (2016) (see Appendix C) presents graphical solutions of these equations for each retardance class and provides details of using Equation 2.20 in solving waterway problems. Chen and Cotton's (1988) relationships for velocity as a function of retardance class, slope, and hydraulic radius. Chin (2013) presents relationships between roughness *n*, velocity *V*, and slope *S* for each retardance class. The downloadables contain spreadsheets for the permissible velocity design of grass waterways, which use the Chin (2013) relationships.

Stability: Apply the basic approach outlined in Equations 2.9 through 2.20.

Select maximum permissible velocity (v_p) based on vegetation and soil conditions (see Tables 3.1–3.3);

Enter the pertinent data into the spreadsheet for the permissible velocity design of grass waterways.

Use the Solver to solve for the bottom width and other elements.

With the triangular channel, $b = 0$, and the side slope z is varied in place of b. With the parabolic channel, we find a topwidth satisfying the stability condition. The downloadables include a separate workbook for the respective geometries. Retardance classes D and E are the usual choices for the stability design.

Design for capacity: Capacity designs require trial and error calculations. The goal is to select a y_c (greater than y) meeting the retardance conditions in retardance class A, B, or possibly C. Retardance class B is the usual choice in the unmowed state. Low maintenance would dictate the roughest condition expected for the vegetation. Frequent, regular mowing may enable the use of a higher retardance class. The following steps define the capacity design using the spreadsheets.

Go to the Capacity sheet of the workbook.

Place a suitable guess value in the depth cell (Other inputs are automatically copied from the Stability page).

Use the Solver to solve the spreadsheet.

Bottom width (if applicable) and side slope z are determined in the Stability design and then utilized in an iterative solution for depth. (Do not open the bioswale sheet for regular waterways with trapezoidal or parabolic spreadsheets). The bioswale is chosen when frequent mowing (or not mowing) is a given. The spreadsheet then computes other elements using the new depth value. Some special consideration for the parabolic channel is required.

Special consideration with parabolic cross sections: The top width at stability cannot be constant. The following relationship provides for computing the new top width of the parabola: T_{cp} guess $= t_s$ * Sqrt(guess y_{cp}/y_s), where the subscripts s and cp refer to stability and capacity. The parabolic geometry is maintained when using the above transformation. As an aside, the above relationship is implemented in the spreadsheets for the design of basic lined channels, even though the parabolic cross section is infrequently used in the usual lined channel design.

Add freeboard: Freeboard accounts for freeze/thaw and wave action. It provides a margin of extra cross section. One does not want incipient overtopping at the usual operating level. For earthen channels, the freeboard factor is usually 0.1 or 0.2 times the design depth.

Several examples that demonstrate hand calculations follow. The Excel folder "PermissibleVelocityGrassedW" contains Excel spreadsheets that address each of the examples below. Likewise, the Mathematica folder "VegetatedChannelsPV" provides a perspective of these problem solutions using Mathematica. The Excel and Mathematica workbooks contain an entry for the freeboard factor in terms of the design depth plus a variable fraction (a factor of 1.2 would give the 20% increase).

Example 3.1 Trapezoidal channel design

We show the Chapter 2 manual solution approach first, then give the results found with the trapezoidal channel permissible velocity spreadsheet.

Given a 500 cfs ($14.17 \, m^3/s$)flow over a Bermuda grass waterway at a 2% slope in erosion-resistant soils. Desire $z = 4$ for crossing ease. $v_p = 2.4 \, m/s$. Find b.

$A = 14.17 \, m^3/2.4 \, m = 5.905 \, m^2$

From the class D retardance, compute $R = 0.38 \, m$ from the Graphs reproduced in Appendix C. Using Equation 2.20, $a_1 = 2$ Sqrt(17) $- 4 = 4.246$; $b_1 = -15.54 \, m$; $c_1 = 5.905 \, m^2$

$y = 0.428 \, m$, $3.23 \, m$; use the shallow depth.

Then from Equation 2.20, $b = [5.905 \, m^2 - 4(0.428 \, m)^2]/0.428 \, m = 12.08 \, m$.

Design for capacity:

$b = 12.08 \, m$ from the stability analysis above.

Add 0.1 m to the depth of 0.43 m found above for $d = 0.53 \, m$.

Then add 0.1 m freeboard to get a depth of $y_f = 0.68$ m, $b = 12.08$ m, $z = 4$, $T_f = 17.52$ m. The spreadsheet gives $b = 11.06$ m, $y_f = 0.68$ m, $z = 4$, $T_f = 17.07$ m. The slight discrepancy is due mainly to errors in reading the design charts and trial and error in the capacity design.

Example 3.2 Trapezoidal channel attempt, radical negative; triangular channels, and other approaches

Given a 50 cfs (1.42 m^3/s) flow over a Bermuda grass waterway at a 2% slope in erosion-resistant soils. Desire $z = 4$ for crossing ease. Find $v_p = 2.4$ m/s for Bermuda (class D) on the 2% slope. Find b using the trapezoidal vegetal waterway spreadsheet.

The permissible velocity spreadsheet would not come to a solution in the stability phase. The hand solution using the charts in Tollner (2016) (see Appendix C) reveals a negative radical. The negative radical provides an opportunity to explore several solution strategies before recommending an approach for dealing with the degenerate trapezoid. One may decrease z (e.g., consider compromising the ability to cross the channel) easily and use triangular geometry. One may increase the channel slope if topography allows, using a triangular cross section. One may increase the channel slope if the situation allows a trapezoidal cross section with a preset bottom width. Finally, one might design only for capacity since the channel is by definition stable, although sedimentation could be problematic. Capacity design with a reduced v_p (not less than 1.4 ft/s or 0.4 m/s) is usually the first choice.

Example 3.3 Parabolic channel design

Given a 500 cfs (14.17 m^3/s) flow over a Bermuda grass waterway at a 2% slope in erosion resistant soils. Design a vegetated waterway having a parabolic cross section. Using the parabolic workbook, $v = 2.4$ m/s and stability is designed using class D. From the class D retardance read $R = 0.39$ m. $A_s = q/v = (14.17$ m^3/s$)/(2.4$ m/s$) = 5.9$ m^2. For the parabolic cross section, $R = 0.39 y_s$ or $y_s = 0.58$ m; $t_s = A_s/(0.67y_s) = 15.18$ m. Design for capacity (class B) gives $y_f = 0.85$ m, $T_f = 18.41$ m using a freeboard factor of 1.2.

In the example above, going to retardance class B from class D changed the velocity from 2.4 to 1.7 m/s. What if one designs for only the unmowed condition and then mows the channel? Such methodology usually leads to a narrower bottom width and deeper channel. Mowing vegetation would reduce roughness leading to excessive velocities. Thus, the channel would be unstable in high flow events.

One may determine the minimum bottom width of the channel based on available construction equipment blade width. Check for stability in the mowed condition. Check for capacity in the unmowed state. If stability is not satisfied, then one must increase the width or decrease the slope. The capacity design is the next step. Warner (1998) found this approach to work well in automated solutions.

The NRCS (2007) provides a practical approach to stream design, including vegetated channels, which includes many nuances to stream design and construction in addition to those presented above. The design manual, cited in the references, is available online.

Bioswales

LID is driving the consideration of bioswales and related vegetated waterways. The bioswale is a stormwater runoff conveyance system that provides an alternative to storm sewers (e.g., see NRCS 2014). NRCS (2005) provides details on bioswale uses. Figure 3.3 shows a typical bioswale,

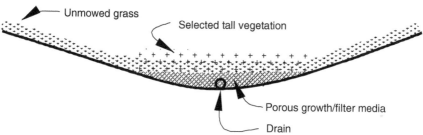

Unmowed grass

Selected tall vegetation

Porous growth/filter media

Drain

Figure 3.3 Photo and definition sketch of a bioswale. *Source:* Photograph courtesy of NRCS.

which is a modified vegetated waterway. Bioswales have only a capacity calculation with a class B or possibly class A retardance condition. The channel is sized to convey a minimum 10-year storm – a trapezoidal cross section with side slopes z values of 3 or higher is recommended. Parabolic sections are frequently employed. Soils should have an infiltration rate of 0.5 in./hour, and excess compaction avoided. Tighter soils may have an amended soil media and a tile drain, as shown in Figure 3.3. Ideally, the modified soil would absorb the equivalent of the water quality storm. The 0.5 in./hour infiltration limit roughly approximates the absorption of the water quality storm. Bioswales may be up to 2500 ft or longer. Long bioswales typically have low slopes (e.g., less than 0.01). Municipalities grant some credit towards treating the water quality storm with bioswales, depending on the gradient, length, and infiltration provision. Clar et al. (2004) provide additional information on the bioswales and other vegetative BMPs.

Bioswale specification involves engineering, ecology, and landscape architecture professionals. The architect conceives an esthetically attractive channel that blends nicely with the surrounding built or natural environment. The Ecologist contributes to the selection of vegetation that can accommodate the anticipated water availability for growth and possibly serve as a habitat for butterflies and other wildlife. Finally, the engineer assesses the hydraulic suitability of the design.

The trapezoidal and parabolic spreadsheets in the downloadables contain bioswale design pages. A single retardance class, usually B–D depending on specified mowing (e.g., frequent mowing or no mowing), forms the design basis. With the bioswale, costing requires some special handling. One may figure drainage costs and media costs per unit length and place this value in the spreadsheet's lining cost cell. After inputting the requested data on the stability page, move to the bioswale page and insert the bottom width (trapezoid) or topwidth (parabolic). Solve and find the required parameters.

Vegetated Filter Strips

The vegetative filter strip is a sheet flow mechanism. Observations of turbid sheet flow moving from a tilled field into a vegetated area inspired the concept of the vegetated filter strip (Tollner 1974). The technique has been further refined and improved by many researchers since 1974, many of whom were protégés of Dr. B.J. Barfield. The practice of removing sands and sand-sized aggregates from sheet flow finds application as a pretreatment for flows coming from parking lots and moving into bioswales or rain gardens. A rain garden is a bioswale acting as a "leaky" impoundment. Clar et al. (2004) provide excellent guidelines for vegetative filter design. Wanielista et al. (1997) provides detailed design approaches for infiltration bioswales that are useful for sandy, well drained soils. The SedCad software (Warner 1998, also a student of B.J. Barfield) provides a vegetated waterway and filter strip design options.

Temporary Linings

Vegetated waterways and bioswales are vulnerable to erosion by storms much smaller than the design event between initial construction and grass establishment. One should divert water from the channel during the construction phase, if possible. Otherwise, temporary linings or nettings, such as those shown in Figure 3.4, are appropriate for protection during the construction period. Protection during the construction phase is necessary for urban settings where adverse esthetic and financial consequences are likely. Straw mulch is sometimes used; however, engineered jute or related fiber mats perform more reliably. Increased reliability is beneficial in urban situations. The selection of vegetation adapted to the region is essential to achieve a stand and minimize maintenance problems throughout the channel life. Biological materials decompose during vegetation establishment, which is desirable. Nonbiological materials remain over a long duration but are generally not environmentally obnoxious if properly installed. A search on the web surfaces a variety of commercially available blankets and nettings.[1]

Haan et al. 1994) provide a design approach attributed to McWhorter (1968) covering a range of temporary linings that can stabilize channels in the interim between construction and vegetation establishment. An empirical equation, Equation 3.1, proposed by Norman (1975), gives the maximum depth over a given lining (see Table 3.4 for linings).

$$d_{max} = mS^n \hspace{4cm} (3.1)$$

---

1 For example, see https://www.erosioncontrol-products.com/erosioncontrolblanket.html

Jute or other nonwoven fiber mat

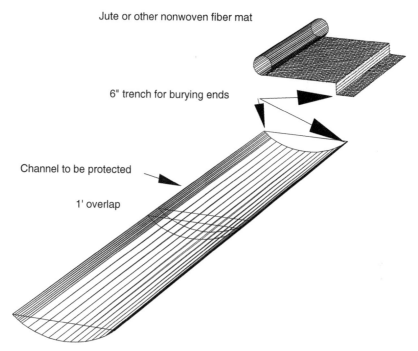

6" trench for burying ends

Channel to be protected

1' overlap

Notes: Soil below mat should be tamped
     Staples on 2' spacing and 2' centers
     Staple with #11 gauge (min) 6" staples (min)

Figure 3.4 A simple procedure for installing a jute mat for stabilizing a newly constructed grassed waterway. *Source:* Adopted from Goldman et al. (1986) © McGraw-Hill.

where d_{max} is the maximum allowable flow depth;

    m, n are empirical parameters available in Table 3.5 for media described in Table 3.4. The m, n coefficients depend on the soil's erodibility, dictated by the Universal Soil Loss Equation (USLE). Table 3.5 gives details on finding the K factor for specific soils.

    S is the channel slope.

    Norman (1975) proposes a modified Manning type equation given in Equation 3.2.

$$V = A * R^B * S^C \tag{3.2}$$

where A, B, and C are empirical constants that vary widely with the media type. Table 3.5 gives values attributed to McWhorter et al. (1968).

    R is the hydraulic radius

    S is the slope

    Downloads provide spreadsheets for the easily erodible case and the erosion-resistant case. It is left for the student to modify the spreadsheet in the soils falling between the resistant and easily erodible cases. The spreadsheets compute the maximum depth, then iterate to find the channel hydraulic elements containing the flow. The general trapezoid and parabolic cross section options are available.

Table 3.4 Temporary lining descriptions.

Excelsior mat	Erosionet
Composed of 0.8 lb/yd² of dried shredded wood covered with a fine paper net covering. Held in place by steel pins or staples with five staples per 6 linear feet of mat	A paper yarn approximately 0.05 in. diameter, woven into a net with openings about 7/8 in × ½ in. The material has little erosion prevention capability and is usually used to hold other lining material in place.
Straw and erosionet	**Fiberglass roving**
Consists of straw applied at a rate of 3 tons per acre and covered with erosionet (see Erosionet discussion)	Fiberglass roving comes as a lightly bound ribbon of continuous glass fibers. The material is applied to the channel bed using a special nozzle driven by an air compressor, which separates the fibers and results in a weblike mat of glass fibers. The glass fibers are often tacked with asphalt for adhesion to each other and the soil. Plants are not able to grow through the roving.
3/8 in Fiberglass mat	**Jute mesh**
A fine, loosely woven glass fiber mat similar to furnace air filter material. It weighs about 0.11 lb per yd³. The material differs from fiberglass roving in that it is not dense enough to block plant growth. Steep pins or staples are used to hold the mat in place.	Jute mesh is a mat lining woven of jute yarn that varies from 1/8 into ¼ in. in diameter. The mat weighs about 0.8 lb/yd² with openings about 3/8 in. × ¾ in. Steel pins or stables are used to hold the jute mesh in place.
½ in. Fiberglass mat	
A fine, loosely woven glass fiber mat, similar to but denser than the 3/8 in. fiberglass mat.	

Source: McWhorter et al. (1968).

The temporary flow resistance is often higher than the resistance offered by the fully established vegetation. In other words, the interim state may dictate the channel size if the channel is to handle the design storm. A designer may propose that a modified design storm of one or two years provide the basis for designing the temporary lining, while the typical design storm of 10 (or more years) provides for the permanent channel design. One may identify the flows expected for various design periods from one year to the maximum return period. Then, evaluate the flow requiring a wetted perimeter of the temporary lining equal to the permanent channel's wetted perimeter.

Example 3.4

Specify a temporary lining for the waterway examined in Example 3.1. This channel conveys a flow of 500 cfs down a 4% slope over an erosion-resistant soil. The channel is trapezoidal ($z = 4$) and conveys a flow of 500 cfs down a 4% slope over an erosion-resistant soil. Straw and Erosinet is available for the lining. Use the temporary lining design spreadsheet to make the recommendation. On the general trapezoidal sheet, complete the design parameter inputs. A bottom width of 105 ft is required, with a depth of 0.67 ft before the freeboard. The bottom width of 12 ft is significantly exceeded, suggesting that one conduct a hydrologic analysis to assess the risk of failure during the construction window.

Table 3.5 Coefficients for temporary lining equation coefficients.

Lining type[a]	Erodible soil (USLE $K \geqslant 0.5$)[b]		Erosion resistant soil (USLE $K \leqslant 0.17$)[b]		Velocity equation coefficient[c]		
	m	n	m	n	A	B	C
Bare soil	0.003	−0.687	0.0084	−0.687	22.81	0.591	0.286
Fiberglass roving with asphalt tack (single layer)	0.0067	−0.960	0.0141	−0.960	42.45	0.667	0.5
Fiberglass roving with asphalt tack (single layer)	0.0143	−1.01	0.027	−1.01	59.2	0.667	0.5
Jute mesh	0.0076	−0.875	0.0202	−0.883	61.53	1.0281	0.431
Excelsior mat	0.0572	−0.585	0.101	−0.585	32.29	1.340	0.351
Straw and Erosinet	0.052	−0.652	0.082	−0.652	70.76	1.455	0.529
Fiberglass (3/8 in. thick)	0.025	−0.670	0.046	−0.670	73.53	1.330	0.512
Fiberglass (1/2 in. thick)	0.048	−0.646	0.083	−0.646	14.84	1.235	0.086
Erosinet	0.049	−0.642	0.084	−0.642	41.45	0.855	0.40

[a] See Table 3.4 for media descriptions.
[b] The USLE K may be found from the Web Soil Survey, located at https://websoilsurvey.sc.egov.usda.gov/App/HomePage.htm. To find K, go to the address, identify the soil, under soil data explorer, go to Soil properties and qualities and go to soil erosion factors and determine the soil (whole soil) factors. For soils with Universal Soil Loss Equation (USLE) K values between 0.17 and 0.5, interpolate to find m and n.
[c] The velocity equation is given by $V = A \, R^B S^C$ where R is the hydraulic radius and S is slope. Units are imperial. The Manning n equivalent is $1.49/A$. Convert to SI units by taking the reciprocal of n.
Source: Adopted from Haan et al. (1994).

Statistical hydrology texts such as Haan (2002) discuss construction window risk assessment. For example, using a binomial distribution, Haan (2002) shows that a designer is taking a 4% chance that a 50-year storm would be exceeded in a two-year construction and vegetation establishment window. The acceptability of such a risk should be conveyed to the client, which is frequently acceptable.

For example, if 1/4 of the flow were allowed over the construction window, a bottom width of 27 ft would suffice. Using a Fiber double tack would reduce the bottom width to 18 ft. One may consider additional linings to be discussed in the Tractive force chapter to find a more suitable temporary liner.

Straw and Erosinet is available for the lining. Use the temporary lining design spreadsheet to make the recommendation. On the general trapezoidal sheet, complete the design parameter inputs. The bottom width of 105 ft is required, with a depth of 0.67 ft before the freeboard. The bottom width of 12 ft is significantly exceeded, suggesting that one conduct a hydrologic analysis to assess the risk of failure during the construction window.

Statistical hydrology texts such as Haan (2002) discuss construction window risk assessment. For example, using a binomial distribution, Haan (2002) shows that a designer is taking a 4% chance that a 50-year storm would be exceeded in a two-year construction and vegetation establishment window. The acceptability of such a risk should be conveyed to the client, which is frequently acceptable.

For example, if 1/4 of the flow were allowed over the construction window, a bottom width of 27 ft would suffice. Using a Fiber double tack would reduce the bottom width to 18 ft. One may consider additional linings to be discussed in the Tractive force chapter to find a more suitable temporary liner.

Summary

Figure 3.5 summarizes the design requirements for vegetated waterways and bioswales. First, create the bottom width of the trapezoidal section, the side slope z of the triangle, or the parabolic cross section's top width, whichever cross section is in use. Stability is now satisfied. Then design for capacity in the high roughness, varying depth, or top width to give the area satisfying the continuity equation. One may bypass the stability step with the bioswale. Finally, provide for freeboard. Conclude the design with a sketch showing key channel elements. Lastly, periodic inspections and maintenance are essential for maintaining functional grassed waterways. An Internet search revealed links to several commercially available software packages for designing channels, including vegetated waterways, in addition to the spreadsheet workbooks available for download, as referenced above and in Appendix A.

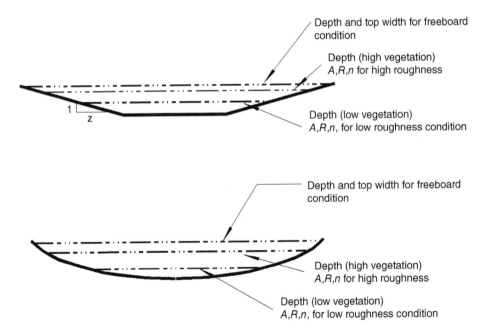

Figure 3.5 Design requirements summary for vegetated trapezoidal and parabolic channels. Note the general similarity with other channels types.

The trapezoidal channel is the "standard" for waterways and bioswales. The side slope z is usually four or more in each case. The triangular waterway is viable when the depth is very shallow and $z > 10$ because it is easily constructed with commonly available equipment. The parabolic waterway is typically not installed in rural areas; however, it serves as an end in the evolution of natural waterways having minimum maintenance. The parabolic channel often finds application in urban areas where esthetics is a high value. Vegetated bioswales are widely used for management devices (WEF 1998; Clar et al. 2004). Other terminology for the grassed waterway includes diversions, which are designed as vegetated waterways or bioswales. The downloadables contain spreadsheets for the permissible velocity design for each of the common cross section types. Commercial software packages for channel design may simplify the design approach to selecting a single n value from a table. Other packages, in effect, implement procedures similar to that of the bioswale method shown above. Others implement the full stability–capacity approach.

As with open channel flow, works such as Harris (1994) provide extensive discussion on earth construction techniques. Associations such as the Land Improvement Contractors Association (LICA, http://ourworld.compuserve.com/homepages/delmarva/staff.htm), and professional associations such as ASABE (http://www.asabe.org) and ASCE (http://www.asce.org) also provide helpful information (e.g., see ASABE 2006). Temporary linings are frequently used to protect the channel during vegetation establishment.

Maintenance of vegetated waterways and bioswales begins with maintaining proper fertility for the chosen grass lining. Apply mulch and surface coverings, as shown in Figure 3.4. Proper application fosters adequate vegetation development and minimal stand injury during establishment. Prolonged droughts or prolonged wet seasons may result in stand deterioration. In the event of damage, reshaping and reseeding may be necessary. Machine traffic, overgrazing, animal traffic, or rodent-induced damages necessitate channel maintenance with vegetated waterways as with other earthen channels. Channels flowing at less than the permissible velocity in the stability phase may be more vulnerable to sedimentation. One should do routine inspections annually and after major storms to assess sediment deposition, scouring, gullying, and traffic damage.

The above channels presume periods of no-flow. If low continuous flows develop during a vegetated waterway's useful life, one may have to put a small lined section in the channel bottom to carry the continuous flow component. If additional protection is needed, one may have to resort to other linings discussed in the Tractive Force chapter. The discussion of maintenance at the end of Chapter 2 applies to grassed waterways as well. NRCS (2007) highlights the reality that grass should only be used as a lining when the channel is ephemeral and that the grass selected is suited to the climate.

An alternative design methodology, the tractive force method, is introduced for unlined channels and grassed waterways in the following chapter. One should understand permissible velocity before using the tractive force method. The permissible velocity method is preferred under mixed vegetation conditions, which often occurs in low maintenance conditions. NRCS (2007) favors the permissible velocity method for stream restoration applications, likely because of mixed vegetation developing over time. The highly managed bioswale (high maintenance typically) is a candidate for the tractive force method discussed in the next chapter.

Problems and Questions

Note: Draw sketches of all completed designs and provide supporting reasons for discussion questions. Show all work and offer a rationale for assumptions made. Remember to use appropriate significant digits. Feel free to use the resources of Appendices B and C. Unless otherwise stated, use a freeboard of 20%.

1   Design a parabolic shaped grassed waterway to carry $1.5\,m^3/s$ in the following conditions. An easily eroded soil with Bermuda grass is in good condition and has a 4% slope.

2   Design a trapezoidal-shaped waterway with a 4:1 side slope to carry $0.6\,m^3/s$ from a terraced field where the soil is resistant to erosion. The channel slope is 4%. The vegetation choice is Bromegrass. Assume the channel must be crossed on an occasional basis.

3   Design a triangular-shaped waterway to carry $1.8\,m^3/s$ from a field resistant to erosion and a 6% channel slope with fescue lining. Can one cross this channel by farm and small construction equipment?

4   Design a trapezoidal channel to carry $25\,m^3/s$ flow over a 1% slope in an easily erodible soil. The producer selects fescue. Farm equipment must traverse the waterway.

5   Design a parabolic channel and a triangular channel to carry $25\,m^3/s$ flow over a 1% slope in an easily erodible soil. Fescue is to be grown. Compare the resulting channels by superimposing a scaled drawing of the channels. Design the channel.

6   The situation to follow is typical in cases where engineered designs from an agricultural era are (mal) functioning in the modern era of urban development. In an easily erodible soil, a trapezoidal channel has been in the unmowed state (fescue) for many years and has accumulated sediment and experienced increased runoff from recent development activity in the watershed. The slope is 2%, 4:1 side slope, 4 m bottom width, and documented flow depth of 0.75 m. Other original design details are unknown. Determine the velocity and flow. Would sediment entrainment occur if the channel were in the mowed condition? Suggest an engineered approach for retrofitting the waterway for the urban situation providing a more active sediment control at the channel discharge.

7   Design a channel to carry $50\,m^3/s$ in a case where the Bermuda grass vegetal lining is a part of a controlled forest burn from time to time. The channel slope is 3%, and the soil is easily erodible. Farm equipment must cross the waterway. Compare this channel to an identical nearby channel with mowed, unburned Bermuda grass.

8   Project: A $100\,m^3/s$ flow is to be conveyed down a 1000 m direct distance with a fall of 100 m. An alternate path of 10000 m is available (e.g., a longer way around). Excavation costs are $2 per m^3 soil and an additional $5 per lineal meter of concrete lining or fescue vegetation lining costing $1 per linear meter. Which lining is the preferred choice based on the initial investment? If annual maintenance is 5% of the initial investment and the project's life is 20 years, what is the choice?

9 What customer requirements would point toward a vegetated waterway recommendation? Refer to Table 2.9.

10 In what situations would one favor a parabolic channel cross section compared to a triangular or trapezoidal?

11 In cases where the slope preset and one initially specifies a trapezoidal channel, what options does one have when meeting the stability condition is impossible?

12 Design a temporary Jute lining for the channel of Problem 4. What flow would give a wetted perimeter equal to the design flow in Problem 4?

13 Design a temporary Excelsior lining for the channel of Problem 4. What flow would give a wetted perimeter equal to the design flow in Problem 5?

14 Design a bioswale that would handle the flow of Problem 4, assuming the channel would never be mowed.

References

ASABE. (2006). Grassed waterway for runoff control. ASABE EP464 DEC06. American Society of Agricultural and Biological Engineers, St. Joseph, MI.

Chen, Y.H. and Cotton, G.K. (1988). Design of roadside channels with flexible linings, FHWA-IP-87-7, Hydraulic Engineering Circular 15. Federal Highway Administration, US Department of Transportation, National Technical Information Service, Springfield, VA.

Chin, D.A. (2013). *Water Resources Engineering*, 3e. Upper Saddle River, NJ: Prentice-Hall.

Clar, M.E., B.J. Barfield, and T.P. O'Connor. 2004. Stormwater best management practice design guide, Vol. 2: vegetative biofilters. EPA/600/R-04/121A. National Risk Management Research Laboratory, Office of Research and Development. US Environmental Protection Agency, Cincinnati, OH.

Goldman, S.J., Jackson, K., and Bursztynsky, T.A. (1986). *Erosion and Sediment Control Handbook*. New York, NY: McGraw-Hill.

Gwinn, W.R. and Ree, W.O. (1980). Maintenance effects on the hydraulic properties of a vegetation lined channel. *Transactions of the ASABE* 23: 636–642.

Haan, C.T. (2002). *Statistical Methods in Hydrology*, 2e. Ames, IO: Iowa State Press.

Haan, C.T., Barfield, B.J., and Hayes, J.C. (1994). *Hydrology and Sedimentology of Small Catchments*. New York, NY: Academic Press.

Harris, F. (1994). *Modern Construction & Ground Engineering Equipment & Methods*, 2e. White Plains, NY: Longman Publishing Group.

Hudson, N. (1971). *Soil Conservation*. Ithaca, NY: Cornell University Press.

Larson, C.L. and Manbeck, D.M. (1960). Improved procedures for vegetated waterways. *Agricultural Engineering* 41: 694–696.

McWhorter, J.C., Carpenter, T.G., and Clark, R.N. (1968). Erosion control criteria for drainage channels. Mississippi State Highway Department and Federal Highway Administration, Department of Agricultural Engineering, Mississippi State University, State College, MS (original not seen, quoted from Haan et al. 1994).

Norman, J.M. (1975). Design of stable channels with flexible linings. Highway engineering circular No. 15, Federal Highway Administration, Washington, DC.

NRCS. (1998). USDA National resources inventory 1997 update. Natural Resources Conservation Service, Washington, DC. www.nrcs.usda.gov (accessed April 2019).

NRCS. (2005). Bioswales. http://www.nrcs.usda.gov/Internet/FSE_DOCUMENTS/ nrcs144p2_029251.pdf (accessed April 2019).

NRCS. (2007). Threshold channel design, Chapter 8 of Part 654. In: *Stream Restoration Design of the National Engineering Handbook*. Washington, DC: USDA National Engineering Handbook. The Stream Restoration Handbook. https://www.nrcs.usda.gov/wps/portal/nrcs/ detail/national/water/manage/restoration/?cid=stelprdb1044707 (accessed April 2019).

NRCS. (2014). Engineering Field Tools (EFT). Washington, DC: Natural Resources Conservation Service, USDA. http://www.nrcs.usda.gov/wps/portal/nrcs/detail/national/ technical/engineering/?cid=stelprdb1186070 (accessed April 2019).

NRCS (2021). Engineering Field Handbook. USDA-NRCS Part 650, Washington, DC.

NRCS (National Resources Conservation Service, formerly Soil Conservation Service, SCS). (1966). Handbook of channel design for soil and water conservation. SCS-TP-61 (metric system revision). Natural Resources Conservation Service, Washington, DC. https://www. wcc.nrcs.usda.gov/ftpref/wntsc/H&H/TRsTPs/TP61.pdf (accessed April 2019).

Ree, W.O. (1949). Hydraulic characteristics of vegetation for vegetated waterways. *Agricultural Engineering* 30 184–187, 189.

Tollner, E.W. (1974). Modeling the sediment filtration capacity of simulated, rigid vegetation. MS Thesis. University of Kentucky, Lexington, KY.

Tollner, E.W. (2016). *Engineering Hydrology for Natural Resources Engineering*, 2e. Cambridge, UK: Wiley.

US Soil Conservation Service (1979). *Engineering Field Manual for Conservation Practices*. Washington, DC: Natural Resources Conservation Service, US Department of Agriculture.

Wanielista, M., Kersten, R., and Eaglin, R. (1997). *Hydrology: Water Quality and Quality Control*. New York, NY: John Wiley & Sons, Inc.

Warner, R.C. (1998). SEDCAD[7] 4 for Windows95/NT. Notes for August 5–7, 1998 short course. Biosystems and Agricultural Engineering Department, University of Kentucky, Lexington, KY.

WEF (Water Environment Federation). (1998). Urban runoff quality management. Water Environment Federation manual of practice No. 23. (ASCE Manual and Report on Engineering Practice No. 87). Water Environment Federation, Alexandria, VA.

4

Tractive Force Methods for Earthen Channels

The tractive force method came about due to more firmly establishing design criteria for earthen channels on first principles rather than on survey results used in establishing the permissible velocity method. Tractive force methods represent what one can call Earthen Channel design II. We conclude the uniform flow discussion with a section on channel costing.

Goals

- To explore the origins of the tractive force and how it works
- To design earthen channels using tractive force methods
- To design grassed waterways using tractive force methods, and discuss when to use Tractive Force and when the permissible velocity method is adequate
- To explore the parabolic cross-section and how it best meets the tractive force criterion across any part of the cross section.
- Channel costing

Riprap-Lined or Earthen Waterways (Earthen II)

Tractive force origins: The tractive force method computes the shear force due to flowing water (see Equation 2.4). It compares the calculated shear to maximum tabulated shear values for the given lining material. We focus on the trapezoidal or triangular cross-section.

To see the similarity in the permissible velocity method and the tractive force method, consider Equation 4.1, which draws upon Equation 2.4. Refer to Equation 2.4 for the definition of nomenclature.

$$v = \left(\frac{\sqrt{\gamma}}{C_d \, \rho} \right) \sqrt{RS} \tag{4.1}$$

Open Channel Design: Fundamentals and Applications, First Edition. Ernest W. Tollner.
© 2022 John Wiley & Sons Ltd. Published 2022 by John Wiley & Sons Ltd.
Companion website: www.wiley.com/go/tollner/openchanneldesign

Now recall the Chezy equation (2.5) and Manning equations, reproduced as Equation 4.2a,b.

$$v = \left(C_{Chezy}\right)\sqrt{RS} \qquad (4.2a)$$

$$v = \left(\frac{\phi}{n}\right) R^{\frac{2}{3}} S^{\frac{1}{2}} \qquad (4.2b)$$

On equating the bracketed terms in Equations 4.1 and 4.2, one finds that the velocity is proportional to $\sqrt{RS}$. The Manning equation includes an extra factor of $R^{1/6}$. Thus, a velocity that causes motion and a tractive force that results in particle motion corresponds to a constant with Chezy hydraulics. With Manning hydraulics, the methods are similar to within $R^{1/6}$ times a constant.

One may modify Equation 2.4 by including parameters that refine the computed tractive force specifically to the channel width and position on the side or bottom. A multiplier provides for including gravity effects on the side slope. Equation 4.3 gives the tractive force of a given flow:

$$\tau = K_s K_t \gamma R S \qquad (4.3)$$

The factor K_t depends on the width/depth ratio and position on the side or bottom. Lane (1955) provided K_t curves for the channel sides and bottom, shown in Figure 4.1. The curve gives the maximum on-the-side force ratio. One frequently approximates the hydraulic radius R with the channel depth y since these are often wide channels. The K_s factor further modifies the relationship on the channel side for noncohesive materials by taking the angle of repose (same as the friction angle, α) of the material into account. Equation 4.4 (derived from a force balance detailed in French 2007 using algebra and application of algebraic and trigonometric identities) defines K_s.

$$K_s = \sqrt{1 - \frac{\sin^2\left(\Gamma\right)}{\sin^2\left(\alpha\right)}} \quad \text{for noncohesive materials and } K_s \equiv 1 \text{ for cohesive materials} \qquad (4.4)$$

where Γ is the angle of the channel side slope ($=\tan^{-1}[1/z]$), and; α is the angle of repose or friction angle of the noncohesive lining (see Figure 4.2).

The parabolic cross-section discussion below offers additional details on the derivation of Equation 4.4.

Figure 4.3 schematically shows the force balance on the side of a trapezoid. These relations apply to a point on any wetted perimeter. The tractive force computed using Equations 4.2 must not exceed critical values for noncohesive and cohesive materials. Figure 4.4,b shows critical tractive forces for noncohesive and cohesive media. For noncohesive material design, one finds a hydraulic radius satisfying the critical tractive force from either Figure 4.4a or b. The Manning equation provides an avenue for computing velocity, given a roughness suitable for the material. For noncohesive materials, Sturm (2010) gives a relationship in Equation 4.5a for computing the Manning n value as a function of roughness element d_{50} and channel hydraulic radius:

$$n = d_{50}^{1/6} \frac{\varphi_{dim}\left(R/d_{50}\right)^{1/6}}{\sqrt{8g}\left(a + b\log\dfrac{R}{d_{50}}\right)} \qquad (4.5a)$$

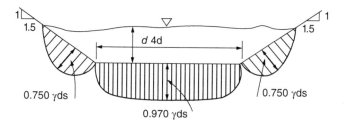

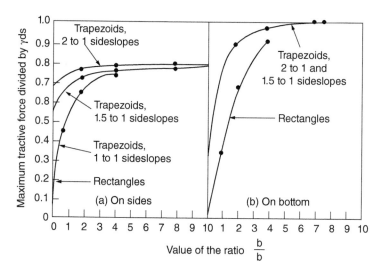

Figure 4.1 Unit tractive force ratio on the channel sides (a) and bottom (b) in the indicated channels as a function of the bottom width (b) to normal depth (d) ratio. *Source:* After Lane (1955); used with permission of the Am. Soc. Civil Engineers.

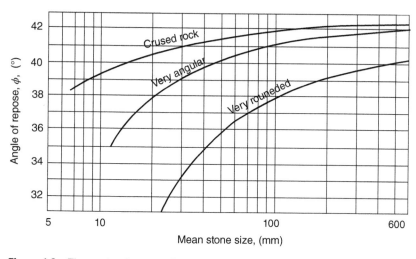

Figure 4.2 The angle of repose of particles as a function of stone geometry and size.

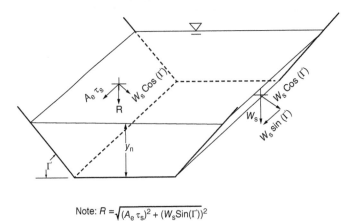

Note: $R = \sqrt{(A_e \tau_s)^2 + (W_s Sin(\Gamma'))^2}$

Figure 4.3 Force balance on a particle on the side of the channel. Flow is along the vector areas.
Source: Adapted following French (2007).

(a)

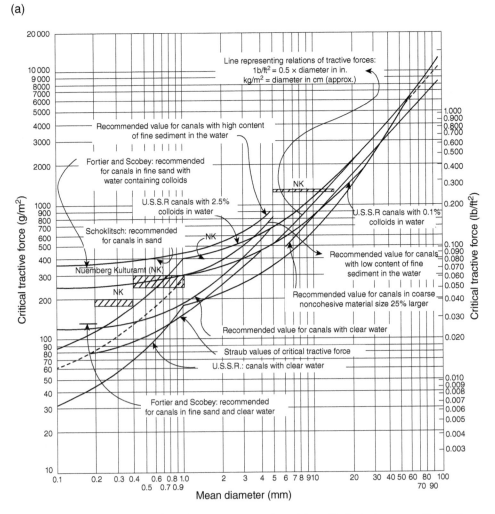

Figure 4.4 (a) Critical tractive force for coarse textured particles of indicated average diameter, showing results of several investigators. One may convert the imperial units (right side) to Pa by lb/ft^2 × 48.57 to get Pa (N/m^2). *Source:* Courtesy of the U.S. Bureau of Reclamation (1973). (b) Critical tractive force vs. void ratio for fine textured materials (Lane and Carlson 1953). One may convert the imperial units to Pa by lb/ft^2 × 48.57 to get Pa (N/m^2).

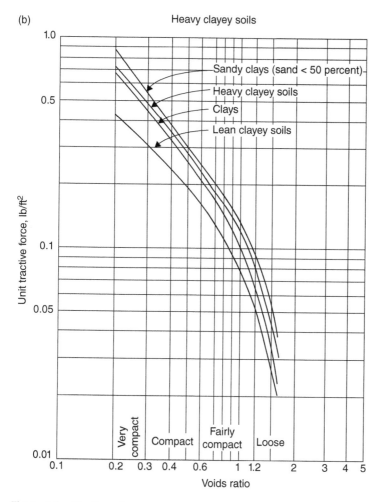

Figure 4.4 (Continued)

In Equation 4.5a, d_{50} is the screen size passing 50% of the noncohesive material, ϕ_{dim} is the units constant, and g is gravity. Constants a and b have values of 0.79 and 1.85, respectively, for noncohesive materials. Equation 4.5a may be simplified, as shown in Equation 4.5b, known as the Strickler equation (Jobson and Froehlich 1988).

$$n = 0.034 d_{50}^{1/6} \tag{4.5b}$$

In Equation 4.5b, d_{50} has dimensions of feet.

For cohesive materials, use Figure 4.4b to estimate the critical tractive force and use tabulated values of the Manning n. Chow (1959) and Jobson and Froehlich (1988) provide additional relationships. One then iteratively seeks a width/depth ratio giving the required design flow rate. As one may gather from considering Figure 4.4b, design in cohesive materials is not precise. The cohesive analysis is applied more toward characterizing natural conditions than for designing new channels.

French (2007) provided a systematic design procedure. The author incorporated French's procedure into an earthen (or riprap-lined) channel tractive force design spreadsheet in the download materials (see the "tractiveearthen" folder under Excel in Appendix A). French (2007) considers a rectangle (considers only the bottom; stabilized sides) and a trapezoid. The spreadsheet determines a depth (corrected for b/y ratio), then determines the needed bottom width to handle the flowrate. The spreadsheet takes some liberties with the side slope, as Lane (1955) provides for 1 : 1 and 1 : 2. The Lane (1955) graphs are modeled in the spreadsheet using regression relationships. The French (2007) process also includes a sinuosity factor, provided in the spreadsheet page's Data tab. In effect, the sinuosity factor decreases the design critical tractive force to give designs accounting for increased erosion forces in curves. Haan et al. (1994) discuss the Federal Highway Administration (FHA) and NRCS procedures for riprap channels. The folder "RiprapDesign" contains spreadsheets implementing these procedures. The sinuosity factor is inversely proportional to the length ratio along the channel to length down the valley. Freeboards are added as before to obtain the final design. The NRCS (2007) recommends that the tractive force method be used to design channels with boundary material larger than sand size. Other methods may provide design checks. We revisit riprap and stable channel design in Chapter 10.

Example 4.1 (Example 2.4 redone with tractive force method)
Given an occasional flow of 100 cfs (2.83 m³/s) over a 2% slope with a natural trapezoidal channel with a $z = 4$. The loamy soil permissible velocity is 2 ft/s (0.61 m/s) and $n = 0.03$. Determine the flow depth. A straight channel is assumed. Compare the depth with that found with the permissible velocity method.

Using a critical tractive force value of 0.15 lb/ft² (7.3 Pa), the spreadsheet gives a bottom width of 99.6 m (320 ft) and a depth of 0.038 m (0.125 ft). The estimated n value is 0.021, as opposed to 0.025. Trying the coarse gravel lining, say 30 mm, the critical tractive force is 0.8 lb/ft² (39 Pa), yielding a depth of 0.203 m (0.66 ft) and bottom width of 11 m (36 ft). One then adds freeboard and computes costs. Costs are more fully considered at the end of this chapter. One could adjust n to account for debris accumulation (not done on the spreadsheet provided), then add freeboard. Hand solution converges rapidly for this case, as it does for wide, shallow channels in general. Rare cases of narrow, deeper channels require several iterations due to the changing coefficients when b/y is low (see Figure 4.4).

The resulting dimensions are similar. Either result would generally be acceptable in practice. The available z values in the Lane (1955) do not include $z = 4$. From the charts, the size slope z of 4 constants approaches those of the bottom. The lack of precise size information of the loamy sand or the coarse gravel also precludes more exact comparisons between the permissible velocity and tractive force methods. In the field, variation in size specifications introduces similar discrepancies.

Lining materials having a higher critical tractive force than natural gravel or cobbles are available. They enable smaller bottom width channels. These new materials are useful for steep slopes. Various transportation departments have published data for modern materials, and Ohio data are featured in the spreadsheet. The following example demonstrates applications.

Example 4.2 Consider the situation of Example 4.1, using a Type 1 Turf Reinforcing Mat with an allowable shear stress of 3 lb/ft² (from the OhioDOT page on the spreadsheet).

Inserting the situation data in the spreadsheet and solving, we find the solver could not find a solution. Lack of solution suggests that the required depth to achieve the critical tractive (shear) stress was not physically possible. Thus, we could back off the critical tractive force to find a solution giving a velocity of 1.4 ft/s (0.6 m/s) to avoid sedimentation. Trying 1 ft²/s gave a $y = 0.254$ m and $b = 7.38$ m (with a velocity of 1.32 m/s) before freeboard was added. Further tractive force reductions may be used as desired. The rock diameter controls the Manning n in the spreadsheet. Manning data for the media are somewhat sparse; however, values ranging from 0.02 to 0.04 seem reasonable. Using a rock diameter entry of 40 mm in the spreadsheet gave a Manning n of around 0.04, which was deemed satisfactory.

Example 4.3   Solve the situation in Example 4.2, except the slope is 10%. The spreadsheet gives $y = 0.152$ m and $b = 7.9$ m before the freeboard. Note that the velocity was 2.1 m/s and that the Froude number was 1.8, indicating supercritical flow. One may consider energy dissipation measures such as piers or abutments at the base of the channel or periodically along the length of long channels. Curves, if any, should be protected, particularly if flows are supercritical.

An advantage of the tractive force method is that, particularly with the new artificial linings, permissible velocity data does not exist. Tractive force data can be collected in a lab setting, whereas permissible velocity data generally requires more expensive field settings. The tractive force design approach compares the older permissible velocity method with traditional materials and designs to new materials where tractive force is relatively easy to measure.

HEC-RAS (Brunner 2016) includes a design tool for the design of stable channels. The tractive force method is demonstrated in the Hydraulic Design Tools option. Using the "DesignFeatures" project, results indicated in Figure 4.5 were observed. Discharge of 50 ft³/s, $z = 4$, $n = 0.02$, and the sediment inputs shown in Figure 4.5 resulted in a depth of 0.6 ft and a bottom width of 18.5 ft. The bolded values of depth D and bottom width W were selected for computation by double-clicking.

Table 4.1 adds to Table 2.6 to provide a step-by-step procedure to the stable channel design tool in HEC-RAS.

Tractive Force for Vegetated Waterways

The tractive force method is available for vegetative channels. This adaptation is ongoing, based on Chin (2013) compared to earlier (e.g. compared with Chin 2006) editions of his text. The NRCS (2007) now includes the tractive force method for vegetated channel design in their Engineering Field Handbook. This text relies on Sturm (2010) and Chin (2013), the former with several worked examples. Seminal work by Kouwen and Li (1980) precipitated the path now being followed. Grasses are classified using the tractive force method as with the permissible velocity method (see Tables 3.1 and 3.2). Instead of a permissible velocity for each class, one uses permissible shear stress, as suggested by Chen and Cotton (1988). Each retardance class has a characteristic stiffness index (MEI), which determines the degree of erectness of the class's vegetation. Table 4.2 summarizes critical data for the various retardance classes. The MEI is the product of the number of stems per unit area times the modulus of elasticity of each stem times the moment of inertia of each plant stem.

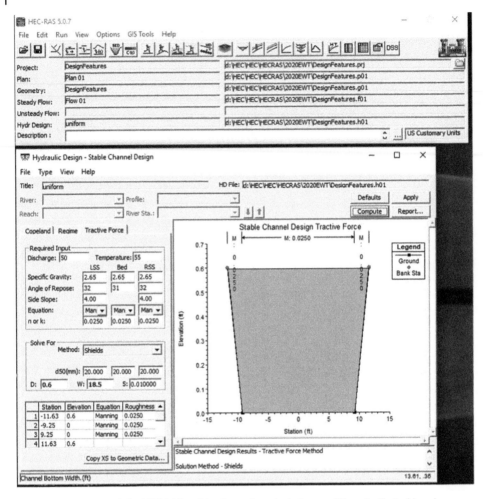

Figure 4.5 Results of the HEC-RAS stable channel analysis for conditions indicated herein. *Source:* HEC-RAS Computer program.

Table 4.1 Using the HEC-RAS stable channel design tool.

1) Create a project file (see Table 2.6) or load an existing HEC-RAS project file.

2) Select a cross section and open the hydraulic design tool (refer to Table 2.6).

3) Enter the target discharge, angle of repose of the aggregate, and side slope.

4) Enter the Strickler k value (or Manning n). A value of 0.034 is a good choice. Strickler is recommended so that a Manning n consistent with aggregate size results.

5) Enter the aggregate size in mm, and enter a value for slope.

6) Use Lane's method. Another available method, that of Shields, is covered in Chapter 10.

7) Apply the data (upper right of the screen).

8) Compute the depth D and width W. Snip the screen. Click on Report and copy to clipboard or file.

Table 4.2 Properties of vegetal retardance classes (adopted from Sturm, 2010).

Vegetal retardance class	Average height, h_s, cm (ft)	Stiffness MEI, N-m^2 (lb-ft^2)	Permissible shear stress τ_p N/m^2 (lb/ft^2)	Computed lining failure minimum criterion ($[k/hs]_{min}$ from a rearranged Equation 4.6)
A	91 (3)	300 (725)	177 (3.7)	0.2
B	61 (2)	20 (50)	100 (2.10)	0.2
C	20 (0.66)	0.5 (1.2)	48 (1.00)	0.3
D	10 (0.33)	0.05 (0.12)	20 (0.60)	0.4
E	4 (0.13)	0.005 (0.012)	17 (0.35)	0.9

Source: Adopted from Sturm (2010) and Chen and Cotton (1988).

Equations 4.6 and 4.7 provide an avenue for computing the Manning n for the flow condition. One first computes an equivalent d_{50} using Equation 4.6 to define the equivalent stand height h_s:

$$k = 0.14h_s \left[\frac{(MEI/\tau_o)^{0.25}}{h_s} \right]^{1.59} \tag{4.6}$$

where MEI is the stiffness index (product of stem density, single stem modulus of elasticity, and stem cross-section moment of inertia), k is the effective stand height (L), h_s is the stand height of the erect vegetation (L), and τ_o is the applied shear stress (F/L^2) in the uniform flow; see Equation 4.3 with the K_s and K_t equal to one.

Minimum values of k/h_s are proposed for each vegetal retardance class. One should test the k/h_s ratio to ensure that it exceeds the stated minimum for the retardance class in question.

Equation 4.7 transforms the roughness height of Equation 4.6 into a Manning roughness value:

$$n = (k)^{1/6} \frac{\varphi_{dim} (R/k)^{1/6}}{\sqrt{8g} \left(a + b \log \dfrac{R}{k} \right)} \tag{4.7}$$

where φ_{dim} is the Manning dimensional constant (1, SI; 1.49, English).

Moreover, a and b are determined using Table 4.3. Kouwen and Li (1980) made these observations of plant responses (classes A–E retardance) to measure flow conditions, with results shown in Tables 4.1 and 4.2. Coefficients a and b reflect the condition (erect or prone) of the vegetation in the flow based on the actual friction velocity $\sqrt{\dfrac{gRS}{\rho}}$ divided by the critical friction velocity $\sqrt{\dfrac{gR_{Crit}S}{\rho}}$. R is the hydraulic radius of the flow, and R_{Crit} is the hydraulic radius that causes a stress condition affecting the vegetation condition in Table 4.3. Class A (high retardance) corresponds to increased resistance and vegetation with a stiff cross-section that remains erect under most flows. On the other hand, class E

Table 4.3 Values of a and b in Equation C.6.

Condition	Classification	Criteria	a	b
1	Erect	$(u_s/u_{sc}) \leq 1.0$	0.15	1.85
2	Prone	$1 < (u_s/u_{sc}) \leq 1.5$	0.20	2.70
3	Prone	$1.5 < (u_s/u_{sc}) \leq 2.5$	0.28	3.08
4	Prone	$2.5 < (u_s/u_{sc})$	0.29	3.50

Source: Adopted from Sturm (2010) and Kouwen and Li (1980).

has a lower resistance and prone (flattened) under many flows. These are similar to the retardance classes used with the permissible velocity method. Differing plant architectures go flat with differing flow exposures and, thus, different n values.

One may occasionally desire to characterize a site with measured stem density, stem modulus of elasticity, stem moment of inertia, and stem height. The critical friction velocity may help one determine the operative retardance class. In this event, one must estimate the critical friction velocity u_{sc} using the minimum of two expressions, as shown in Equation 4.8 (see Sturm 2010). One may review the fundamentals in Chapter 2 to calculate the actual friction velocity. Knowledge of friction velocity (see Equation 4.8) enables the estimation of the actual retardance class.

$$u_{sc} = \min\left(0.28 + 6.33\text{MEI}^2, 0.23\text{MEI}^{0.106}\right) \tag{4.8}$$

Sources such as Persson (1987) and references cited therein provide additional details on various vegetation species' stem moduli. The above approach is not generally used for routine design work.

The following steps summarize the tractive force method for vegetated channels:

1) Choose a vegetal retardance class for the vegetation in question. Select the class associated with the shortest condition. Determine the critical shear from Table 4.2.
2) Estimate the flow depth using $y = \tau_p/\gamma S$ (recall from Equation 2.4 that $\tau = \gamma y S$, where we are taking $R \approx y$. Select a guess value of bottom width b and compute the hydraulic elements A, P, and R).
3) Using Equations 4.7 and 4.8, along with Tables 4.1 and 4.2, calculate the Manning n.
4) Compute the flow using the Manning equation. Adjust the bottom width to reach an agreement between the computed and target flows.
5) Check the k/h_s ratio to ensure it is above the minimum for the retardance class.
6) Now repeat steps 3–5 for the tall condition, holding the bottom width constant. Vary the depth until the flow equals the design target. The final depth should not exceed the maximum critical shear stress for the retardance class.
7) In the case of the bioswale, one chooses a single retardance class consistent with the vegetation management. One may have to manually iterate to find a, b, z, or T that meets minimum critical tractive force requirements.

The downloadables contain an Excel spreadsheet (see the Excel "TractiveForceGrass W" folder) for implementing the above procedure for the trapezoidal channel (SI units only;

imperial units are not supported). There is a Stability page, a Capacity page, and a Bioswale page with the corresponding permissible velocity spreadsheet. Parameters for the selected retardance class are inserted using if–then logic. The Manning roughness n is computed with Equation 4.7. One then chooses a depth corresponding to the critical tractive force. The bottom width is then computed to carry the required flow. The bottom width is maintained in the capacity section to calculate the depth required to handle the flow with the increased n value arising from higher retardance. One checks velocity then adds a freeboard for a final design. The rationale for the bioswale is similar to that used in the permissible velocity method. Bioswale designs entail some manual iterations to ensure that the chosen retardance class's critical tractive forces are not exceeded. The following examples apply the spreadsheets.

Example 4.4 Rework Example 3.1, using the tractive force design approach. Entering the $S = 0.02$, $Q = 14.17\,\text{m}^3/\text{s}$, Stability $= D$, Capacity $= B$, $z = 4$, and freeboard factor of 1.2, and solving the Stability page gives a bottom width $= 177\,\text{m}$. The permissible velocity stability sheet gives $b = 11.6\,\text{m}$. The velocity of the tractive force is $0.54\,\text{m/s}$. In contrast, the permissible velocity of the erosion-resistant soil was $2.4\,\text{m/s}$. The roughness n of the tractive force in the stability phase was 0.07, whereas for permissible velocity $n = 0.031$. Moving to the Capacity design, depth before freeboard for the tractive force analysis $= 0.24\,\text{m}$ compared to $0.57\,\text{m}$ for permissible velocity. The roughness n was 0.15 from the tractive force analysis compared to 0.048 for permissible velocity.

Example 4.5 Rework Example 3.2, using the tractive force method. Inserting the required information into the tractive force spreadsheet's stability page and solving find the bottom width to be $18\,\text{m}$. The velocity is $0.51\,\text{m/s}$, and the roughness n is 0.074. In contrast, the permissible velocity stability sheet would not solve. Reducing permissible velocity to $1\,\text{m/s}$ resulted in a bottom width of $16\,\text{m}$ and depth of $0.085\,\text{m}$. The roughness n was 0.044 with permissible velocity and 0.16 with tractive force. Now, move to the capacity design phase. Tractive force gives a depth of $0.23\,\text{m}$ before the freeboard, whereas the permissible velocity method gave a depth of $0.11\,\text{m}$.

Example 4.6 Rework Example 3.3. Load the parabolic cross-section grass waterway tractive force spreadsheet and insert the required information from the problem statement. The tractive force procedure gives a top width of $539\,\text{m}$. Solving for capacity, the depth before the freeboard is $0.25\,\text{m}$. The roughness is 0.36, and the velocity is $0.12\,\text{m/s}$. Permissible velocity results are as follows: the top width is $17\,\text{m}$, the depth is $0.71\,\text{m}$, the roughness at capacity was 0.03, and the velocity was $1.57\,\text{m/s}$.

Examples 4.3–4.6 suggest substantial differences in the tractive force design results vs. permissible velocity design for grass waterways. Note that the flows in these examples are large, and they highlight differences in the design approaches. Designs using the tractive force method generally appear more conservative, requiring higher bottom widths compared to the permissible velocity method. Velocities appear lower (even below the minimum of $0.6\,\text{m/s}$) in the tractive force method. Manning roughness values appear to be higher. Runoff that was free of sediment is preferred. The tractive force data appear to be reflective of erodible soils. They make no specific allowance for nonerodable soils. Class D's relatively low critical tractive force values seem to force the tractive force method to result in wide channels. Stem densities (e.g. planting density) increases for D and E classes would help (see Table 4.3). Tables 4.2 and 4.3 do not seem to reflect the variation in stem parameters found in the field.

Further field investigations regarding stem count, stem condition, and mixed-species may shed additional light. Designing as an unmowed bioswale, where one sets the width such that the tractive force is less than critical, results in a smaller channel. The less conservative permissible velocity method may suffice in rural situations. The tractive force method may be more appropriate in urban settings, particularly for well-managed bioswales. The tractive force method (which allows no direct control of velocity) can result in velocities less than 0.6 m/s (1.4 ft/s) that facilitate sediment deposition.

Further research seems desirable to better reconcile the results of the two design approaches. Factors such as vegetation age and particular varieties may influence the parameters of Tables 4.1 and 4.2. NRCS (2007) provides an extended discussion of the tractive force method applied to vegetated waterways.

Chapter 2 examined the strengths of the permissible velocity method. The power of the tractive force design methods stands out when vegetation is highly specified and managed. The bioswale appears to be the most likely application of the tractive force method to grassed waterway design.

Temporary liners are needed to help in vegetation establishment as with permissible velocity designs. Haan et al. (1994) discuss the details of a temporary lining specification.

Design method recommendations: When precision in inputs is relatively low (e.g. the vegetation may not be frequently maintained, cover density varies), and the principle velocity method is probably sufficient. In situations where the cover is precisely maintained, the Tractive Force method could be justified.

Details and Origins of the Parabolic Cross-section[1]

By now, you may be curious about why the parabolic cross section has been featured along with easily constructed or manufactured cross sections such as trapezoids or circles. The parabolic approximation occurs in nature. We now explore the reasons for parabolic type occurrences. We first derive the hydraulic elements of the area and the wetted perimeter of the parabola. We use the Mathematica notebook, ParabolaWettedPerim, available in the ParabolicChannels folder of the Mathematica folder. Figure 4.6 provides a definition sketch.

Standard parabolic cross-section area: Follow along with the discussion in the ParabolaWettedPerim Mathematica notebook. We define the parabola from the center of the channel as $y = a*x^2$. Set the depth of the channel as d. Then, the area of interest is the area above the parabola. Taking symmetry into account, we integrate $2*(d - a*x^2)$ over the top width $t/2$. The resulting area is $2\left(\dfrac{dT}{2} - \dfrac{aT^3}{24}\right)$. Now, substitute d for y and x at $t/2$ in

---

1 This section may be skipped without loss in continuity.

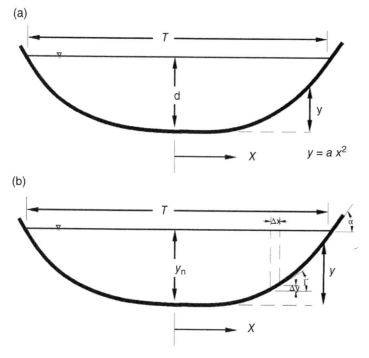

Figure 4.6 (a) Definition sketch of the basic parabolic cross section. (b) Definition sketch of the general parabolic cross section with nomenclature used in the basic analysis relationship.

$y = a*x^2$, Solve for coefficient a, resulting in $a = d/(t/2)^2$. Substituting this result into the area equation results in Equation 4.9.

$$\text{Area} = 2*d*\frac{t}{3}$$ (4.9)

Standard parabolic wetted perimeter: Wetted perimeter evaluation requires integrating along the line defined by the parabola. The integrand for a line integral is Sqrt[$1-f'(x)$], where $f(x) = a*x^2$. Integrating from 0 to $T/2$ and multiplying by 2 gives

$$P = \frac{T^2\left(\dfrac{4d\sqrt{1+\dfrac{16d^2}{T^2}}}{T} + \text{ArcSinh}\left[\dfrac{4d}{T}\right]\right)}{8d} = \frac{T^2\left(\dfrac{4d\sqrt{1+\dfrac{16d^2}{T^2}}}{T} + \text{ArcSinh}\left[\dfrac{4d}{T}\right]\right)}{8d}$$ (4.10)

Now substitute for $16d^2/T^2$ for X^2 and then $4d/T$ for X, giving the following.

$$P = \frac{1}{2}T\sqrt{1+X^2} + \frac{T^2\text{ArcSinh}[X]}{8d}$$ (4.11)

The real part of a complex expansion of the ArcSinh term shows one can write ArchSinh(x) as

$$\text{ArcSinh}(X) = \frac{1}{2}\text{Log}\left[\left(X + \sqrt{1 + X^2}\right)^2\right]$$

(4.12)

Substituting Equation 4.12 into Equation 4.11 results in the long-form expansion of wetted perimeter for a trapezoid (see Chow 1959):

$$P = \frac{1}{2}T\sqrt{1 + X^2} + \frac{T^2\text{Log}\left[X + \sqrt{1 + X^2}\right]}{8d} \text{ where } X = 4 * \frac{d}{T}$$

(4.13)

Taking three terms of a McLaurin series expansion of Equation 4.13 gives

$$P \approx \frac{8d^2}{3T} + T$$

(4.14)

Equation 4.14 is the often used short form of the wetted perimeter equation often cited for the parabolic cross section. Equation 4.14 is generally accepted when $X = 4*d/T$ is less than one, implying a wide, shallow channel. Other hydraulic elements for the parabolic section shown in Table 2.1 are found by applying definitions given in Chapter 2.

The stable hydraulic section and its relation to the parabolic cross section French (2007): We first consider the noncohesive particle's forces in a flowing channel section. Figure 4.3 shows the forces on the side of a trapezoidal section. Note that both tractive force (based on distance from the surface to the point) and gravity is in play on the side, leading to the resultant force R in Figure 4.3. Tractive force alone moves particles along the bottom (the sine of the low channel slope ≈ 0). At the point of incipient motion, we can relate the tractive force on a particle to the weight of the particle times its coefficient of friction, $\tan(\alpha)$, leading to Equation 4.15.

$$A_e\tau_S = W_s \tan(\alpha)$$

(4.15)

Solving Equation 4.15 for τ_s yields Equation 4.16.

$$\tau_S = \frac{W_s}{A_a}\tan(\alpha)$$

(4.16)

The particle moves when the resultant force of Figure 4.3 is equal to the weight of the particle times the coefficient of friction times the cosine of the side angle, as shown in Equation 14.17.

$$W_s\cos(\Gamma)\tan(\alpha) = R = \sqrt{\left(A_e\tau_s\right)^2 + \left(W_s\sin(\Gamma)\right)^2}$$

(4.17)

Substituting the resultant in Equation 4.17 for R and solving for τ_s yields Equation 4.18.

$$\tau_s = \frac{W_S}{A_e}\cos(\Gamma)\tan(\alpha)\sqrt{1 - \frac{\tan^2(\Gamma)}{\tan^2(\alpha)}}$$

(4.18)

The ratio of τ_s / τ_L, where $\tau_L = \gamma\, y_n\, S$, processing through trigonometric identities, and defining the ratio as K, leads to Equation 4.19.

$$K = \frac{\tau_s}{\tau_L}\cos\left(\Gamma\right)\sqrt{1 - \frac{\tan^2\left(\Gamma\right)}{\tan^2\left(\alpha\right)}} = \sqrt{1 - \frac{\sin^2\left(\Gamma\right)}{\sin^2\left(\alpha\right)}} \tag{4.19}$$

Equation 4.19 enables the computation of tractive force on the channel side based on geometric and sediment properties. Equation 4.4 stems directly from the development leading to Equation 4.19. This analysis applies only to noncohesive channel linings.

Considering Figure 4.6b, the tractive force, τ_s, at any point on the side of the curved surface, can be expressed differentially as Equation 4.20.

$$\tau = \frac{\gamma\, y\, S\, dx}{\sqrt{dx^2 + dy^2}} = \gamma\, y\, S\cos\left(\Gamma\right) \tag{4.20}$$

In Equation 4.20, we use the approximation of tractive force with depth above the point of interest, S is the slope, and y is the distance from the bottom to a point on the side. Figure 4.6b provides other nomenclature. We substitute Equation 4.19 into Equation 4.20 yields Equation 4.21.

$$\tau_s = K\tau_B = \gamma\, y_n\, S\cos\left(\Gamma\right)\sqrt{1 - \frac{\tan^2\left(\Gamma\right)}{\tan^2\left(\alpha\right)}} \tag{4.21}$$

where $\tau_L = \gamma\, y_n\, S$ as before. Combining Equations 4.20 and 4.21 and solving for y gives Equation 4.22.

$$y = \frac{y_n}{\tan\left(\alpha\right)}\sqrt{\tan^2\left(\alpha\right) - \tan^2\left(\Gamma\right)} \tag{4.22}$$

Setting $dy/dx = \tan(\Gamma)$ and isolating dy/dx enables Equation 4.23.

$$\left(\frac{dy}{dx}\right)^2 + \left(\frac{y}{y_n}\right)^2 \tan^2\left(\alpha\right) - \tan^2\left(\alpha\right) = 0 \tag{4.23}$$

The solution of Equation 4.23 defines the shape of the profile giving a resultant force R that is constant regardless of position on the cross-section profile. In brief, in the bottom, the tractive force of the fluid controls. As one moves up the side and the tractive force lessens, the gravitational component increases to compensate. Figure 4.6b looks like the parabola in Figure 4.6a and is a very close approximation, as we show.

Equation 4.23 is solved in the Mathematica notebook, ParabolicAnalysisDesign, which is located in the Mathematica, Parabola folder of the downloadables. The initial solution of Equation 4.23 is given in Equation 4.24.

Soln = DSolve$\left[\left\{\left((y'[x])^{\wedge}2+\left(((y[x]/yn)^{\wedge}2)*\tan[\alpha]^{\wedge}2\right)-\tan[\alpha]^{\wedge}2=0\right),\right.\right.$
$\left.\left.(y[T/2]=0)\right\},y[x],x\right]$

$\cdots$**Solve** : Inverse functions are being used by Solve, so some solutions may not be found; use Reduce for complete solution information.

$\cdots$**Solve** : Inverse functions are being used by Solve, so some solutions may not be found; use Reduce for complete solution information.

$\cdots$**Solve** : Inverse functions are being used by Solve, so some solutions may not be found; use Reduce for complete solution information.

$\cdots$**General** : Further output of Solve :: ifun will be suppressed during this calculation.

$$\left\{\left\{\left\{y[x]\rightarrow-\dfrac{ynTan\left[\dfrac{\dfrac{1}{2}T\tan[\alpha]-x\tan[\alpha]}{yn}\right]}{\sqrt{1+Tan\left[\dfrac{\dfrac{1}{2}T\tan[\alpha]-x\tan[\alpha]}{yn}\right]^{2}}}\right\},\left\{y[x]\rightarrow\dfrac{ynTan\left[\dfrac{\dfrac{1}{2}T\tan[\alpha]-x\tan[\alpha]}{yn}\right]}{\sqrt{1+Tan\left[\dfrac{\dfrac{1}{2}T\tan[\alpha]-x\tan[\alpha]}{yn}\right]^{2}}}\right\}\right\}\right\}$$

FullSimplify$[\%]$

$$\left\{\left\{\left\{y[x]\rightarrow-\dfrac{ynTan\left[\dfrac{(T-2x)\tan[\alpha]}{2yn}\right]}{\sqrt{Sec\left[\dfrac{(T-2x)\tan[\alpha]}{2yn}\right]^{2}}}\right\},\left\{y[x]\rightarrow-\dfrac{ynTan\left[\dfrac{(T-2x)\tan[\alpha]}{2yn}\right]}{\sqrt{Sec\left[\dfrac{(T-2x)\tan[\alpha]}{2yn}\right]^{2}}}\right\}\right\}\right\}$$

Soln1 = Soln$[[2,1]]$

$$y[x]\rightarrow\dfrac{ynTan\left[\dfrac{\dfrac{1}{2}T\tan[\alpha]-x\tan[\alpha]}{yn}\right]}{\sqrt{1+Tan\left[\dfrac{\dfrac{1}{2}T\tan[\alpha]-x\tan[\alpha]}{yn}\right]^{2}}} \qquad (4.24)$$

Several simplifications and substitutions allow writing Equation 4.24, as shown in Equation 4.25.

$$:=Soln4 = yn*Simplify\left[Soln3\right]$$

$$y=ynCos\left[\dfrac{\pi x}{T}\right] \qquad (4.25)$$

A McLaurin's series expansion of the right-hand side of Equation 4.25 gives:

$$\text{Series}\left[\text{yn}\cos\left[\frac{\pi\,\text{xx}}{T}\right],\{\text{xx},0,9\}\right]$$

$$\text{yn}-\frac{\left(\pi^2\,\text{yn}\right)\text{xx}^2}{2\,T^2}+\frac{\pi^4\,\text{yn}\,\text{xx}^4}{24\,T^4}-\frac{\left(\pi^6\,\text{yn}\right)\text{xx}^6}{720\,T^6}+\frac{\pi^8\,\text{yn}\,\text{xx}^8}{40320\,T^8}+0\big[\text{xx}\big]^{10}$$

Note that higher-order terms of the series decay rapidly after the second term (xx is an introduced variable for x). Thus, the parabola is approximated rather well, which justifies the hydraulic elements derived earlier.

We conclude the discussion by deriving the hydraulic elements for the exact representation. The Mathematica notebook contains the details and highlights are summarized as follows.

$$\text{In}\big[21\big]:=\text{Wperfkernel}=\text{Sqrt}\left[1+\left(-\big(y\big[z\big]/\text{yn}\big)^\wedge2*\text{Tan}\big[\alpha\big]^\wedge2+\text{Tan}\big[\alpha\big]^\wedge2\right)\right]$$

$$\text{Out}\big[21\big]=\sqrt{1+\text{Tan}\big[\alpha\big]^2-\cos\left[\frac{z\,\text{Tan}\big[\alpha\big]}{\text{yn}}\right]^2\text{Tan}\big[\alpha\big]^2}$$

$$\text{In}\big[22\big]:=\text{Wperfkernel1}=\text{Wperfkernel}/.\text{yn}\rightarrow T*\text{Tan}\big[\alpha\big]/\pi$$

$$\text{Out}\big[22\big]=\sqrt{1+\text{Tan}\big[\alpha\big]^2-\cos\left[\frac{\pi z}{T}\right]^2\text{Tan}\big[\alpha\big]^2}$$

$$\text{In}\big[23\big]:=\text{WP1}=\text{Assuming}\left[\text{Re}[\text{Tan}\big[\alpha\big]^\wedge2]\geq-1\|\big[\text{Tan}\big[\alpha\big]^\wedge2\not\in\text{Reals},\right.$$
$$\left.2*\text{Integrate}\left[\text{Wperfkernel1},\{z,0,T/2\}\right]\right]$$

$$\text{Out}\big[23\big]=\text{Conditional Expression}\left[\frac{2\,T\,\text{EllipticE}\big[-\text{Tan}\big[\alpha\big]^2\big]}{\pi},\ T\in\mathbb{R}\right]$$

$$\text{In}\big[24\big]:=\text{WP1}$$

$$\text{Out}\big[24\big]=\text{Conditional Expression}\left[\frac{2\,T\,\text{EllipticE}\big[-\text{Tan}\big[\alpha\big]^2\big]}{\pi},\ T\in\mathbb{R}\right]\quad(4.26)$$

The conditional expression in Equation 4.26 is the wetted perimeter. The conditions for validity are satisfied. Equation 4.27 is the Area. We note that French (2007) has sin(α) in the elliptical integral in Equation 4.26 instead of $-\tan(\alpha)^2$. The approximation may lead to a small difference in solutions because $\sin(\alpha)\approx-\tan(\alpha)^2$ and elliptical integral tables are not readily available for Elliptical($-\tan(\alpha)^2$). Numerical approximations in the hand solution may also explain the small discrepancies.

$$\text{AreaParabola2}=2*\text{Integrate}\left[\text{Soln4},\{\text{xx},0,T/2\}\right]$$
$$\frac{2\,T\,\text{yn}}{\pi}\quad(4.27)$$

Other hydraulic elements follow from the element definitions.

The design of the stable hydraulic channel proceeds as follows. One first determines the base flow that the specified sediment is at the point of incipient motion. Substitute Equations 4.26 and 4.27 into the Manning equation, then substituting the critical tractive force relation for yn gives the following.

$$\text{Qbase} = (1/n)*\big((\text{AreaParabola}/\text{WP1})^\wedge 0.67\big)*(\text{Slope}^\wedge.5)*\text{AreaParabola}$$

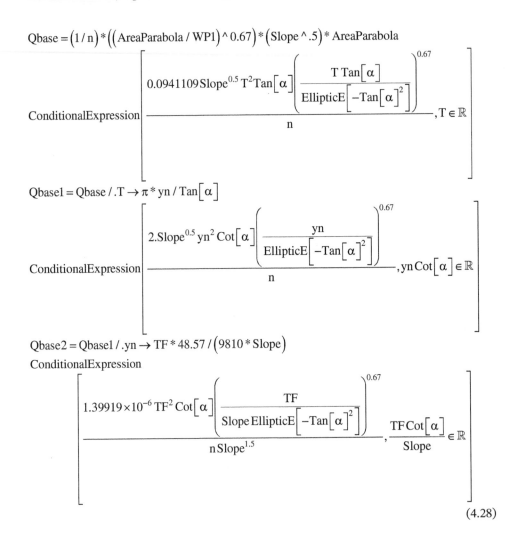

$$\text{ConditionalExpression}\left[\dfrac{0.0941109\,\text{Slope}^{0.5}\,T^2\text{Tan}[\alpha]\left(\dfrac{T\,\text{Tan}[\alpha]}{\text{EllipticE}\left[-\text{Tan}[\alpha]^2\right]}\right)^{0.67}}{n},\,T\in\mathbb{R}\right]$$

$$\text{Qbase1} = \text{Qbase}/.T\to\pi*yn/\text{Tan}[\alpha]$$

$$\text{ConditionalExpression}\left[\dfrac{2.\text{Slope}^{0.5}\,yn^2\,\text{Cot}[\alpha]\left(\dfrac{yn}{\text{EllipticE}\left[-\text{Tan}[\alpha]^2\right]}\right)^{0.67}}{n},\,yn\,\text{Cot}[\alpha]\in\mathbb{R}\right]$$

$$\text{Qbase2} = \text{Qbase1}/.yn\to TF*48.57/(9810*\text{Slope})$$

$$\text{ConditionalExpression}$$

$$\left[\dfrac{1.39919\times10^{-6}\,TF^2\,\text{Cot}[\alpha]\left(\dfrac{TF}{\text{Slope EllipticE}\left[-\text{Tan}[\alpha]^2\right]}\right)^{0.67}}{n\,\text{Slope}^{1.5}},\,\dfrac{TF\,\text{Cot}[\alpha]}{\text{Slope}}\in\mathbb{R}\right]$$

$$(4.28)$$

All the conditionals are real numbers and are thus satisfied. Substituting the slope, n, angle of repose, and critical tractive force into Equation 4.28 gives the base flow. We plot a typical plot in Figure 4.7.

The design flow is usually larger or smaller than the base flow. In the design flow case, we insert a rectangular section in the profile where the bottom of the rectangle experiences the same tractive force as the parabolic section's original bottom. We adjust the wetted

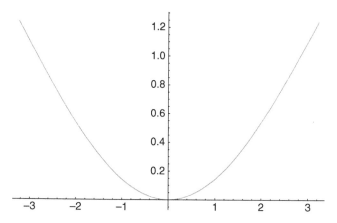

Figure 4.7 Plot of a typical base parabola for the hydraulically stable channel, with scales on *x* and *y* being different.

perimeter by adding a constant T and adjust the area by adding the depth $\times$ T. Substituting these modifications into the Manning equation and iteratively solving for T gives the extension. Equations 4.29 summarize the computations.

$= \text{Wperlarge} = \text{WPbase} + \text{Tprime}$

$= 7.02176 + \text{Tprime}$

$=$

$= \text{AreaLarge} = \text{AreaParabolaBase} + \text{Tprime} * \text{NormDepth}$

$= 5.10004 + 1.23777\,\text{Tprime}$

$= \text{QdesignLarge} = \text{QlargeTarget} = (1/n) * (\text{AreaLarge} / \text{Wperlarge})\wedge.67) * (\text{Slope}\wedge 5)$
$\qquad\qquad * \text{AreaLarge} == 0$

$= \text{QlargeTarget} - 1.\left(5.10004 + 1.23777\,\text{Tprime}\right)\left(\dfrac{5.10004 + 1.23777\,\text{Tprime}}{7.02176 + \text{Tprime}}\right)^{0.67} == 0$

$= \text{QlargeTarget} = 8.5$

$= 8.5$

$= \text{QdesignLarge}$

$= 8.5 - 1.\left(5.10004 + 1.23777\,\text{Tprime}\right)\left(\dfrac{5.10004 + 1.23777\,\text{Tprime}}{7.02176 + \text{Tprime}}\right)^{0.67} == 0$

$= \text{val} = \text{FindRoot}\left[\text{QdesignLarge}, \{\text{Tprime}, .1\}\right]$

$= \{\text{Tprime} \rightarrow 3.30457\}$ \hfill (4.29)

Thus, to accommodate the design flow, in this instance, we add 3.30 m to the top width, giving the channel profile shown in Figure 4.8.

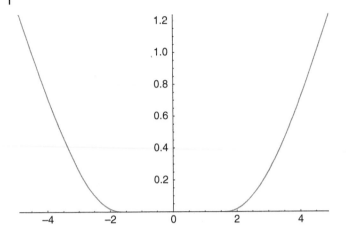

Figure 4.8 Plot of a typical base parabola for the hydraulically stable channel, expanded to accommodate a larger design flow compared to the base flow. Scales on x and y are different.

If the design flow is smaller than the base flow, one must subtract part of the base parabolic cross section. We use the same approach as that used for scaling parabolic cross-sections earlier. The computations appear as follows in Equation 4.30.

$$:= \text{TsmallAdjust} = \text{Topwidthbase} * \left(1 - \text{Sqrt}\left[\text{QsmallTarget/Qbase2}\right]\right)$$
$$\models 1.9609$$
$$:= \text{Tsmall} = \text{Topwidthbase} - \text{TsmallAdjust}$$
$$\models 4.51133$$
$$:=$$
$$:= \text{Ynsmall} = \text{Tsmall} * \text{Tan}\left[\alpha\right]/\pi$$
$$\models 0.862759 \tag{4.30}$$

Spreading the topwidth adjustment over both sides, one arrives at a plot similar to Figure 4.9.

Here are the details of problems solved to develop Figures 4.7–4.9.

Example 4.7 Find the design cross-section to convey a flow of 8.5 m³/s down a slope of 0.0004. The critical tractive force is 0.1*48.57 N/m² = 4.857 N/m², and the roughness is 0.02. From the Mathematica notebook, Qbase is 4.1165 m. The normal depth is 1.23 m (using the tractive force equation). The base topwidth is 6.47 m, the base wetted perimeter is 7.02 m, the base area is 5.1 m². Figure 4.7 shows the base cross section for this problem. Since the design flow of 8.5 m³/s exceeds the base flow, additional topwidth is needed. Modified area is base area plus yn * tprime = 5.1 + 1.23*tprime. Modified wetted perimeter is base wetted perimeter + tprime = 7.02 + tprime. The design flow of 8.5 m³/s = Manning with the modified elements substituted therein. One then numerically solves for tprime for tprime = 3.3 m. Put half of tprime on each side of Figure 4.6 to arrive at Figure 4.7. French (2007) solved this problem by hand and found results similar to those shown here.

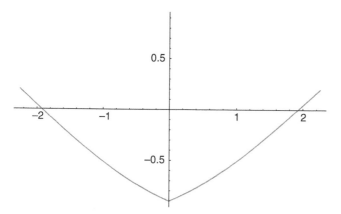

Figure 4.9 Plot of a typical base parabola for the hydraulically stable channel, contracted to accommodate a smaller design flow compared to the base flow. Scales on *x* and *y* are different.

Suppose the design flow was $2\,m^3/s$. Find tsmalladjust by taking the base topwidth times (1 − sqrt(Qsmall/Qbase)). Tsmall then equals base topwidth − tsmall adjust. In this case, we find tsmalladjust = 1.96 m and Tsmall = 4.51 m. The depth of the small channel is Tsmall * tan(α)/π = 0.86. Figure 4.8 shows the final plot.

In practice, one could use the d_{35} to represent the sediment for determining the critical tractive force. The larger particles mostly hide smaller sediment – the stable hydraulic section of more theoretical interest than practical interest. The stable hydraulic design approach requires precision in inputs that is not often available in field applications.

The tractive force method is gaining popularity because it provides a structure suitable for serving as an umbrella method for living and temporary nonliving linings, along with earthen channels. The NRCS (2007) recommends the tractive force method when the aggregate lining is large sand or larger. Permissible velocity is suggested for fine sands and smaller aggregate. The NRCS (2007) now recommends the tractive force method for well-managed vegetation situations such as the urban bioswale. The tractive force method readily extends to other lining media, such as temporary lining materials used to stabilize channels during the construction phase. Kilgore and Cotton (2005) provide the needed parameters for tractive force temporary lining design. Simplifications of the critical tractive force methods are sometimes used (e.g. see Fifield 2004). One may also use a smaller design storm, such as a two- or five-year storm. Fifield (2004) gives additional critical tractive force values and design roughness values for various linings. Refer to texts and references such as Graf (1971); Chin (2013); French (2007); and NRCS (2007).

A method related to tractive force, tractive power, is recommended by the NRCS (2007) when the boundary material is not discrete particles. Tractive power is a product of velocity times the tractive force. The NRCS (2007) method bases design on the unconfined compressive strength of the soil. High unconfined compressive strength requires more power to cause erosion. The NRCS (2007) provides step by step methodology for completing tractive power designs. The method is generally presented in the context of alluvial channel sediment transport. Alluvial channels are unlined channels carrying substantial sediment loads. Alluvial channels have bedforms that add roughness depending on sediment load and flowrate. They are more fully addressed in Chapter 10.

Costing Channel Designs

Each of the design spreadsheets referenced thus far has a section on cost calculation, assuming that a lining cost per square foot of lining over the wetted perimeter is constant. An excavation cost per unit volume of material removed is known. The cost of the channel based on excavation and lining may then be computed. The simple accounting ignores factors such as machinery mobilization, job-specific labor costs, and site-specific elements such as trees and stones hindering excavation. Appendix C provides some cost data for various lining materials. Cost decisions should be made collaboratively, with much input from senior engineers. The NRCS field office in the area also can provide experience-based recommendations.

Thus far, we have presented optimum cross sections, sub-optimum cross sections where soil conditions constrain z, and arbitrary cross sections where other factors determine bottom width. We now consider the effects of a cost differential of the sides and bottom.

Ideally, the minimum cost for a channel cross section motivated the search for the minimum hydraulic sections discussed in Chapter 2. With the trapezoidal section, the implied assumption is that the unit lining cost of the sides equals that of the bottom. Substituting the hydraulic elements into Equation 2.9 results in Equation 4.31.

$$AR^{2/3} = \frac{\left(by + zy^2\right)^{5/3}}{\left(b + 2y\sqrt{1+z^2}\right)^{2/3}} = \frac{Qn}{\Phi\sqrt{S}} \qquad (4.31)$$

Rearranging Equation 4.31 for an implicit equation for flow depth y gives Equation 4.32 (French 2007, drawing upon Trout's work 1982).

$$y = \frac{\left(\dfrac{b}{y} + 2\sqrt{1+z^2}\right)^{1/4}}{\left(\dfrac{b}{y} + z\right)^{5/8}} \left(\frac{Qn}{\phi S}\right)^{3/8} \qquad (4.32)$$

Figure 4.10 shows elements contributing to the cost of a lining for a general trapezoidal cross section. Equations 4.33 accumulate the costs as follows.

$$C_b = \mu_b t_b \left(b + b'\right) = Bb + k \qquad (4.33a)$$

$$C_s = \mu_s t_s \left(2E + 2E'\right) = 2\Gamma\left(y + F\right)\sqrt{1+z^2} \qquad (4.33b)$$

$$C = C_b + C_s = Bb + k + 2\Gamma\left(y + F\right)\sqrt{1+z^2} \qquad (4.33c)$$

The nomenclature for Equations 4.41 is as follows.

C total lining materials cost,
C_b material cost for channel base per unit length,
C_s material cost of sides per unit channel length,

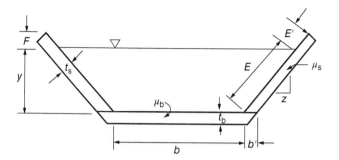

Figure 4.10 Schematic diagram of cost elements in a general trapezoidal channel cross section used in cost optimization with differing bottom and side lining costs.

b' bottom corner width,

t_b channel base-lining thickness,

t_s channel side-lining thickness,

E wetted side length,

E' freeboard side length,

μ_b cost of the base-lining material per unit volume,

μ_s cost of side-lining material per unit volume,

B cost of the base-lining material for specified thickness per unit area,

K cost of corner materials per unit length,

Γ cost of side-lining material for specified thickness per unit area, and

F vertical freeboard requirement.

Marginal changes in cost with respect to bottom width b and depth y can be equated as follows in Equation 4.33d.

$$\frac{\partial \left(AR^{\frac{2}{3}} \right)}{\partial b} = \frac{\frac{\partial C}{\partial b}}{\frac{\partial C}{\partial Y}} \quad (4.33d)$$

Substituting Equations 4.31 and 4.33c into Equation 4.33, applying equalities, resulted in the optimal solution given in Equations 4.34.

$$K_1 \left(\frac{y}{b} \right)^2 + K_2 \frac{y}{b} + K_3 = 0 \quad (4.34a)$$

where

$$K_1 = 20\left(z^2 + 1 \right) - \left(1 + 4\frac{B}{\Gamma} \right) 4z\sqrt{z^2 + 1} \quad (4.34b)$$

$$K_2 = \left(1 - \frac{B}{\Gamma}\right)6\sqrt{z^2 + 1} - 10z\frac{B}{\Gamma} \qquad (4.34c)$$

$$K_3 = -5\frac{B}{\Gamma} \qquad (4.34d)$$

Solving Equations 4.34 results in the following optimal solution for b/y shown in Equation 4.35.

$$\frac{b}{y} = \frac{2K_1}{-K_2 z + \left(K_2^2 + 20K_1\frac{B}{\Gamma}\right)^{1/2}} \qquad (4.35)$$

Given Q, S, n, z, B, and Γ, values for K_1, K_2, and K_3 are determined knowing the cost factors in Equations 4.33. The minimum cost value of the b/y ratio is then estimated using Equation 4.35. One then computes the depth using Equation 4.33. Then, multiply the b/y by the depth to get the bottom width. The Excel spreadsheet, AdvancedLiningCosts, in the Optimums folder computes the costs and final design for Trout's (1982) example. This spreadsheet computes the optimal b/y ratio given a unit b following a somewhat different procedure from Equations 4.35 and 4.35. One then introduces flow conditions using the computed b/y ratio to establish channel dimensions.

Interestingly, when an optimum $z = 0.5445$ and equal cost of the bottom and side materials were similar, the optimum $b/y = 1.15$ was recovered. Likewise, if z was constrained to a non-optimum value, and the ratio of side cost to bottom cost was unity, the expected b/y value was recovered. The spreadsheet addresses rectangular and trapezoidal cross sections. It is not relevant to the triangular, circular, or parabolical cross sections.

Steady Uniform Flow Conclusion

Our discussion of various uniform flow channel design is now mostly complete. We have discussed lined and unlined channels of a variety of standard cross-sections. The trapezoid is the most common cross-section, close to the minimum optimum cross-section, the semi-circle from the wetted perimeter, and area considerations. We have discussed a variety of linings ranging from artificial to vegetation to earthen linings. We have considered the permissible velocity method and the tractive force method. We argue why natural channels often take on an approximate parabolic cross section shape based on tractive force.

Earthen and other natural linings designed using permissible velocity or tractive force methods often result in far from optimum channels from a cost viewpoint. Cost optimality does not address shear-related issues in earthen channels.

Comments regarding maintenance in past chapters apply with tractive force designs as well. Tractive force attempts to be more precise than permissible velocity; however, the difference is small. The NRCS (2007) suggests using the tractive force method when the lining aggregate size is small gravel or larger. On the other hand, the NRCS (2007) recommends the permissible velocity method when aggregate sizes are fine sands and smaller. Critical tractive force suffers from the lack of precision that also is characteristic of permissible velocity. We revisit tractive force in our discussion of alluvial channels and sediment transport in Chapter 10, called Earthen Channel design III. Appendix D contains a table summarizing the design approaches discussed thus far.

We now transition to varied flow situations. The energy equation and momentum equations supplement the Manning and continuity equations, providing the capability to analyze additional relevant problems. Unlined channels generally transport sediment, which influences the roughness. A comprehensive examination of the interactions between sediment transport and flow is covered in Chapter10.

Problems and Questions

1   Redo Examples 2.5 with a flow of 1 cfs. Compare and contrast results with results found from the permissible velocity method. Discuss the differences in the two design methodologies and suggest the pros and cons.

2   Redo Example , assuming a flow of $0.5 \, m^3/s$. Compare and contrast results with results found from the corresponding examples using the permissible velocity approach. Discuss the differences in the two design methodologies and suggest the pros and cons.

3   Redo Example 4.3, assuming a flow of $0.5 \, m^3/s$. Compare and contrast results with results found from the corresponding examples using the permissible velocity approach. Discuss the differences in the two design methodologies and suggest the pros and cons.

4   Compare the tractive force solution to the permissible velocity solution for problem 3.2. Discuss the pros and cons.

5   Redo Example 4.2, assuming a flow of $0.1 \, m^3/s$. Compare and contrast results with results found from the corresponding example using the permissible velocity approach. Discuss the differences in the two design methodologies and suggest the pros and cons.

6   Take the results of Example 4.2 with the $0.1 \, m^3/s$ flow, and suggest a temporary lining that would suffice.

7   Suggest a temporary lining from the OhioDOT spreadsheet page for the channel of Example 4.1 as designed using the tractive force method.

8   Suggest a temporary lining design strategy for Example 4.3 as designed using the tractive force method.

9   Find the cost of constructing 500 ft of the channel of Example 4.1, where the Excavation cost is $10/ft^3$, the lining cost (equal side and bottom cost) is $2.00/ft^2$ of the lining. Assume $z = 4$ as the channel must be crossed. Compare with the case where the side-lining cost (Γ) is \$4 per unit area, and the bottom cost (B) is \$2 per unit area. Assume other associated lining costs are the same as the Trout (1982) data of the spreadsheet example.

10 Develop a chart of the various types of channels we have discussed thus far and list the pros and cons or strengths and weaknesses of the different approaches.

11 List the pros and cons of the standard cross-sections used in channels.

12 Project (revisit of a project in Chapter 2): A $10\,\text{m}^3/\text{s}$ flow is conveyed down a $1000\,\text{m}$ direct distance with a fall of $1\,\text{m}$. Flow events are occasional. An alternate path of $12\,000$ m is available (e.g. a longer way around). The soil is a shale. If the channel runs along the longer route, assume it must be crossed and built with a small dozer. Further, assume that if the riprap, if used (the size of the riprap), a side slope no steeper than $z = 6$ is used. The crossing is not necessary along the steeper route. A dozer is available for constructing the short channel also. Excavation costs are $\$200/\text{m}^3$ soil and an additional $\$10.50$ per linear m per unit wetted perimeter of concrete lining or $\$1.25$ per linear foot per unit wetted perimeter for a rock/cobble lining. Identify a feasible and most economical channel design. A grassed lining of Fescue having a cost of $\$1.75$ per square meter is available. A mat lining is available with a tractive force of $4\,\text{lb/ft}^2$ costs $\$20$ per square meter. Identify a feasible and most economical channel design. Consider additional options given those covered since Chapter 2.

13 Project (revisit of project in Chapter 2): A $100\,\text{m}^3/\text{s}$ flow is conveyed down a $1000\,\text{m}$ direct distance with a fall of $100\,\text{m}$. Flow events are occasional. An alternate path of $10\,000\,\text{m}$ is available (e.g. a longer way around). The soil is a shale. If the channel runs along the longer route, assume it must be crossed and built with a small dozer. Further, assume that if the riprap, if used, the side slope no steeper than $z = 6$ is used. Crossing is not necessary along the steeper route. A dozer is available for constructing the short channel also. Excavation costs are $\$200/\text{m}^3$ soil and an additional $\$10.50$ per linear m per unit wetted perimeter of concrete lining or $\$1.25$ per linear foot per unit wetted perimeter for a rock/cobble lining. Identify a feasible and most economical channel design. Consider additional options given those covered since Chapter 2.

14 Using HEC-RAS design tools, specify the flow depth of a stable channel that conducts a flow of $100\,\text{cfs}$ down a slope of 0.5% in aggregate of $1''$ gravel. Use Lane's method.

15 Construct a QFD table using factors included in Tables 2.9 and 2.10, having the tractive force option for one of the above projects. What additional factors would you add to the customer requirements or engineering requirements?

16 Using Equations 4.32 and Appendix C (or other available cost data), estimate the cost data per square feet (or meter) of a grass-lined channel.

17 Consult a public works department to recreate a current table of costs for excavating and lining a concrete channel.

18 Consult a public works department to recreate a current table of cost for excavating and lining a grassed waterway. Include a temporary lining.

<u>The following problems require Mathematica</u>

19 Apply the stable hydraulic design approach to the channel of Example 4.1. Compare with the parabolic-based tractive force method.

20 Apply the stable hydraulic design approach to the channel of Example 4.3. Compare with the parabolic-based tractive force method.

References

Brunner, G.W. (2016). *HEC-RAS River Analysis System: User's Manual*. Davis, CA: US Army Corps of Engineers, Hydrologic Engineering Center.

Chen, Y.H. and Cotton, G.K. (1988). *Design of Roadside Channels with Flexible Linings*. FHWA-IP-87-7, HEC 15. Federal Highway Administration, USDOT, National Tech. Information Service, Springfield, VA.

Chin, D.A. (2006). *Water Resources Engineering*, 2e. Upper Saddle River, NJ: Prentice-Hall Publishing Co.

Chin, D.A. (2013). *Water Resources Engineering*, 3e. Upper Saddle River, NJ: Prentice-Hall Publishing Co.

Chow, V. T. (1959). *Open Channel Hydraulics*, New York, NY: McGraw-Hill.

Fifield, J.S. (2004). *Designing for Effective Sediment and Erosion Control on Construction Sites*. Santa Barbara, CA: Forester Press.

French, R.H. (2007). *Open Channel Hydraulics*, 2e. Highlands Ranch, CO: Water Resources Publications.

Graf, W.H. (1971). *Hydraulics of Sediment Transport*. New York, NY: McGraw-Hill.

Haan, C.T., Barfield, B.J., and Hayes, J.C. (1994). *Hydrology and Sedimentology of Small Catchments*. New York, NY: Academic Press.

Jobson, H.E. and Froehlich, D.C. (1988). Basic hydraulic principles of open-channel hydraulics. USGS Open-File Report 88-707, Reston, VA.

Kilgore, R.T. and Cotton, G.K. (2005). *Design of Roadside Channels with Flexible Linings*. HEC-15, 3e. Arlington, VA: FHWA.

Kouwen, N. and Li, R.M. (1980). Biomechanics of vegetated channel linings. *J. Hydr. Div, ASCE* 106 (6): 1085–1103.

Lane, E.W. (1955). Design of stable channels. *Trans. ASCE* 120: 1234–1279.

Lane, E.W. and Carlson, E.J. (1953). Some factors affecting the stability of canals constructed in coarse granular materials. Proceedings, Minnesota International Hydraulic Convention.

NRCS (2007). Threshold channel design, Chapter 8 of Part 654. In: *Stream Restoration Design National Engineering Handbook*. Washington, DC: USDA-NRCS https://directives.sc.egov.usda.gov/OpenNonWebContent.aspx?content=17784.wba (accessed 15 March 2021).

Persson, S. (1987). *Mechanics of Cutting Plant Material*. ASAE Monograph #7. St Joseph, MI: American Society of Agricultural & Biological Engineers.

Sturm, T.W. (2010). *Open Channel Hydraulics*, 2e. New York, NY: McGraw-Hill.

Trout, T.J. (1982). Channel design to minimize lining material costs. *Proc. ASCE* 108 (IR4): 242–249.

US Bureau of Reclamation (1973). *Design of Small Dams*, 2e. Washington, DC: US Dept. of the Interior – Bureau of Reclamation.

US Bureau of Reclamation (1987). *Design of Small Dams*. Denver, CO.

5

The Energy Equation and Gradually Varied Flows

Channel design work has thus far involved the mean velocity. Understanding depths in channel cross-section, elevation, or slope transitions requires the introduction of energy concepts. Being able to account for flow depths through transition situations provides the opportunity to cost transition structures more appropriately. The effects of infrastructure, such as bridges on flows in natural channels, can inform structure designs. The energy equation is a primary tool in transition analysis. Figure 5.1 shows a general schematic of gradually varied flow at a channel transition.

Goals

To identify and quantify the differences between laminar and turbulent flows.

- To develop a relationship between velocity and depth in laminar and turbulent flows.
- To analyze and compute flow profiles away from various channel transitions ranging from expansions, contractions, and slope changes.
- To analyze and compute flows from a reservoir into a channel

Energy Preliminaries – Velocity Profiles and Boundary Effects

Velocity profiles: Laminar flows conform to Equation 5.1.

$$\tau = \mu \frac{du}{dy} \tag{5.1}$$

where τ is the shear stress (Pa or lb/ft$^{2)}$, u is the horizontal velocity at a point y away from the bottom; and, μ is the dynamic viscosity (Pa-s or lb/ft^2-s).

A similar equation characterizes turbulent flows with one other term, as shown in Equation 5.2.

$$\tau = \left(\mu + \eta\right)\frac{du}{dy} \tag{5.2a}$$

Open Channel Design: Fundamentals and Applications, First Edition. Ernest W. Tollner.
© 2022 John Wiley & Sons Ltd. Published 2022 by John Wiley & Sons Ltd.
Companion website: www.wiley.com/go/tollner/openchanneldesign

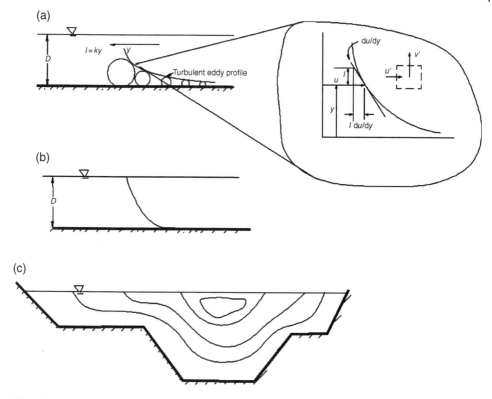

Figure 5.1 Velocity profile development in a wide channel (a and b), and hypothetical measured profile in a compound channel (c).

where η is a turbulent eddy viscosity of the same units as μ.

The dynamic viscosity μ is a fluid property that depends on temperature (we generally use the 10 °C or 20 °C values, which are similar). There is a weak dependence on the pressure that is often ignored. On the other hand, η depends only on the fluid density, the velocity gradient, and a term to be defined called the "mixing length." The term η goes to zero as the Reynolds number falls below 2000.

Prandtl (quoted by Streeter 1971) developed a useful turbulence theory called the mixing length theory (MLT). In the MLT in open channels, the turbulent shear stress is related to velocity fluctuations u' and v'. The turbulent velocity u', if averaged separately over a brief time, approaches zero (see Figure 5.1a).

Turbulent eddies are not periodic. Eddies have a stochastic character. The velocity at any instant is probabilistic. Instantaneous velocities are expressed as the sum of the mean velocity $\bar{u}$ + an instantaneous deviation u'. The mean of u' is zero. Turbulence may be isotropic, wherein measurements in one dimension are representative of measures in all three dimensions. Turbulence is frequently anisotropic wherein the instantaneous velocities are different in one or more dimensions. The fluctuating velocity u' has implications for sediment transport, which are more fully developed in Chapter 10.

The turbulent shear stress in the Prandtl theory is expressed as follows.

$$\tau = \rho u' v' \tag{5.2b}$$

Based on water being incompressible, u' has to be like v' as shown in Figure 5.1a. Therefore, Prandtl argued from an analogy with the kinetic theory of gases that

$$u' \approx v' \approx l\frac{du}{dy} \tag{5.2c}$$

where l is the mixing length with dimensions of L. The mixing length is proportional to the size of an eddy, as shown in Figure 5.1a. Another influential contributor to turbulent flow theory, von Karmen, also made significant advances in flow characterization. The eddy size increases as one moves away from the channel bottom.

Because of Equation 5.2a, b, and c, one can write the turbulent shear as follows.

$$\tau = \rho l^2 \left(\frac{du}{dy}\right)^2 \tag{5.2d}$$

In the laminar zone, one can write

$$\frac{u}{u^*} = \frac{\mu}{\rho}\frac{u}{y} = v\frac{u}{y} \tag{5.3}$$

where v is the kinematic eddy viscosity (L/T^2).

The laminar zone is usually very thin once the flow develops.

$$u^* = \sqrt{\frac{\tau}{\rho}} \tag{5.4}$$

Equation 5.4 enables writing Equation 5.2d in the following manner.

$$u^* = l * \frac{du}{dy} \tag{5.5}$$

The term u^* is referred to as the friction velocity and has units of L/T. Substituting $l = k$ y into Equation 5.5 and rearranging gives the following.

$$\int du = \int_{\varepsilon}^{D} \frac{1}{k}\frac{dy}{y} \tag{5.6}$$

Integrating Equation 5.6 results in the following.

$$u = \frac{u^*}{k} Ln\left(\frac{D}{\varepsilon}\right) + constant \tag{5.7a}$$

Equation 5.7a gives the velocity profile of the turbulent zone of the velocity profile,

In Equation 5.7, ε is the top of the laminar sublayer, D is the channel depth, k is the von Karmen coefficient, equal to 0.4, D is the maximum flow depth, and u is the velocity at a point in the vertical profile.

Several variations of Equation 5.7 are available. For smooth and rough boundaries, the following variations are often used.

$$v = 5.75u * \log\left(\frac{9 * y * \text{ustar}}{v}\right)(\text{smooth}) \tag{5.7b}$$

$$v = 5.75u * \log\left(\frac{30 * y}{\varepsilon}\right)(\text{rough}) \tag{5.7c}$$

Rough vs. smooth is defined in Figure 5.2 and computed using Equation 5.7d.

$$\varepsilon < \frac{5v}{\sqrt{gRS}} = \frac{5v}{u*} = \frac{5v}{\sqrt{\frac{\tau}{\rho}}} \tag{5.7d}$$

The laminar sublayer boundary layer thickness is given as $\frac{5v}{u*} < 5$, which suggests that ε is small. If $\frac{5v}{u*} < 70$, the boundary is classified as rough. Between 5 and 70, the flow is in transition. Equation 5.7 assumes a wide channel where the depth approximates the hydraulic radius. Side effects would introduce a transverse logarithmic function, which we ignore. We reiterate that boundaries can take a considerable distance to form (see Equation 2.20).

The logarithmic function has significant implications for expressing the flow's kinetic energy ($V^2/2g$). Figure 5.1b shows a logarithmic velocity profile. To find the total kinetic energy, one must integrate V^2, weighted by V, over the depth from the top of the laminar sublayer to the maximum depth. The downloads include a Mathematica folder,

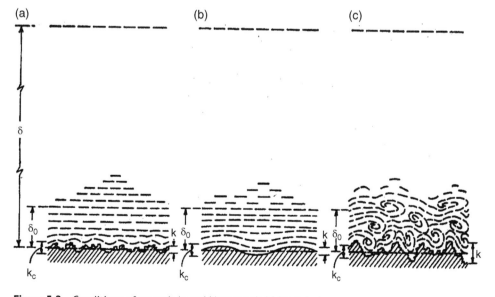

Figure 5.2 Conditions of smooth (a and b) vs rough (c) boundaries.

"VelocityProfiles." This folder consists of a logarithmic profile, among other profiles relevant in wide channels. We divide the energy summation by the cube of the mean velocity to obtain a ratio α. One finds that α varies from 1.01 to 1.05 or so in simple prismatic channels with turbulent flow. We discuss the coefficient β in conjunction with the momentum equation to be covered later. Thus, one may express kinetic energy by multiplying that found by substituting mean velocity by a coefficient α. The programs in the "VelocityProfiles" folder analyze several profiles. The coefficient α is generally close to one in prismatic channels, which often justifies ignoring the effect. It can approach two in compound (natural) channels and when any channel is covered with ice. Transverse boundary effects in narrow channels also result in larger α values; however, we usually ignore these effects except in natural channels. Figure 5.1c shows a velocity profile in a compound channel. Example 5.1 demonstrates the α coefficient computation in a natural channel using an approach involving conveyance put forward by Jobson and Froehlich (1988).

Example 5.1   Calculate the kinetic energy coefficient α for the channel of Example 2.1, using the single slope assumption.

Recall that conveyance $K = (\phi/n)*A*R^{2/3}$ and $V = K\,S^{1/2}$. Thus, one may compute the α coefficient based on conveyance and area (Jobson and Froehlich 1988). Henderson (1966) suggests that an equivalent slope S_o may be computed using

$$S_f = \left(\frac{\sum_{i=1}^{n} Q_i}{\sum_{i=1}^{n} K_i} \right)^2$$

$$\alpha = \frac{\sum_{i=1}^{n}\left(K_i\right)^3 / A_i^2}{\dfrac{\sum_{i=1}^{n}\left(K_i\right)^3}{\sum_{i=1}^{n}\left(A_i\right)^2}}$$

Here is the table computed for Example 2.1.

Channel section	Slope	Hydraulic Radius* (m) ≈d in over banks	Area (m²)	Conveyance (m³/s; see Equation 2.8)	Velocity	Q (m³/s)
Single slope of the main channel						
Left	0.02	6	600	79 721	18.79	11 274
Right	0.02	6	750	71 179	13.42	10 066
Main	0.02	12.6	1200	328 192	38.7	46 413
Totals	–		–	479 092		67 753

A. Sum of $Ki^3/Ai^2 = 2.4545\ E10$
B. Sum of $K^3 = 1.1\ E17$
C. Sum of $Ai^2 = 6\,502\,500$

$\alpha = A/(B/C) = 1.45$

$S_f = 0.02$ as one would expect since valley slope and stream slope were considered equal. Had the valley slope been used for the right and left banks and the stream slope used for the main channel, the calculated S_f would be a sufficient average slope for subsequent computations. The kinetic energy term warrants correction, as indicated by the magnitude of α. For prismatic channels, the correction α is close enough to unity.

Macroscopic energy equation: The energy equation describes the conservation and dissipation of energy in the flow. Jain (2001) provides a detailed development of the differential energy equation, including using the α energy correction coefficient. The macroscopic form of the energy equation is sufficient for most design work, and that is our focus. Flow energy arises from three components. Elevation energy, energy due to depth or pressure, and energy due to motion (kinetic energy) describe energy at a point. Other forms of energy due to chemical composition and temperature are deemed negligible. The units of the energy equation are L-F/F or L. Thus, fluid energy is expressed as length units or "head." Equation 5.8a shows how the kinetic energy term has units of L.

$$\frac{\text{kinetic energy}}{\text{unit weight}} \rightarrow \frac{ke/\text{Vol}}{wt/\text{Vol}} \rightarrow \frac{\rho V^2}{2\gamma} \rightarrow \frac{V^2}{2g} \rightarrow \frac{(ft/\sec)^2}{ft/\sec^2} \rightarrow ft \tag{5.8a}$$

"Refer to a fluid mechanics text such as Streeter (1971) for an in-depth derivation of the energy equation. The energy equation is the Bernoulli equation. One may relate energy loss to the surface slope, enabling the energy relationships for a constant channel geometry shown in Equation 5.8b–d:

$$E_i = E_{i+1} + \text{losses}_{\Delta i} \tag{5.8b}$$

$$E_i = \frac{\alpha v_i^2}{2g} + y_i + z_i \tag{5.8c}$$

where (E_i represents the kinetic, depth, and elevation energy components.), y_i is the depth of flow at i; z_i is the absolute elevation of the point i above some datum, and $\text{losses}_{\Delta i}$ is the change in energy per unit travel length, the surface slope per unit length, as shown in Equations 8.10a through 8.12.

The energy equation may be written, as shown in Equation 5.8d:

$$e_1 + z_1 = e_2 + z_2 + \text{losses}_{1-2} \tag{5.8d}$$

where e_i is kinetic and pressure depth energy summation. One sometimes calls the parameter e, the specific energy. Specific energy, e, measures energy from the bottom of the channel. Note that the specific energy does not include absolute elevation. Absolute elevation differences move the flow down a channel in a uniform flow where energy dissipation is, in effect, the dissipation of elevation energy. Losses due to slope changes, channel geometry changes, roughness, or other obstructions are measured from the channel bottom when flow quantity is constant with distance. One may estimate losses by multiplying the downstream kinetic energy term by a small constant (see King and Brater (1963) for typical values). The constant value approaches zero if corners and edges are well rounded. It may be as high as 0.1 when edges are sharp (more details of the constant to be provided in the reservoir to channel discussion). Figure 5.3 shows such a channel having a geometric transition.

Froude number and critical depth: The Froude number (Fr) is defined in Equation 2.3[1]:

$$Fr = \frac{v}{\sqrt{gD_h}} = \frac{q}{A\sqrt{gD_h}} \tag{2.3}$$

where g is gravity (L/T^2); and, D_h is the hydraulic depth.

The formulation for the Froude number in Equation 2.3 is the typical form.

Froude numbers greater than one indicate supercritical flows, with high velocities and low depths. Conversely, Froude numbers less than one indicate subcritical flows, with comparatively low velocities and more considerable depths. Maintaining α in the equation and adjusting D for steep slopes by including $D\cos(\theta)$ results in Equation 5.9, a generalized Froude number.

$$F = \frac{V}{\sqrt{gD\cos(\theta)/\alpha}} \tag{5.9}$$

where θ is the slope angle, which is near one at typical slopes.

The hydraulic depth D_h is a cross-sectional area divided by the topwidth, as discussed previously. The hydraulic depth is y for rectangular cross-sections. Flows with a Froude number exceeding 1 are supercritical. Flows with Froude numbers less than 1 are deemed subcritical or tranquil. Flows with Froude number $= 1 (\pm 0.1)$ are considered critical. Critical flow is the basis of flow measuring devices such as weirs. A simple test for assessing sub- or

---

1 One may write $E = \dfrac{V^2}{2g} + d$. Substituting flow Q into the relationship results in $E = \dfrac{Q^2}{A^2 2g} + d$. Take the derivative with respect to d, using the chain rule with area A, and find $\dfrac{dE}{dd} = \dfrac{-2Q^2}{2gA^3}\dfrac{dA}{dd} + 1$. Now,

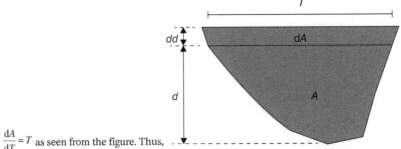

$\dfrac{dA}{dT} = T$ as seen from the figure. Thus,

$\dfrac{dE}{dd} = \dfrac{-2Q^2}{2gA^3}\dfrac{dA}{dd} + 1$. Setting $\dfrac{dE}{dd} = 0$ at the specific energy curve inflection point, and substituting

$D = A/T$ results in $\dfrac{-Q^2}{gA^2D} + 1 = 0$. Now, substituting $Q/A = V$ leads to $\dfrac{V}{\sqrt{gD}} = 1$, which is the Froude number at the inflection point.

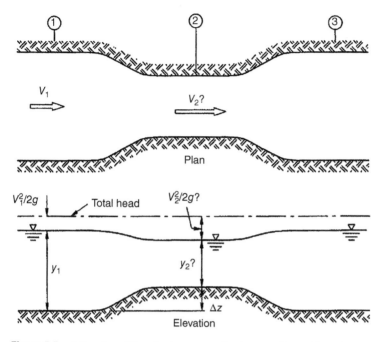

Figure 5.3 A flow transition due to geometric changes. *Source:* From Jain (2001); used with permission of John Wiley & Sons.

supercritical flow is to test whether a wave propagates upstream. The flow is subcritical if a wave propagates upstream. If the flow shoots up the sampling device, the flow is supercritical. Critical and supercritical flows require additional channel engineering to route flows through contractions, bends, and obstacles due to the jetting nature. Such flows exert high erosive forces and potential for standing waves (critical flows) or hydraulic jumps (see Sturm 2010). One may readily show that hydraulic depth is equal or nearly equal to the actual depth of a rectangular or shallow trapezoidal channel. Slopes resulting in supercritical flows are designated "steep," and slopes supporting subcritical flows are designated "mild."

Consider a channel having a uniform flow in conditions of a varying bottom slope. Flows of 200 ft^3/s in a rectangular channel with a 10 ft bottom width and roughness n of 0.015 were evaluated. We then change the bottom width to 3 ft, holding all other parameters constant. The normal depth with slopes ranging from 0.1 to 2.5% is computed using the "TrapezoidalChannelNormalCritical" Excel sheet in the downloadables (see the Example Specific Energy Curve page). The specific energy (depth + $V^2/2g$) is plotted as a function of depth for each flow, as shown in Figure 5.4. The specific energy curves are asymptotic to the 1:1 line on the upper side, denoting subcritical flow. The other extreme is the supercritical flow regime, which runs near the x-axis. The Froude number is one at the inflection point, indicating critical flow (see the previous footnote).

The specific energy curves shown in Figure 5.4 are typical for any simple prismatic channel. The compound channel can result in more complicated specific energy curves, as shown in Figure 5.5. The lower rectangular portion contains the 100 m^3/s flow. Therefore, the specific energy curve looks like Figure 5.5. The 220 m^3/s flow extends into the left and right

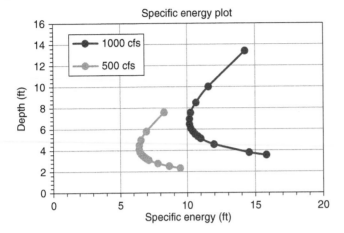

Figure 5.4 Specific energy curves for two different flows.

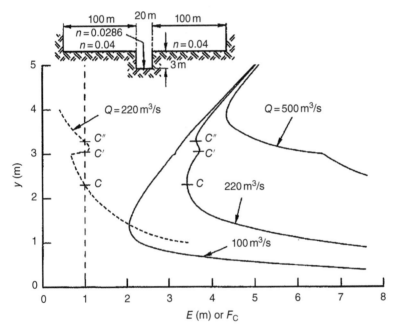

Figure 5.5 Specific energy plots over a range of flows in the indicated compound channel. The dashed line plots the Froude number for the $220\,m^3/s$. *Source:* From Jain (2001); used with permission of John Wiley & Sons.

sections to a shallow extent. Interestingly, the Froude number may not be minimum at the inflection points in compound channels. The dashed line plots the Froude number for the intermediate flowrate. As the flow further increases, the effect persists but is not as dramatic.

Short transitions – chokes and gates

Cross-section changes: A transition zone is generally short when the boundary layer development (see Equation 2.14) is not well developed. The energy equation, along with the continuity equation, enables the analysis of flow-through situations such as slope changes, geometric and elevation transitions, through an orifice, and under a sluice gate. Figure 5.6 shows such situations. Case a in Figure 5.6 indicates the equality of the energy equation pictorially at points A and B, which is also true for the other cases. Cross-section changes, slope changes, and steps (up or down) create interesting possibilities, as indicated in Figure 5.6. Example 5.2 demonstrates using the energy equation with a small elevation change with a constant cross-section geometry to assess a transition.

Example 5.2 You are given a rectangular channel with a bottom width of 10 ft, roughness of $n = 0.015$, and a flow of 200 cfs down a 1% slope. A transition consisting of a smooth (e.g., assume no energy loss) elevation change of +0.25 ft with no geometry changes or slope changes occurs. Find the normal depths, critical depths, and energy values upstream and downstream of the transition. Figure 5.6c indicates this scenario.

We use the spreadsheet, "EnergyTrapezoidChoke" in the "TransitionChokes" folder to assess the transition. Figure 5.7 provides relevant portions of the spreadsheet pages. The critical flows are estimated, as shown below, using the CriticalDepth page. They are equal, as one would expect, given no change in the cross-section area. The Goalseeker is used to compute the respective critical depths.

The "EnergyBalance" page computes the normal depth following the elevation change. Assume no energy losses unless otherwise stated. The sheet computes the normal entering depth using the goal seeker, then calculates the normal exit depth with the energy needed to balance the energy equation using the solver. Computations and a summary of results are shown in Figure 5.7b.

The energy balance computes the specific energy above and following the transition. If there is an elevation change, the exit's specific energy is adjusted to account for the energy loss (positive step) or gain (negative step). Energy losses at the transition are also included in the exit energy adjustment. This example had equal critical depths, normal depths, and similar energy values because there was no cross section or slope change.

One should check two conditions in the summary section: First, does the Froude number pass through one as the flow moves through the transition? Secondly, does the specific energy of the flow move below the specific energy at the critical condition? Either condition suggests a choke. In this example, the Froude number was less than one in the entrance and exit. The specific energy of the exit remained above the critical specific energy here, so no choke occurred. The normal depth on the top of the step elevation would propagate upstream as a backwater effect, creating a nonuniform flow situation discussed in the gradually varied flow section below. If there were an elevation change of, say, 0.5 ft, then the energy exit would be less than the critical energy. A choke would occur (verify that finding). The critical flow depth on the step would propagate upstream as a backwater profile, as discussed above. The critical flow situation would create a weir, discussed in the hydraulic structures chapter.

The specific energy equation is needed to complete a short hump analysis, as shown in Figure 5.8. The specific energy relation determines the actual depth on the hump surface. In contrast, the computed uniform flow is the limiting case. On a short step where the elevation would decrease back to the original level with no losses, the results above would mirror those with a positive hump. The scenario illustrated with the solid line in Figure 5.8

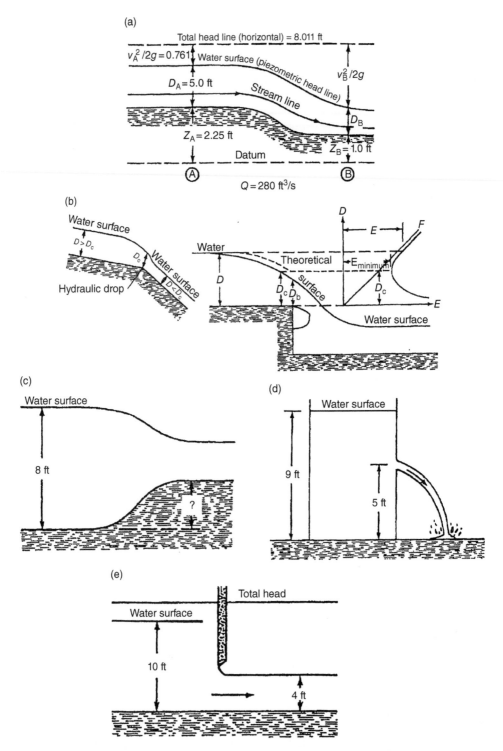

Figure 5.6 Flow transition situations: (a) flow transition down a gradual transition showing energy relations; (b) slope change transition and a drop showing the specific energy curve; (c) flow transition having a gradual elevation change; (d) flow through an orifice; and, (e) flow transition under a sluice gate. *Source:* From Jobson and Froehlich (1988).

(a)

Energy balance and loss analysis

Trapezoidal channels
Critical entering side

b	z	Q	n	S	Rho	Phi	Qtarg		Color codes
10	0	200	0.015	0.0015	64.4	1.49	200		

g	A	P	R	T	Dhy	V	F1	QcritCalc	E1crit
32.2	23.1597	14.6319	1.58282	10	2.31597	8.63568	1	199.999	3.47397

dcrit ->   2.31597   Force F1 to 1 using goal seek while varying dcrit

Critical exit side

b	z2	Q2	n2	S2	Rho	Phi	Qtarg2	QcritCalc	E2crit
10	0	200	0.015	0.0015	64.4	1.49	200	200	3.47397

g	A2	P2	R2	T2	Dhy2	V2	F2
32.2	23.1598	14.632	1.58282	10	2.31598	8.63565	1

dcrit2 ->   2.31598   Force F2 to 1 using goal seek while varying dcrit2

(b)

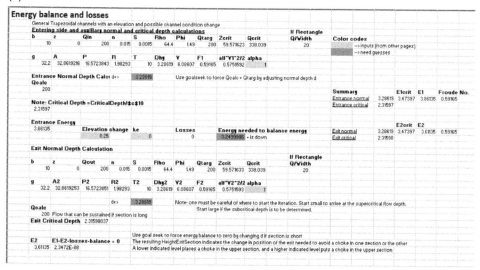

Figure 5.7 (a) Critical depths computation page of the EnergyTrapezoidChoke spreadsheet showing computations for Example 5.2. *Source:* Used with permission from Microsoft Corporation. (b) Exit entrance depth and energy balance computation page, showing the summary of results for the choke analysis. *Source:* Used with permission from Microsoft Corporation.

represents the case analyzed above with the mirror image included. In the above example, both y_1 and y_2 were above the critical depth. The situation moved from A to B and back to A on the energy curve. Had the hump been higher, the status would have moved to C, and a choke would have occurred. In the case of supercritical flow, the dashed line is operative. Had the hump forced the supercritical energy to C, a choke would occur. The introduction

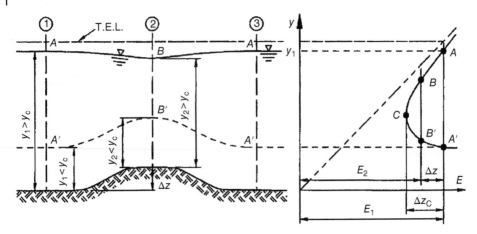

Figure 5.8 Flow situation with a 0.1 ft short hump and with a converging section from 10 to 9 ft. *Source*: From Jain (2001); used with permission of John Wiley & Sons.

of losses due to dissipation at the hump in practice favors choking. A long hump would favor choking due to losses arising due to the roughness of the surface.

The orifice and sluice gates shown in Figure 5.6 are straightforward applications of the energy and continuity equations. One may derive the orifice equation using the energy equation. Referring to Figure 5.6c, energy in the open tank is simply the depth from the center of the orifice to the surface. The flow energy in the orifice is $\alpha V^2/2g$. Neglecting losses and solving yields $V = \sqrt{2gh}$ or $Q = A\sqrt{2gh}$. The sluice gate shown in Figure 5.6e is somewhat more involved. It is an interesting application of the energy equation in a rectangular channel. The Mathematica notebook, "GeneralSluiceGate" in the "Energy" folder, sets up the energy and continuity equations to solve for flowrate and velocity in a rectangular channel with a gate at a set height in a channel with known width. The Excel folder "SluiceGateForce" contains spreadsheets for typical cross-sections that calculate flow through and velocity downstream from a sluice gate.

Another practical application of a short geometric transition is the bridge pier, as shown in Figure 5.9. Flow transitions from a natural channel profile into one or more rectangular sections (depending on the number of support piers in the flow), then back to the natural channel. Bridge piers cause small but finite energy loss, computed using a constant K multiplied by the kinetic energy inside the pier section. Constant K varies from 0.9 (semicircular nose and tail) to 1.25 (square nose and tail). Jobson and Froehlich (1988) and French (1985) provide details on hand calculating the transition's flow profile. Software packages such as HEC-RAS (Brunner 2016) include bridges and piers analysis options. This package is the go-to option for bridge pier analysis. The downloads include a "WeirBridgeCuvChan" project that demonstrates the bridge pier effect.

Reservoir-channel transition: A flow transition occurs when flows move from a reservoir to a channel, channel to culvert, or from one channel geometry to another. We discuss two types of transitions. Transitions cause energy loss, expressed as $K_{tran}v_2^2/2g$, where v_2 is the exit velocity. Table 5.1 shows the values of K_{tran} for several transition types. These coefficients are useful for any analysis where energy loss is computed as a function of kinetic energy. Texts on open channel flow texts such as French (1985) cover flow transitions in additional detail.

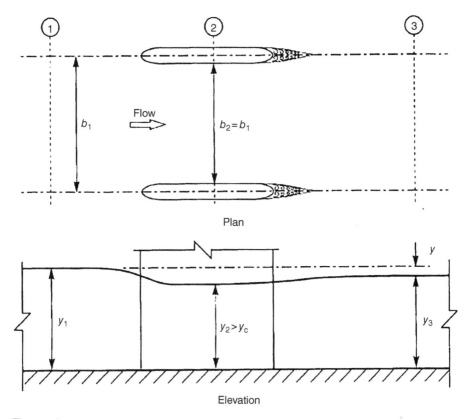

Figure 5.9 Flow between bridge piers, showing the effects on the flow profile. *Source*: From Jain (2001); used with permission of John Wiley & Sons.

Table 5.1 Suggested energy loss coefficients (K_{tran}) associated with selected channel transitions.

Flow situation	"K_{tran}" value
Sharp transition from reservoir to outlet channel	1.0
Flow from channel into culvert	0.7–1.0
Smooth transition from reservoir to outlet channel	0.5–0.6
Smooth transition from one geometry to a narrow geometry (flow convergence)	0.1
Sharp transition from narrow geometry to a wide geometry (diverging flow)	$0.3\,(v_u^2 - v_d^2)/2g$
Flow around smooth barrier	0.05–0.2

Source: Used with permission from Jain (2001).

The transition from a reservoir to a spillway employs the energy equation to describe energy conservation between the reservoir and the spillway. The Manning equation provides the uniform flow relationship in the spillway. Figure 5.10 shows a transition from a reservoir to a channel. Example 5.3 sets up the computation to determine the channel bottom's elevation at the reservoir that provides the desired flow quantity and depth in the discharge channel.

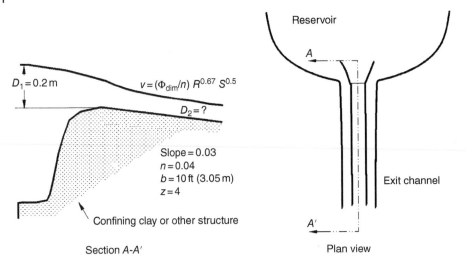

Figure 5.10 Schematic of a flow transition problem.

Example 5.3 Suppose a flow is entering into a 0.6 m wide rectangular channel at a 5% slope from an impoundment. The depth puts the surface level at 0.2 m. Losses are $K_{tran}v_2^2/2g$ with K_{tran} value of 0.75 (see Table 5.2). Velocity v_1 can is zero. Velocity $v_{chan} = v_2$ is the uniform velocity expected in the channel (e.g., use the Manning equation). Depth $y_1 = 0.2$ m and y_{chan} (y_2) is required. We can sketch the problem, as shown in Figure 5.10. The energy equation, written in terms of specific energy (recall specific energy is constant with uniform flow), appears as shown in Equation 5.10:

$$y_1 = \frac{v_2^2}{2g} + y_{chan} + \frac{K_{tran}v_2^2}{2g};$$ (5.10a)

$$v_2 = v_{chan} = \frac{\Phi_{dim}}{n_{chan}} R_{chan}^{2/3}\sqrt{S_o}$$ (5.10b)

Substitute the expression for channel hydraulic radius, slope, and roughness into the Manning equation. Then substitute the Manning equation for velocity v_{chan} in the energy equation. Refer to the nomenclature in Figure 5.10.

One can iteratively solve Equation 5.10 for y_2. Substitute values for y_2 into the Manning equation expressed using hydraulic elements of the channel to solve for flow depth. Substitute this depth back into the Manning equation to get the velocity. Calculate the flow by applying the continuity equation. For the above condition, the depth is 0.12 m (before freeboard), and the flow is 0.40 m³/s. One may increase or decrease the channel bottom width to meet flow requirements. The downloadables contain spreadsheets for solving transition problems from a reservoir to several different channel geometries.

The Excel "ReservoirTransitions" folder contains spreadsheets for several different channel geometries that compute the flowrate given the depth of the entrance bottom below the reservoir surface. Conversely, a sheet is provided for calculating the flowrate given a

Table 5.2 Image of part of a spreadsheet implementing the direct step method for assessing the profile behind a structure such as a spillway.

1. Calculate normal depth using the goal seeker to force test normal to zero by varying yn.

Channel elements

Botwtom W	Z	Manning	Slope	Q	yn	Phi	ycrit	Gravity		Color codes
20	2	0.025	0.001	1000	6.226991	1.49	3.74013	32.2		→Inputs
										→Subsequent input

Area	WettedP	R	T	Hyd Dpth	Vel Norm					→Needs guess
202.0906	47.847 949 69	4.223 601	44.907 96	4.500 107	4.948 275					

Pred Q	Test normal	Froude normal
1000	−1.3139E−05	0.411 069

2. Critical depth prediction – force TestQcrit to zero by varying ycrit.

Z								

176.2269				Acrit	Vcrit	Tcrit	Dcrit	Fcrit
Qcrit	TestQtrit			102.7797	9.729 545	34.960 52	2.939 88	0.999 999
1000.001	−0.000 532 647							

Note: Do not erase the first two or three lines of the table below.

One may fill in y, then drag the second line down to reconstruct the tabl

y	Area	Wp	T	R	D	Vel	Frict term	F^2	Frict bar	F^2 bar	So-frictba	DeltaE	DeltaX	Sum X	Bot El	Total EL
8	288	55.77708764	52	5.163411	5.538462	3.472222	0.00038	0.067604					0	0	0	8
7.9	282.82	55.32987404	51.6	5.111524	5.481008	3.535818	0.0004	0.070838	0.00039	0.069221	0.00061	−0.09308	−152.591	−152.591	0.152591	7.9
7.8	277.68	54.88266045	51.2	5.059521	5.423438	3.601268	0.00042	0.074264	0.00041	0.072551	0.00059	−0.09274	−157.203	−309.794	0.309794	7.952591
7.7	272.58	54.43544685	50.8	5.007399	5.365748	3.668648	0.000442	0.077898	0.000431	0.076081	0.000569	−0.09239	−162.466	−472.26	0.47226	8.009794
7.6	267.52	53.98823326	50.4	4.955154	5.307937	3.738038	0.000466	0.081753	0.000454	0.079826	0.000546	−0.09202	−168.519	−640.779	0.640779	8.07226
7.5	262.5	53.54101966	50	4.902783	5.25	3.809524	0.000491	0.085847	0.000478	0.0838	0.000522	−0.09162	−175.545	−816.323	0.816323	8.140779
7.4	257.52	53.09380607	49.6	4.850283	5.191935	3.883194	0.000517	0.090197	0.000504	0.088022	0.000496	−0.0912	−183.787	−1000.11	1.00011	8.216323
7.3	252.58	52.64659247	49.2	4.797651	5.13374	3.959142	0.000545	0.094823	0.000531	0.09251	0.000469	−0.09075	−193.575	−1193.69	1.193685	8.30011
7.2	247.68	52.19937888	48.8	4.744884	5.07541	4.037468	0.000576	0.099745	0.00056	0.097284	0.00044	−0.09027	−205.374	−1399.06	1.39906	8.393685
7.1	242.82	51.75216528	48.4	4.691978	5.016942	4.118277	0.000608	0.104987	0.000592	0.102366	0.000408	−0.08976	−219.85	−1618.91	1.61891	8.49906
7	238	51.30495168	48	4.638928	4.958333	4.201681	0.000642	0.110574	0.000625	0.107781	0.000375	−0.08922	−238.001	−1856.91	1.85691	8.61891
6.9	233.22	50.85773809	47.6	4.585733	4.89958	4.287797	0.000679	0.116534	0.000661	0.113554	0.000339	−0.08864	−261.389	−2118.3	2.1183	8.75691
6.8	228.48	50.41052449	47.2	4.532387	4.840678	4.376751	0.000719	0.122897	0.000699	0.119716	0.000301	−0.08803	−292.61	−2410.91	2.41091	8.9183
6.7	223.78	49.9633109	46.8	4.478887	4.781624	4.468675	0.000761	0.129696	0.00074	0.126296	0.00026	−0.08737	−336.305	−2747.21	2.747215	9.11091
6.6	219.12	49.5160973	46.4	4.425228	4.722414	4.563709	0.000807	0.136967	0.000784	0.133331	0.000216	−0.08667	−401.687	−3148.9	3.148902	9.347215
6.5	214.5	49.06888371	46	4.371406	4.663043	4.662005	0.000856	0.14475	0.000832	0.140859	0.000168	−0.08591	−509.992	−3658.89	3.658894	9.648902
6.4	209.92	48.62167011	45.6	4.317416	4.603509	4.76372	0.000909	0.15309	0.000882	0.14892	0.000118	−0.08511	−723.597	−4382.49		

The initial depth was defined as 8 ft. The normal depth was nearly reached over 4000 ft upstream.

channel placed at a given depth below the surface. One gives a trial with a flow depth in the channel and a trial flowrate if there is a known Y_1 (same as D_1 in Figure 5.10). When Q is known, provide a trial Y_1 value and trial depth in the channel, using the known flow spreadsheet. Solvers are programmed to solve the respective sheets to solve for either Q and channel depth or Y_1 and channel depth.

Longer Transitions – Gradually Varied Flow Analyses

Flow profiles behind structures or slope changes may be defined for mild and steep slopes. Figure 5.10 shows examples of flow profiles found with longer transitions. Focusing first in Figure 5.11a, the flow profiles are organized around the normal depth (y_n) and critical depth (y_c) lines. Steep slopes have a normal depth less than critical depth, while mild slopes have a normal depth greater than critical depth. The critical slope occurs when normal and critical depths are the same. The horizontal slope has no normal depth (e.g., is infinitely deep), and the adverse gradient has a horizontal asymptote. Most attention is devoted to the mild slope cases. The M_1 curve occurs in the case of a significant constriction of flow. M_2 occurs with the flow over a free discharge, and M_3 frequently results from a flow out of a sluice gate. Flows in a steep slope occur at slope changes (S_2), flows under a gate at a slope transition (S_3), or flows down a gradient that impacts an impounded water body (S_1).

The situation in compound channels is complicated with multiple critical flow depths. The lower set of curves in Figure 5.11b mirrors those in Figure 5.11a. The upper group of curves brings in the fully compound channel. Table 5.3 gives details of Froude number, the relation of friction slope S_f to bottom slope S_o, and dy/dx of profiles in Figure 5.11b. Jain (2001) and Sturm (2010) provide an expanded discussion of flow dynamics in compound channels.

Direct step analysis: One may compute simple flow profiles over mild slopes using numerical integration of the energy equation. Figure 5.12 shows a schematic of flow through a free discharge (M_2 in Figure 5.10a). Equation 5.11 discretizes the energy equation as follows (see the previous footnote also).

$$\frac{\Delta E}{\Delta d} = 1 - \alpha \frac{Q^2}{gA^3} \frac{dA}{dd} = 0 \tag{5.11}$$

Substituting $dA/dd = T$, $D = A/T$, $\alpha = 1$, rewriting in terms of the Froude number, and discretizing leads to the following.

$$\Delta E = \Delta d \left(1 - F^2 \right) \tag{5.12a}$$

Energy changes as one moves upstream following Equation 5.12b.

$$\Delta E = \Delta x \left(S_o - \overline{S_f} \right) \tag{5.12b}$$

One may then equate the above equations and solve for Δx to have the following.

$$\Delta x = \frac{\Delta d \left(1 - F^2 \right)}{S_o - \overline{S_f}} \tag{5.12c}$$

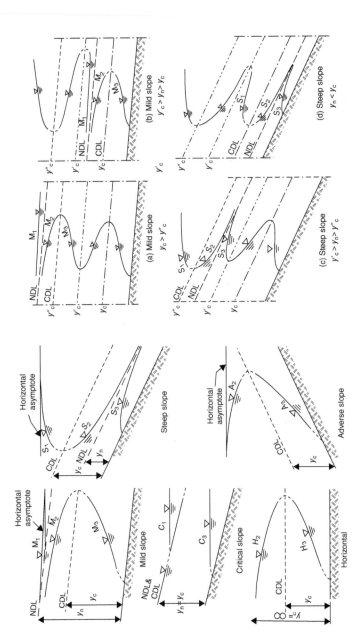

Figure 5.11 Flow profiles in simple prismatic channels (a) and compound channels (b) with mild and steep slopes. *Source:* From Jain (2001); used with permission of John Wiley & Sons.

Table 5.3 Flow characteristics in compound channels.

Depth	F	S_f	dy/dx
(a) Mild slope; $y_n > y_{c''}$			
$y > y_n$	<1	$<S_0$	>0
$y_n > y > y_{c''}$	<1	$>S_0$	<0
$y_{c''} > y > y_{c'}$	>1	$>S_0$	>0
$y_{c'} > y > y_c$	<1	$>S_0$	<0
$y_c > y > 0$	>1	$>S_0$	>0
(b) Mild slope; $y_{c'} > y_n > y_c$			
$y > y_{c''}$	<1	$<S_0$	>0
$y_{c''} > y > y_{c'}$	>1	$<S_0$	>0
$y_{c'} > y > y_n$	<1	$<S_0$	>0
$y_n > y > y_c$	<1	$>S_0$	<0
$y_c > y > 0$	>1	$>S_0$	>0
(c) Steep slope; $y_{c'} < y_n < y_{c''}$			
$y > y_{c''}$	<1	$<S_0$	>0
$y_{c''} > y > y_n$	>1	$<S_0$	<0
$y_n > y > y_{c'}$	>1	$>S_0$	>0
$y_{c'} > y > y_c$	<1	$>S_0$	<0
$y_c > y > 0$	>1	$>S_0$	>0
(d) Steep slope; $y_n \lessgtr y_c$			
$y > y_{c''}$	<1	$<S_0$	>0
$y_{c''} > y > y_c$	<1	$<S_0$	>0
$y_{c'} > y > y_c$	<1	$<S_0$	>0
$y_c > y > y_n$	>1	$<S_0$	<0
$y_n > y > 0$	>1	$<S_0$	>0

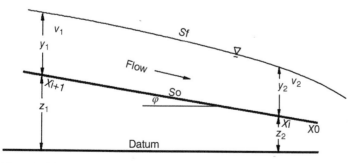

Figure 5.12 Basic nomenclature for open channel flow energy computation.

The Manning equation solved for slope leads to an expression for $\overline{S}_f$ as follows.

$$S_f = \frac{n^2 V^2}{\phi^2 R^{4/3}}$$ (5.13)

One may compute the mean surface slope between any two points by taking a simple average, harmonic average, or geometric average. Sturm (2010) reports that the simple average slope has a slight advantage for the M_1 profile. The harmonic average slightly improves predictions for M_2 profiles (see Figure 5.11a). HEC-RAS, by default, divides the square of the average conveyance into the flowrate squared to arrive at an average energy slope. Another popular model, WSPRO, uses the geometric mean. Jain (2001) suggests that the simple average is generally adequate for most cases. The simple average form appears as follows.

$$S_f = \frac{\left(\dfrac{n^2 v_1^2}{\phi_{dim}^2 R_1^{4/3}}\right)_1 + \left(\dfrac{n^2 v_2^2}{\phi_{dim}^2 R_2^{4/3}}\right)_2}{2}$$ (5.14)

We summarize the nomenclature for the above relationships below.

S_f is the surface slope of the water surface (−); n is the Manning roughness (discussed in detail in uniform flow analysis); v_i is the velocity (L/T); R_i is the hydraulic radius (L); and ϕ is a dimensions-related constant (SI, 1; imperial, 1.49).

The losses are equal to the elevation fall per horizontal unit length with uniform flow. The bottom and water surface slopes provide a way to visualize losses. Although total energy changes due to elevation changes, *the specific energy does not vary with uniform flow.* Thus, knowing the conditions at the outfall of a channel, one can integrate upstream until $\Delta e/\Delta x$ approximates $= 0$ (uniform flow, relative to the bottom). Upstream integration also applies to a backwater curve (M_1), given the depth at a point behind a dam. Upstream integration is applicable in any subcritical flow situation. Downstream integration applies to supercritical flows such as that in M_3. These integration directions enable one to progress upstream of a control point or, with M_3, downstream of the control point. The control point is critical flow (M_2), flow height over a dam (M_1), or sluice gate opening (M_3).

Spreadsheets are available in the Excel "GraduallySpatVariedFlow" folder that implements the direct step method in several different cross-section types. For example, the "DirectStepTrapezoid" sheet first computes critical flow and normal flow depths using the goal seek function. For flow over a free discharge, one inserts the critical depth at the first y value. One may then fill the y column with points at some interval such that y increases towards the normal depth. The sheet computes the hydraulic element values, velocity, friction slope term, and the square of the Froude number. At the next y value, repeat the above computations. The average friction term between the previous and current friction slope values and the mean Froude number squared are completed. The Δx is computed that defines the distance between the pre-targeted y values. One may then drag and drop all rows (excluding the y column) to get a cumulative sum of x and the elevations. The computation scheme was modeled after that presented by French (1985). The same sheet may be used to compute flows over a spillway. In this case, ignore critical flow. The y values start at the depth at the exit, then decrease to approach (but not become equal to) the normal depth. Table 5.2 shows an image of the spreadsheet. Figure 5.13a plots the data.

Standard step analysis: The direct step method presumed that hydraulic elements could be easily computed at a distance where an arbitrary value of depth was assigned. Natural and compound channels have changing cross sections. Thus, one needs to specify *x* values where measured values of hydraulic elements may be inserted. A trial and error solution, known as the standard step method, solves the depth at each of the known cross sections is required. The standard step method includes energy losses associated with cross-section changes. Several

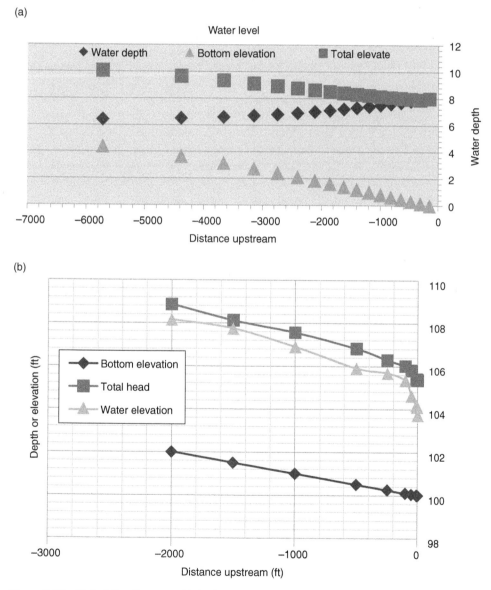

Figure 5.13 Plot of the direct step M_3 backwater curve in the conditions portrayed in Table 5.1.

cross-section changes were introduced in the problem analyzed in Table 5.4. To solve, one assumes a value for depth, computes the total head based on presumed depth and bottom elevation and actual total head based upon energy loss from the head of the previous step. Iterate each step until the depth, d, is found to equate the head values. Iterate on depth. The iteration is performed using the solver. The Excel Goalseek tool is used to solve normal and critical depths. The evidence of channel perturbations is evident on the curve for the data in Figure 5.12b.

The energy equation is useful for describing transitions from reservoirs to channels, culverts to channels, or related situations. As is the custom in pipe flow analyses, one may express losses as $K_{tran}v_2^2/2g$. The transition geometry determines K_{tran}, the loss coefficient.

HEC-RAS provides an excellent platform for analyzing gradually varied flows via an advanced integration technique for the energy equation. We demonstrate the use of HEC-RAS for a prismatic channel using the "GradVariedFlow" project. The primary difference in the gradually varied flow and the uniform flow analysis are (i) a need for finer interpolation and (ii) selecting a free discharge or a defined boundary elevation. We demonstrate free water discharge. Figure 5.14 shows results and menus for a single prismatic channel with a flow of 20 cfs in the 2000 ft channel with a slope of 0.0005, a roughness of $n = 0.025$, side slope $z = 4$, and bottom width of 10 ft. This channel reached normal depth within 100 ft from the outfall. Table 5.5 provides a step-by-step procedure for analyzing a free discharge from the channel developed in Table 2.6.

Analytical solutions[2]: Before the age of computers, engineers developed some workable solutions to a more extended transition problem. Many of the solution approaches involve tables of integration functions. The text will not attempt to cover all the available solution approaches. Chow (1959), French (1985), and Graf and Altinakar (1998) give a survey of known techniques. The derivations below provide only the high points; consult these references for more details. The analytical solutions fell into declining use when computers and numerical methods became available. Analytical techniques often give insights to solutions that are not as obvious when using numerical methods. The advent of tools such as Mathematica causes one to revisit analytical tools. The hydraulic exponents method is briefly described herein. We use Mathematica to solve the distance upstream required to approach the normal depth to within 5%.

We begin by defining the section factor for critical flow, Z. following Chow (1959) in Equation 5.15.

$$A * \sqrt{D} = A * \sqrt{\frac{A}{T}} = Z = \frac{Q_{crit}}{\sqrt{g/\alpha}} \tag{5.15}$$

where Q_{crit} = critical flow and other nomenclature are as previously defined.

One can write based on Equation 5.15, the following function of Z and y.

$$Z^2 = Cy^2 \tag{5.16}$$

Where c is a coefficient of proportionality, and M is the hydraulic exponent for critical flow. To find M as a function of hydraulic elements, one may write

$$M = 2\frac{d\ln Z}{d\ln y} = 2\frac{3T}{2A}y - \frac{y}{2T}\frac{dT}{dy} \tag{5.17}$$

---

2 This section may be skipped without loss in continuity.

Table 5.4 Partial image of standard step spreadsheet for computing a free discharge.

1. Calculate normal depth using the goal seeker if prismatic

Channel elements

Bottom W	Z	Manning	Slope	Qtarg	yn	Phi	ycrit	Gravity	Color codes:
20	2	0.025	0.001	1000	6.226991	1.49	3.74013	32.2	—

(If the channel is nonprismatic, measure the flow rate, depth, area, and wetted P at the initial, final, or other relevant point)

Area	Wetted P	R	T	Hyd Dpth	Vel Norm
202.0907	47.84795	4.223601	44.90796	4.500107	4.948274

Pred Q	Test normal	Froude normal
1000	−7.8E−05	0.411069

2. Critical Depth Prediction if needed or prismatic

Z

		Acrit	Vcrit	Tcrit	Dcrit	Fcrit	Notes
176.2269							
Qcrit	TestQcrit	102.7797	9.729545	34.96052	2.93988	0.999999	Allways kee
1000.001	−0.00053						O

Fill station	Fill Nonprism BottomEl	Assume y	Water El	Fill if Nonprism area	Vel	Check area against survey	Fill alpha	alfv^2/2g	Tot Head	Check WP against survey Wp	R	Frict term	Frict bar	DeltaX	hf	he	Head	Test
−0.1	100	3.74	103.74	102.7752	9.729974		1.1	1.617075	105.3571	36.72579	2.798448	0.006758	—	—	—	—	105.35707	—
−5	100.005	4.113018	104.118	116.0942	3.613695		1.1	1.267319	105.3853	38.39398	3.023761	0.004777	0.005768	4.9	0.028262	0	105.38534	0
−10	100.01	4.191792	104.2018	118.9781	8.404909		1.1	1.206627	105.4084	38.74626	3.070698	0.004456	0.004616	5	0.023082	0	105.40842	0

−50	100.05	4.583054	104.6331	120	8.333333	1.1	1.186163	105.8192	50	2.4	0.006084	0.00527	40	0.210798	0.2	105.81922	0
−100	100.1	5.25491	105.3549	160.3264	6.237277	1.1	0.664503	106.0194	43.50067	3.685607	0.001924	0.004004	50	0.200197	0	106.01941	−1.8E−08
−250	100.25	5.433713	105.6837	167.7247	5.96215	1.1	0.607173	106.2909	44.3003	3.786085	0.001696	0.00181	150	0.271473	0	106.29089	1.02E−08
−500	100.5	5.409516	105.9095	160	6.25	1.5	0.909841	106.8194	40	4	0.001732	0.001714	250	0.42847	0.1	106.81936	1.37E−08
−1000	101	5.900995	106.901	187.6634	5.32869	1.5	0.661373	107.5624	46.39005	4.045337	0.00124	0.001486	500	0.743011	0	107.56237	1.35E−08
−1500	101.5	6.244854	107.7449	202.8935	4.928694	1	0.377205	108.1221	47.92784	4.233312	0.000999	0.001119	500	0.559692	0	108.12206	1.55E−08
−2000	102	6.14521	108.1452	180	5.555556	1.5	0.718887	108.8641	40	4.5	0.00117	0.001084	500	0.542037	0.2	108.8641	1.58E−08
																	5.05E−08

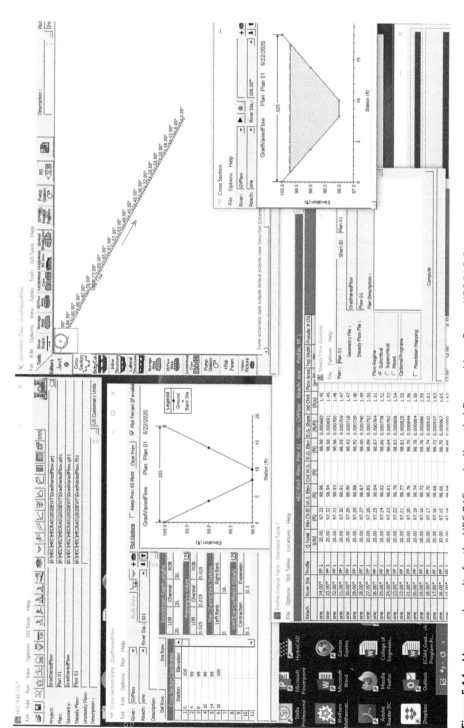

Figure 5.14 Menus and results for the HEC-RAS gradually varied flow problem. *Source:* HEC-RAS Computer program.

Table 5.5 Steps for using HEC-RAS for a free discharge analysis from the channel developed in Table 2.6.

---

1) Load the project created using Table 2.6

2) On the main HEC-RAS menu, go to Edit->steady flow data

3) In the cell under "PF 1," insert the flowrate of interest.

4) In the upper right corner, press "Apply Data."

5) Press "Reach Boundary Conditions" in the steady flow panel. This brings up the Reach Flow boundary Conditions panel

6) Click "Downstream." For a free discharge, press "Critical Depth." One would press "Known WS" Other options explained in the documentation for a backwater curve with known depth.

7) Click "Upstream." Press "Normal Depth" and enter the channel slope. Click "Apply Data." Press OK

8) Go back to the main HEC-RAS panel and press "Run" -> "Steady Flow Analysis."

9) In the flow regime block, press "Mixed," (Subcritical would also work)

10) Press "Compute."

11) On the main menu, press the [image] to see a channel cross-section. Press [image] to see a flow profile. The [image] button brings up a table of results.

The derivative of T vs. y can be found analytically for all the standard cross-sections. Numerical methods are required for natural channels.

A similar argument holds for uniform flow. If conveyance K is written for uniform flow, one has Equation 5.18.

$$K = \frac{\phi}{n} AR^{2/3}.$$ (5.18)

One may write the following approximation.

$$K^2 = Cy^N$$ (5.19)

In Equation 5.19, C is a proportionality constant, and N is the hydraulic exponent for uniform flow. As we did with critical flow, one may write Equation 5.20.

$$N = 2\frac{d\ln K}{d\ln y} = \frac{2y}{3A}\frac{dA}{dy} + \frac{2y}{3R}\frac{dR}{dy} = \frac{2y}{3A}\left(5T - 2R\frac{dP}{dy}\right)$$ (5.20)

In the right-hand side of Equation 5.20, we substituted $dA/dy = T$ and $R = A/P$. One can find dP/dy for any prismatic cross section

Coming back to gradually varied flow, after several substitutions (see Chow 1959; French 1985), one may write the following

$$S_f = \frac{Q^2}{K^2} \quad \text{and} \quad S_0 = \frac{Q^2}{K_n^2} \quad \rightarrow \quad \frac{S_f}{S_0} = \frac{K_n^2}{K^2} \tag{5.21}$$

Now the basic equation for gradually varied flow is Equation 5.22 in terms of the hydraulic exponents appears as follows more substitutions (see Chow 1959):

$$\frac{dx}{dy} = \frac{1}{S_0} \frac{1 - \left(\dfrac{y_n}{y}\right)^N}{1 - \left(\dfrac{y_c}{y}\right)^M} \tag{5.22}$$

Equation 5.22 is the routing equation in terms of hydraulic exponents. Table 5.6 summarizes hydraulic exponents for some typical cross sections.

Writing Equation 5.22 in terms of the Froude number (see Equation 5.12) and inverting to solve for x, we have Equation 5.23, which is the differential form of Equation 5.12c.

$$\frac{dx}{dy} = \frac{1 - F^2}{S_0 - S_f} \tag{5.23}$$

Equations 5.22 and 5.23 are solvable with prismatic channels. See the "Mathematica, GraduallyVariedFlow" folder for several solutions. The solution computes the normal depth, critical depth, and then integrate concerning x. We compare the result of integrating Equations 5.22 and 5.23 and find for a specific case in the Mathematica notebook to be $x = -2331$ ft vs. -2283 ft. A similar range was found with the circular cross section.

Table 5.6 Values of M and N for indicated cross-sections.

Cross-section	M	N
Rectangular	3.0	3.33
Triangular	5	5.33
Trapezoidal	$\dfrac{3\left(1+2z\dfrac{y}{b}\right)^2 - 2z\dfrac{y}{b}\left(1+z\dfrac{y}{b}\right)}{\left(1+2z\dfrac{y}{b}\right)\left(1+z\dfrac{y}{b}\right)}$	$\dfrac{10\left(1+2z\dfrac{y}{b}\right)}{3\left(1+z\dfrac{y}{b}\right)} - \dfrac{8\dfrac{y}{b}\sqrt{1+z^2}}{3\left(1+2\dfrac{y}{b}\sqrt{1+z^2}\right)}$

Source: Adapted from French (1985).

Conclusions

The energy equation, coupled with a flow equation such as the Manning equation, provides the primary relationships used to analyze transition flows such as channel transitions from reservoirs. The energy equation is the primary relationship used in developing backwater curves. *Main Point: In uniform flow, the fluid loses elevation potential at the same rate the flow dissipates energy. Open channel flows dissipate energy via friction at the wetted perimeter and turbulence within the flow.* Mathematica provides some theoretical analysis of Gradually varied flows in prismatic channels. Excel is a convenient tool for prismatic channels. HEC-RAS is the recommended tool to analyze flows in non-prismatic channels. Cross-section data for natural channels in a GIS platform (ESRI 2020 or CIVIL3d from AutoDesk 2020) interface with HEC-RAS. From a practical viewpoint, gradually varied flow techniques over short transitions and longer transitions give one a basis for more accurate infrastructure pricing involving these situations.

Problems and Questions

1 A 500 ft^3/s flow that is 20 ft wide, a slope of 0.01 in a rectangular channel encounters a transition where there is a step elevation change of -0.5 ft and a narrowing to 15 ft. Assume there are no losses. What is the depth in the channel above the transition? Is the flow choked?

2 There is a step change of 0.5 ft in problem 1. Is the flow choked?

3 The flow of problem 1 goes through a slope transition from 0.01 to slope $= 0.02$ and a drop of -1.21 ft. The bottom width is 20 ft above and below the transition. Is the flow choked?

4 A 10 cfs flow in a trapezoidal channel ($z = 1$) converges from a 2 ft bottom width to a 1 ft bottom width. The slope changes from 1 to 0.5% over the convergence. There are no step changes in the channel bottom. Assume the transition is smooth. The channel is concrete, with a Manning n value of 0.018. Does the channel choke the flow?

5 A 10 cfs flow in a trapezoidal channel ($z = 1$) steps up from zero to 0.5 ft abruptly. The bottom width is 2 ft above and below the step and has a slope of 1%. Use a loss k of 0.2. The channel is concrete, with a Manning n value of 0.018. Does the channel choke the flow?

6 A reservoir serves as a source of water for an irrigation canal. The static height of the reservoir is 0.5 m about the concrete channel bottom entrance to the reservoir. The opening is smooth and rounded. If the channel has a 1% slope and is trapezoidal, $Z = 1$, $b = 1.5$, what is the normal flow depth? What is the flowrate? If the flow were 5 m^3/s, find the required entrance height? If the channel were optimum, the entrance rounded, and the flow was 5 m^3/s; what is the size of the opening needed?

7 A concrete-lined channel conveying 5 ft^3/s with a slope $= 0.001$, $z = 4$, and $b = 1$ ft freely discharges. How far upstream should we expect the flow to be within 1% of the normal depth? Use standard step and direct step methods.

8 Use HEC-RAS to solve problem 7.

9 Use HEC-RAS to solve the following problem. You have a finished concrete channel that is 2000 ft long, conveying a flow of 400 cfs. The channel bottom is at 1000 ft at the inlet and 998 ft at the outlet. The channel is a concrete trapezoid that is 7 ft deep and has a bottom width of 6 ft and a side slope of 3:1. The discharge is free. Work out the station-depth profile. Analyze the slope. Assume that the left and right overbank stations are the same as the initial and final stations. Solve and Report (Profile view, Froude no plot, Geometric editor showing a profile).

10 Redo the above problem with an optimum side slope.

11 Suppose the channel of problem 7 had an obstruction 5 ft above the channel bottom at the outfall. How far upstream would the backwater effect be noticed (e.g., within 1% of the normal depth)?

12 Develop the specific energy curve for a concrete-lined channel conveying $15\,\text{ft}^3/\text{s}$ with a slope $= 0.001$, $z = 2$, and $b = 1$ ft. Compare this with the specific energy curve with an optimum z, concrete-lined channel, slope of 0.001, and

13 A sluice gate in a 10 ft wide rectangular channel is open 4 ft. The depth behind the gate is 10 ft. Assume a loss coefficient of 0.1 and an alpha of unity. Compute the discharge through the gate.

14 Using the direct step method, estimate the affected distance upstream due to the backwater curve caused by the gate of the problem. Assume the slope is 0.005, the channel is rectangular and concrete-lined, and has a bottom width of 2 ft.

References

Autodesk (2020). *CIVIL3D Reference Manual*. San Rafael, CA: 111 McInnis Parkway.

Brunner, G.W. (2016). *HEC-RAS River Analysis System User' Manual, Version 5*. Davis, CA: Hydrologic Engineering Center.

Chow, V.T. (1959). *Open-Channel Hydraulics*. New York, NY: McGraw-Hill.

ESRI (2020). *ArcGIS Mapping and Analytics Platform*. Redlands, CA: Esri Headquarters.

French, R.H. (1985). *Open Channel Hydraulics*. New York, NY: McGraw-Hill.

Graf, W.H. and Altinakar, M.S. (1998). *Fluvial Hydraulics: Flow and Transport Processes in Channels of Simple Geometry*. Chichester, UK: Wiley.

Henderson, F.M. (1966). *Open Channel Hydraulics*. New York, NY: MacMillan Publishing.

Jain, S.W. (2001). *Open Channel Hydraulics*. New York, NY: Wiley.

Jobson, H.E. and Froehlich, D.C. (1988). Basic hydraulic principles of open-channel hydraulics. USGS Open File Report 88-707, Reston, VA.

King, H.W. and Brater, E.F. (1963). *Handbook of Hydraulics*. New York, NY: McGraw-Hill.

Streeter, V.L. (1971). *Fluid Mechanics*, 5e. New York, NY: McGraw-Hill.

Sturm, T.W. (2010). *Open Channel Hydraulics*, 2e. New York, NY: McGraw-Hill.

6

Momentum Equation for Analyzing Varied Steady Flows and Spatially Varied Increasing Flows

The energy equation has distinct advantages. Energy is not a vector, which leads to a straightforward analysis of problems for which energy loss is predictable. Hydraulic jumps and channel junction analysis exemplify problem types where energy loss is difficult to predict. Momentum analysis is the choice where forces related to drag are either negligible or possibly foreseen (e.g., drag force due to bridge pier). The hydraulic jump, the channel junction, and a spatially varied *increasing* flow represent three applications that may be analyzed using a momentum balance. The *decreasing* spatially varied flows topic is presented for completeness; however, an energy balance is sufficient, and momentum is not used for the decreasing flow case.

Goals

- To develop the one-dimensional momentum equation
- To compare and contrast the momentum curve with the specific energy curve
- To apply momentum to an analysis of the hydraulic jump
- To analyze a channel junction using the momentum equation
- To consider spatially varied flow (dQ/dx not zero, dy/dx not zero)

Rapidly Varying Steady Flows ($dQ/dt = 0$, $dQ/dx = 0$, dy/dx varies)

We provide a brief survey of selected rapidly varying flows herein. Chow (1959), French (2007), Graf and Altinakar (1998), Jain (2001), and Sturm (2010) provide additional topics and additional treatment of issues presented.

Theoretical background: The momentum equation comes directly from the application of Newton's second law. One equates the sum of external forces to the fluid momentum change to arrive at the momentum equation. Equation 6.1a shows the momentum balance. Figure 6.1a and b shows selected momentum balance.

Open Channel Design: Fundamentals and Applications, First Edition. Ernest W. Tollner.
© 2022 John Wiley & Sons Ltd. Published 2022 by John Wiley & Sons Ltd.
Companion website: www.wiley.com/go/tollner/openchanneldesign

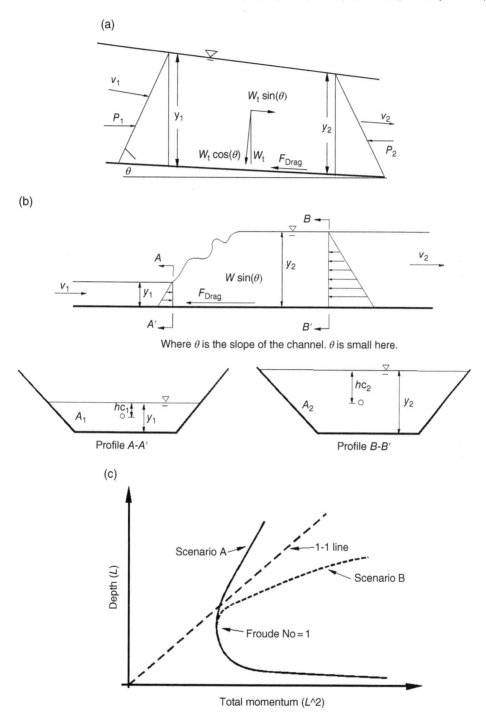

Figure 6.1 (a) Schematic of a channel showing the hydrostatic and velocity components of the momentum balance along the flow direction. (b) Schematic of a channel with a hydraulic jump showing the application point of the equivalent hydrostatic force. (c) Example momentum function plot for a rectangular channel, with indicated scenarios A and B suggesting a lack of relation to the 1–1 line as was the case with the energy function.

$$\sum \text{force} = \text{mass} \frac{\Delta v}{\Delta t} = \frac{v_1 \gamma v_1 A_1}{g \, \text{unit time}} - \frac{v_2 \gamma v_2 A_2}{g \, \text{unit time}} = M_1 - M_2 \qquad (6.1a)$$

Substituting $q = A_1 v_1 = A_2 v_2$, and inserting the momentum coefficient β, one can also write Equation 6.1a as follows:

$$\sum \text{force} = \text{mass} \frac{\Delta v}{\Delta t} = \frac{q\gamma}{g}(v_1 - v_2) = \frac{\beta q^2 \gamma}{g A_1} - \frac{\beta q^2 \gamma}{g A_2} = M_1 - M_2 \qquad (6.1b)$$

where pressure force P_1 is $\gamma y_1^2/2$ (F); pressure force P_2 is $\gamma y_2^2/2$ (F); $W_t \sin(\theta)$ is the horizontal weight component (F); F_{drag} is the drag force induced by channel bottom and wall resistance (F); $\Sigma \text{force} = P_1 - P_2 + F_{drag} + W_t \sin\theta \, (\gamma/g) \, q \, (v_1-v_2)$ is the momentum change (F); q is the flowrate (L^3/T); v_i is velocity at locations $i = 1$ or 2 (L/T); A_i is the cross-sectional area at $i = 1$ or 2 (L^2); γ is the unit weight of the fluid (F/L^3); g is the acceleration of gravity (L/T^2); and β is the momentum correction coefficient discussed below (–).

In words, the momentum equation states:

Pressure force$_1$ + Horizontal component of weight – Pressure force$_2$ – drag = change in momentum.

This statement of the momentum equation assumes a small slope and negligible drag forces. We now use the momentum equation to develop a relationship describing the hydraulic jump in a rectangular channel. Starting with Equation 6.1b, we have:

$$M_1 - M_2 = \Sigma \text{force} \qquad (6.1c)$$

where $M_1 = \gamma \beta q^2/gA_1$; $M_2 = \gamma \beta q^2/gA_2$; M_1 and M_2 represent the fluid dynamics part of momentum here.

Jain (2001) provides a detailed development of the differential momentum equation, including the momentum coefficient, β. We focus on the macroscopic form, which is sufficient for most design work.

Forces that cannot be ignored are the pressure forces. The pressure force is computed by multiplying the centroid's cross-sectional area as measured from the centroid's water surface. With the rectangular cross section, the centroid is simply $y/2$. Thus, the pressure component for the rectangular section is by $3/2$. Looking at Equation 6.1b, each side has a fluid dynamics component, q^2/gA_i. Combining the fluid dynamics momentum term and the pressure term gives what one may call the momentum function for various cross sections (Sturm 2010). Table 6.1 includes tabulated Momentum functions. The momentum functions in Table 6.1 assumes negligible drag and small channel slopes.

Momentum coefficient β: Momentum of a flowing fluid based on the mean velocity requires correction for reasons paralleling those discussed with energy in Chapter 4. We developed a coefficient α that would correct the energy when computed with the mean velocity. Similarly, one may calculate a weighted average of momentum over the velocity profile by integrating the velocity profile equation's square from roughness height to depth and dividing by the mean velocity squared times the depth. The Mathematica notebook "Veldistlogalphbeta.nb" in the "VelocityProfiles" folder makes this computation. With laminar flows, both α and β are higher than one and should be considered.

In a compound channel, the momentum β could be significantly higher than one for the same reasons that the energy α was greater than one. The momentum β is computed similarly to the energy α in Example 5.1. We reduce the exponents on all terms in Equation 5.1 by one to give Equation 6.2.

$$\beta = \frac{\dfrac{\sum_{i=1}^{n}\left(K_i\right)^2 / A_i}{\sum_{i=1}^{n}\left(K_i\right)^2}}{\sum_{i=1}^{n} A_i} \tag{6.2}$$

Since momentum involves the square of velocity and energy uses the cube of velocity, the β tends to be somewhat less than α for given channel velocity profiles. For prismatic cross sections, we routinely use β less frequently than α. Thus, our momentum computations usually do not include the β coefficient.

Momentum curve: A graph of the momentum function can be constructed from the momentum function. They appear similar to the specific energy function graph when plotting total momentum (*x*-axis) from Table 6.1 against depth (*y*-axis).

Figure 6.1c presents a momentum curve plot for a rectangular channel. The upper portion is not asymptotic to the one-to-one line because the momentum function's pressure term is not depth alone. Graphs may or may not cross the one-to-one line in the plotted range of total momentum. Unlike with specific energy, the dimension of momentum changes with the channel cross section. Multiplying the depth by the centroid of the depth explains the dimension change of the hydrostatic component. Similar factors at play in the flow portion of total momentum likewise explain the dimension variation with cross section. The Froude number is one at the curve inflection point, as it was on the energy equation. The momentum curve does not convey more information than does the energy curve. For that reason, energy curves suffice for demonstrating key concepts.

Sequent depths: A hydraulic jump is shown in Figure 6.1b. With a rectangular channel of width b, substituting $q = V_1 A_1 = V_2 A_2$, $A_1 = by_1$, $A_2 = by_2$, entering the pressure forces on the righthand side, canceling the γ, and neglecting drag results in Equation 6.3:

$$\frac{q_{uw}^2}{g}\left(\frac{1}{y_1} - \frac{1}{y_2}\right) = \frac{1}{2}\left(y_2^2 - y_1^2\right) \tag{6.3}$$

In Equation 6.3, q_{uw} is the flow per foot of width ($=q/b$); units are L^2/T

g is gravity $\left(L / T^2\right)$.

Substituting the definition of the Froude number $Fr = q_{uw}/y(gy)^2$ for the entering flow allows us to write Equation 6.4:

$$\frac{y_2}{y_1} = \frac{1}{2}\left(\sqrt{1 + 8Fr^2} - 1\right) \tag{6.4}$$

The Mathematica notebook "SequentDepthsRectangularTheoretical" in the momentum folder gives the derivation of Equation 6.4. Only for the rectangular section can one arrive

Table 6.1 Momentum functions for standard channel cross sections.

Cross section	Momentum function
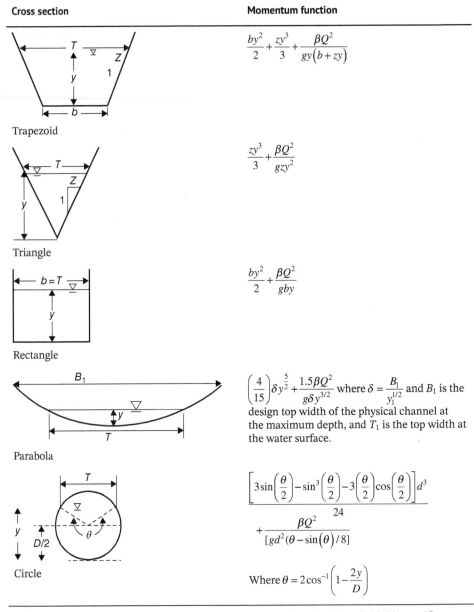 Trapezoid	$\dfrac{by^2}{2} + \dfrac{zy^3}{3} + \dfrac{\beta Q^2}{gy(b+zy)}$
Triangle	$\dfrac{zy^3}{3} + \dfrac{\beta Q^2}{gzy^2}$
Rectangle	$\dfrac{by^2}{2} + \dfrac{\beta Q^2}{gby}$
Parabola	$\left(\dfrac{4}{15}\right)\delta y^{\frac{5}{2}} + \dfrac{1.5\beta Q^2}{g\delta y^{3/2}}$ where $\delta = \dfrac{B_1}{y_1^{1/2}}$ and B_1 is the design top width of the physical channel at the maximum depth, and T_1 is the top width at the water surface.
Circle	$\dfrac{\left[3\sin\left(\dfrac{\theta}{2}\right) - \sin^3\left(\dfrac{\theta}{2}\right) - 3\left(\dfrac{\theta}{2}\right)\cos\left(\dfrac{\theta}{2}\right)\right]d^3}{24}$ $+ \dfrac{\beta Q^2}{[gd^2(\theta - \sin(\theta))/8]}$ Where $\theta = 2\cos^{-1}\left(1 - \dfrac{2y}{D}\right)$

Source: Adapted from Sturm (2010). The momentum functions above have been divided by specific weight γ.

at a closed-form solution for the ratio of depth. Froude numbers passing through the supercritical to subcritical regimes indicate the presence of a hydraulic jump. Passing from subcritical to supercritical flow suggests a weir (further discussed in Chapter 7). Both instances indicate a choke (Chapter 5).

HEC-RAS (Brunner 2016) can demonstrate hydraulic jumps. The HEC-RAS folder (HydraulicJmp.prj) contains a hydraulic jump demonstration. The demonstration comprises the following. A simple channel with a supercritical slope was created following the procedure of Table 2.6. A proscribed depth above the critical depth was established at the outlet. A profile plot clearly shows the hydraulic jump.

A closed-form solution for energy loss is possible by substituting Equation 6.4 into the energy equation. The Mathematica notebook referenced above contains the derivation of the energy loss equation, given as Equation 6.5.

$$\text{Energy loss} = \frac{\left(y_1 - y_2\right)^2}{4 y_1 y_2} \tag{6.5}$$

Where Energy loss is in $N–M/N$ or $Ft–lb/lb$ One may compute the temperature change across the jump by equating the energy loss with flow * density* specific heat* ΔTemperature and solving for temperature. Temperature changes across the typical jump are on the order of 0.01 °C because of the high flowrate and high specific heat of the water.

Using the momentum functions in Table 6.1, one may analyze the general trapezoid, parabolic, and circular cross-sections by simply setting $M_1 = M_2$ and solving. Closed-form solutions are not forthcoming, but one may numerically solve for y_2/y_1. Excel sheets for the trapezoid, circle, and parabolic cross sections are available in the Excel "SequentDepths" folder. The numerical solutions may be substituted into the energy equation to compute energy loss, if desired.

The momentum principle is useful when energy dissipation is substantial over a short distance, and drag at the channel bottom is negligible. The typical hydraulic jump is one such example. Therefore, the momentum before and after the jump is similar. The small drag force is reflected in the actual chord in Figure 6.1c. Sturm (2010) and French (2010), among others, discuss the effects of friction loss in some specific types of hydraulic jumps. For nonrectangular channels, use the appropriate hydraulic element expressions and use the hydraulic depth D in the Froude number. The hydraulic jump equations for nonrectangular are more involved, as shown on the Mathematica files for sequent depth in other cross sections. Refer to works such as French (2007) and Sturm (2010) for more details. The hydraulic jump is significant for energy dissipation, implying that a theoretical chord on the specific energy curve deviates substantially from the theoretical vertical. Energy dissipaters find an application at the foot of spillways and other transitions from steep to mild slopes.

Example 6.1   A flow of 80 cfs in a 3 ft wide rectangular channel down a steep slope and the depth happens to be 2 ft. The slope makes a sudden transition to a less steep value. Will a hydraulic jump occur?

The channel cross section area is $3 \times 2\,\text{ft} = 6\,\text{ft}^2$. The hydraulic depth for a square channel is $A/t = 6\,\text{ft}^2/3\,\text{ft} = 2\,\text{ft}$; $v_1 = q/A_1 = (80\,\text{ft/s}^3)/6\,\text{ft}^2 = 13.33\,\text{ft/s}$.

The Froude number $Fr_1 = v/\text{Sqrt}(gy) = (13.33\,\text{ft/s})/\text{Sqrt}(32.2\,\text{ft/s}^2\,(2\,\text{ft})) = 1.66$ and flow is supercritical.

From Equation 6.4, $y_2 = 2\,\text{ft}/2\,(\text{Sqrt}(1 - 8(1.66^2) - 1) = 3.8\,\text{ft}$ with conservation of momentum. The depth would be somewhat less in practice due to some momentum being lost due to drag forces on the channel bottom and sides.

From continuity, $v_2 = 80\,\text{ft}^3/\text{s}/(3.8\,\text{ft}*3\,\text{ft}) = 7.01\,\text{ft/s}$. It is assumed the channel is steep enough to maintain this velocity. $Fr_2 = (7.01\,\text{ft/s})/\text{Sqrt}(32.2\,\text{ft/s}^2\,(3.8\,\text{ft})) = 0.63$ (subcritical). Expect a jump.

Example 6.2   How much energy is lost in the hydraulic jump of Example 6.1? The kinetic energy is:

$$v_1^2/2g = \left(13.33\,\text{ft/s}\right)^2/\left(\left(2\,\text{ft}\right)\left(32.2\,\text{ft/s}^2\right)\right) = 2.759\,\text{ft};$$

$$v_2^2/2g = \left(7.01\,\text{ft/s}\right)^2/\left(2\left(32.2\,\text{ft/s}^2\right)\right) = 0.767\,\text{ft}$$

The static energy is:
$y_1 = 2\,\text{ft}\ y_2 = 3.8\,\text{ft}$
Combining static and kinetic energy leads to:
$e_1 = v_1^2/2g + y_1 = 4.759\,\text{ft}\ e_2 = v_2^2/2g + y_2 = 4.567\,\text{ft}$
The lost specific energy was $e_1 - e_2 = 4.759 - 4.567\,\text{ft} = 0.192\,\text{ft-lb/lb}$ or ft.

The specific energy difference is not surprising since the hydraulic jump is an energy dissipater. The hydraulic jump provides energy dissipation in structures, further discussed in Chapter 8.

In cases where one simply analyzes a hydraulic jump by setting $M_1 = M_2$, the jump positions itself based on flowrate and phenomena above or downstream of the jump. One can exert some control over the jump by placing a drag force at the desired location. The equation to be solved then changes from $M_1 = M_2$ to $M_1 = M_2 - F_d$. If the slope were steep and no longer ignorable, one could insert a term to include the fluid component's weight. Steep slope scenarios complicate the analysis because the jump's length is a function of the sequent depth. The jump region's length is approximately $6\,y_2$ for a smooth channel and $3.6\,y_2$ to $4.1\,y_2$ for rough channels. The incoming Froude number ranged from $4.5 < F < 9$ in this study. Jain (2001) and Sturm (2010) provide additional relationships for predicting jump length. Sturm (2010) indicates that with hydraulic jumps occurring in rectangular channels on slopes (θ relative to the horizontal) and depths d_1 and d_2 measured normal to the channel bottom, one may correct the incoming flow Froude number ($F_{incoming}$) by dividing it by the factor shown in Equation 6.6.

$$\sqrt{\cos(\Theta) - \frac{KL\sin(\theta)}{d_2 - d_1}} \tag{6.6}$$

Equation 6.6 depends on the jump (L) length, as modified by a correction factor (K). The product of KL is the length of the jump. Thus, an iterative process becomes involved with jumps on a steep slope angle. Kindsvater (1944) gives detailed discussion and relationships for hydraulic jumps on sloping channels. One may use the guidance provided in the above paragraph and consult Chow (1959), Jain (2001), and French (2007) for additional details.

One way to incorporate drag forces to locate a jump is to insert vertical rods imparting a drag force $= (0.5 * Cd * y1 * \rho * \text{Diam} * V^2)/(\gamma)$. The force term was divided by specific weight γ to be consistent with the momentum functions in Table 6.1. The Excel sequent depth spreadsheets allow one to insert several piers with a drag coefficient to simulate an imparted drag force.

The US Bureau of Reclamation (1987) presents several spillway designs that, in effect, locate jumps by imparting various drag forces. Abrupt drops, abrupt rises, and sills can all be used to locate the hydraulic jump. Jain (2001) illustrates hydraulic jump control by baffle piers, as shown in Figure 6.2. The elevated depth within the structure (point 2 in Figure 6.2) indicates an energy loss that exceeds the channel's normal depth (point 3).

Hydraulic jumps frequently occur at steep to flatter slope changes or with cross-section constrictions. Our hydraulic jumps presentation does not consider oblique jumps (where a supercritical flow encounters a channel curve). Graf and Altinakar (1998) provide a lucid discussion of oblique jump analysis.

Channel junction analysis (90-degree junction): Momentum analysis enables the computation of an incremental backsplash panel opposite a joining channel (see Figure 6.3). With a 90-degree channel junction, the fluid dynamic component of joining channel momentum goes to zero. Thus, the joining momentum converts to the pressure force term. The joining flow momentum is changed 90-degrees, which also contributes to the momentum increment. The momentum increment is the end goal

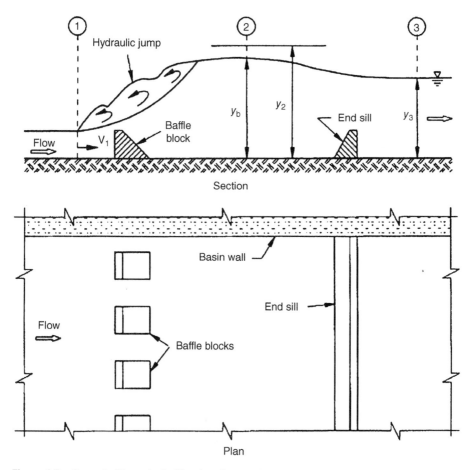

Figure 6.2 Control of jump by baffle piers *Source:* From Jain (2001); used with permission of John Wiley & Sons.

of this analysis. The momentum increment is added to the primary channel depth in the junction's vicinity to avoid splash-over across from the incoming channel.

In the "channelJunctions" folder, there is a workbook called "Trapezoidchannel Junction90Modn" that analyses a 90-degree junction. There is a page for each of the entering main, joining, and exiting channels. The normal depth and critical depth is computed for each channel using the goal seek function. The momentum sheet summarizes the critical and normal depths and has a solver set-up for the momentum depth increment. The momentum increment is added to the normal depth of the exiting flow. Suppose the joining channel's flow is supercritical. In that case, one may consider placing pier structures just before the entrance (see Figure 6.3) to provide drag to dissipate books from the joining flow. The junction analysis presented herein is a demonstration version indicating rough effects and is intended only for approximate designs.

Most junctions are not at 90-degrees. In this case, one must vectorially add the momentum components. The analysis becomes complicated. The 90-degree intersection gives a conservative value. Channel transitions and curves handling supercritical flows require an analysis of waveforms and their interaction with channel boundaries. Graf and Altinakar (1998) and other references cited herein provide more details.

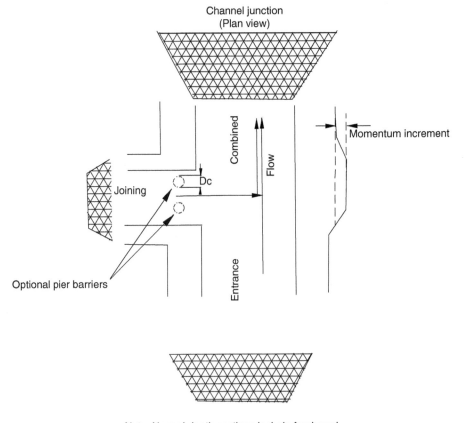

Note: Normal depth sections include freeboard

Figure 6.3 A 90-degree channel junction.

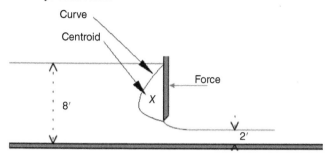

Approximate hydrostatic force

Figure 6.4 Approximate hydrostatic pressure forces on a partially opened sluice gate.

Sluice gate forces: A closed gate may be analyzed as a simple hydrostatic force loading as there is no flow under this gate. A momentum balance provides an approach for analyzing a partially opened gate. The essential force balance is simply the momentum function above the sluice gate subtracted from the momentum function downstream of the gate equals the force on the gate. The Excel spreadsheets in the "SluiceGateForce" folder contain a "Momentum Force" page that computes the gate force. The force diagram curves back to near zero at the gate's bottom edge when there is flow under the gate (see Figure 6.4). The pressure distribution may be quantified using computational fluid dynamics numerical approaches beyond the scope of this text. The point of action of the force is the centroid of the pressure force. This gate force locates approximately down two-thirds of the submerged portion of the gate.

Spatially Varying Steady Flow (dQ/dt = 0, dQ/dx varies, dy/dx varies)

Spatially (gradually) varied flow has a nonuniform discharge resulting from the addition or subtraction of water along the channel. Added or diminished flow causes disturbance of the flow energy and, in limited cases, momentum. Thus, the hydraulic behavior of a spatially varied flow is more complicated. Added flow affects both energy and momentum. Diminished flow is, in effect, a flow diversion where the diverted water does not affect the energy head. Thus, increasing and decreasing discharge cases are handled separately.

Increasing discharge: In this type of spatially varied flow, an appreciable portion of the energy loss is due to the turbulent mixing of the added water and the water flowing in the channel. Added water must undergo a momentum adjustment because of the flow direction change. Figure 6.5 provides a definition sketch of the increasing discharge scenario. The continuity equation may be written as follows.

$$\frac{dQ}{dx} = q^*$$

(6.7)

Parameter Q is the channel flowrate (L³/s), and q^* is the lateral inflow rate per unit length (ft³/s-ft). Knowing that $Q = AV$, the left side of Equation 6.7 may be expanded using the chain rule as follows in Equation 6.8.

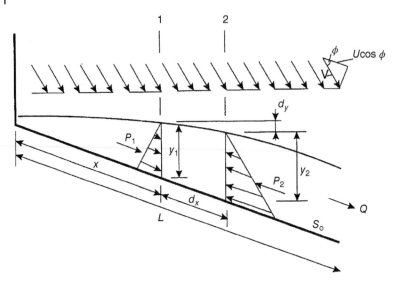

Figure 6.5 Flow in a spatially varied flow with increasing discharge. *Source:* French (2007); used with permission of Water Resources Publications.

$$\bar{u}\frac{dA}{dx}+A\frac{d\bar{u}}{dx}=q^* \tag{6.8}$$

For the control volume, the momentum flux in the x-direction is given as Equations 6.9

$$\Delta M = M_2 - \left(M_1 + M_L\right) \tag{6.9a}$$

where $M_2 - M_1 = \iint \rho u^2 \, d\iota$ \hfill (6.9b)

and $M_L = \rho q^* \Delta x V \cos(\varphi)$ \hfill (6.9c)

Bringing in the fluid's pressure forces and weight, one may write Equations 6.9 in the x-direction as Equation 6.10.

$$\frac{d}{dx}\iint \rho u_x^2 dA - \rho u^* V \cos(\varphi) = \gamma S_o A - \tau_b P - \frac{d}{dx}\iint p dA \tag{6.10}$$

Where $S_o = \sin\theta = $ bottom slope $(-)$, $p = $ pressure (F/L^2), $P = $ wetted perimeter (L), $\gamma = $ fluid unit weight (F/L^3), and $\tau_b = $ average boundary shear stress (F/L^2).

French (2007), following work of Yen and Wenzel (1970) provides simplifications and manipulations of Equation 6.10 to give the following Equation 6.11 of water surface slope. Equation 6.11 is the most general representation of the momentum balance on one dimension.

$$\frac{dy}{dx} = \frac{S_o - S_f + \dfrac{q^*}{gA}(V\cos\theta - 2\beta\bar{u}) - \dfrac{\overline{u^2}}{g}\dfrac{d\beta}{dx} - y\dfrac{d\left(\alpha'\cos\theta\right)}{dx}}{\alpha'\cos\theta\left(1+\dfrac{y}{D}\right) - \dfrac{\beta\overline{u^2}}{gD}} \tag{6.11}$$

where β is the momentum correction factor, α' is the pressure correction coefficient (allows for non-hydrostatic pressure distribution (−), D is the hydraulic Depth (L), and S_f is the friction slope (−).

Let us assume that $\beta=1$, slope θ near zero, $V\cos\phi=0$, $\alpha'=1$ (hydrostatic pressure), we have the most common form of the spatially varied flow with increasing discharge (Equation 6.12).

$$\frac{dy}{dx} = \frac{S_o - S_f - \dfrac{2q^*V}{gA}}{1 - \dfrac{u^2}{gD}} = \frac{S_o - S_f - \dfrac{2Q}{gA^2}\dfrac{dQ}{dx}}{1 - \dfrac{Q^2}{gA^2D}} \tag{6.12}$$

Henderson (1966) concluded that critical flow occurs if the numerator of 6.12 is zero. One integrates upstream in subcritical flow regimes. Conversely, one integrates downstream in a supercritical flow regime. Thus, knowledge of the critical flow zone is essential to solve spatially varied flow problems. Henderson (1966) was credited by French (2007) for presenting Equation 6.13 for locating the critical flow section.

$$x = \frac{8Q_x^2}{gT^2\left(S_o - \dfrac{gP}{C^2T}\right)^3} \tag{6.13}$$

where $Q_x = dQ/dx$ and C is the Chezy coefficient, one could go back to Equation 2.6 and write $C=(\phi/n)*R^{1/6}$.

If x is greater than the channel's length, then a critical flow section does not exist. Suppose there is a critical flow section, and beyond this section, there is a downstream control. In that case, the critical section may be submerged.

French (2007) works an example set up in a convenient spreadsheet form. The "Momentum" folder of the Excel software contains a program, "SpatiallyVariedExample," that is available for experimentation with a decreasing discharge scenario.

Decreasing discharge: Side weirs and bottom racks exemplify this type of spatially varied flow. The energy equation is directly applicable because there is no significant energy loss. On the other hand, momentum changes are difficult to quantify due to uncertainty in the flow departure angle. We, therefore, use the energy principle to analyze decreasing charge. One may write energy at a point as follows in Equation 6.14.

$$H = z + y + \frac{\alpha Q^2}{2gA^2} \tag{6.14}$$

Differentiating concerning the longitudinal coordinate x yields Equation 6.15.

$$\frac{dH}{dx} = \frac{dz}{dx} + \frac{dy}{dx} + \frac{dy}{dx} + \frac{1}{2g}\left(\frac{\alpha 2Q}{A^2}\frac{dQ}{dx} - \frac{2\alpha Q^2}{A^3}\frac{dA}{dx}\right) \tag{6.15a}$$

$$\frac{dH}{dx} = -S_f \tag{6.15b}$$

$$\frac{dz}{dx} = -S_o \tag{6.15c}$$

$$\frac{dA}{dy} = \frac{dA}{dy}\frac{dy}{dx} = T\frac{dy}{dx} \tag{6.15d}$$

Substituting Equations 6.15b–d into Equation 6.15a and rearranging to isolate dy/dx gives Equation 6.16.

$$\frac{dy}{dx} = \frac{S_o - S_f - \dfrac{\alpha Q}{gA^2}\dfrac{dQ}{dx}}{1 - \dfrac{\alpha Q^2}{gA^2 D}} \tag{6.16}$$

Equation 6.16 is the dynamic equation for spatially varied flow with decreasing discharge. The Excel spreadsheet "SpatiallyVariedExample" mentioned above has a page for decreasing discharge scenarios.

Consider the case of water withdrawal through a rack in the bottom of a rectangular channel, as shown in Figure 6.6, an equation for the water surface profile by French (2007). The velocity through the rack is normal to that of the channel. The channel flow energy E may be considered constant. If we take $\alpha = 1$ and $\theta \approx 0$, then the specific energy at any section of a rectangular channel is shown in Equation 6.17.

$$E = y + \frac{Q^2}{2g(by)^2} \tag{6.17}$$

Chow (1959) maintained that specific energy along the channel was constant, which provides a basis for solving the problem. French (2007) shows that the rate of change of y with respect to x can be written as Equation 6.18.

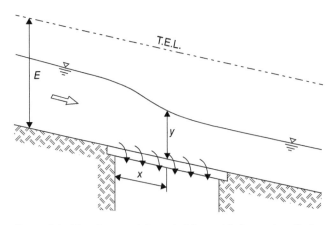

Figure 6.6 Flow in a spatially varied flow with decreasing discharge via a bottom screen. *Source:* Jain (2001); used with permission of John Wiley & Sons.

$$\frac{dy}{dx} = -\frac{Qy\dfrac{dQ}{dx}}{gb^2 y^3 - Q^2}$$

(6.18)

The water loss may be described as Equation 6.19.

$$-\frac{dQ}{dx} = \varepsilon C_D b \sqrt{2gE}$$

(6.19)

where C_D is the coefficient of discharge (typically 0.435 (1:5 slope) to 0.497 (horizontal gradient) and ε is the ratio of open area in the rack to rack surface area. One may solve Equation 6.17 to give a solution for Q, shown in Equation 6.20.

$$Q = by\sqrt{2g(E - y)}$$

(6.20)

Substitution of 6.20 and 6.19 into Equation 6.18 results in an equation for the profile along the rack (Equation 6.21).

$$\frac{dy}{dx} = \frac{2\varepsilon C_D \sqrt{E(E - y)}}{3y - 2E}$$

(6.21)

Integrating Equation 6.21 where $x = 0$ at $y = y_1$, gives Equation 6.22 for x as a function of y along the discharging rack. Depth y_1 may be taken as the channel normal depth before the rack.

$$x = \frac{E}{\varepsilon C_D}\left(\frac{y_1}{E}\sqrt{1 - \frac{y_1}{E}} - \frac{y_1}{E}\sqrt{1 - \frac{y}{E}}\right)$$

(6.22)

The length of rack required to remove all flow in the channel is given by Equation 6.22, setting $y = 0$. The rack length is measured from the initial encounter point and is negative.

Example 6.3 A flowing rectangular channel carrying 1.5 cfs with a bottom width of 0.5 ft, $n = 0.015$, $S = 0.005$, has a rack with a $C_D = 0.45$ and $\varepsilon = 0.3$. Assuming the rack has the same width as the channel, how long must the rack be to capture all the flow?

One must first compute the normal depth in the channel before the discharge initiation. Using a spreadsheet, find that $y = 0.855$ ft. One can then compute the energy using Equation 6.17, which is 0.89 ft-lb/lb. Substituting into Equation 6.22, letting $y = 0$, result in $x = -5$ ft.

French (2007) also presents a solution for drainage through a system where the exit velocity is not normal to the channel flow. A perforated screen is a case in point. The distinctive difference between the screen and the rack is that energy loss with a screen exit is not negligible because the exit velocity has a significant component in the channel flow direction. Instead of writing Equation 6.19 as a function of Energy E, we write it in terms of y. French (2007) provides the solution details and give a relationship between depth y and distance x. Sturm (2010) offers a solution for a side discharge weir.

HEC-RAS can numerically accommodate increasing and decreasing flow profile analyses. The example set contains example analyses with lateral discharging weirs (decreasing flows). Increasing flow is simulated with a series of stream junctions, which is supported by HEC-RAS.

Conclusions

This chapter developed the one-dimensional momentum equation, which was useful in analyzing hydraulic jumps and channel junction analysis. Supercritical flow behavior in curves and channel transitions requires analyses beyond the scope of this text.

Momentum analysis is essential for spatially varied flows with increasing discharge. We present some spreadsheet solutions for hydraulic jumps in standard cross sections and spatially varied flows in rectangular channels. The decreasing discharge problem can be handled with energy analysis. A simple solution was given for predicting the length of a grate for discharging flow in a channel.

Problems and Questions

Note: Draw sketches of all completed designs and provide supporting reasons for discussion questions. Show all work and offer a rationale for assumptions made. Remember to use appropriate significant digits. Feel free to use the software resources. Assume freeboards of 20% and smooth concrete lining unless otherwise stated.

1 A flow of 100 cfs is moving down a steep slope in a trapezoidal channel ($b = 10$, $z = 3$, $n = 0.015$, $S = 0.05$) and transitions to a nearly flat channel ($S = 0.001$) with similar geometry and roughness. What is the depth expected before and after the transition? What is the Froude number before and after the transition?

2 Compute the energy dissipated in problem 1. What is the temperature change of the water, assuming no heat is otherwise lost?

3 Determine the sequent depths for a hydraulic jump in a 7 ft diameter storm sewer, $n = 0.015$, entering slope $= 0.08$, exiting slope $= 0.005$, $Q = 100$ cfs. Note: If the exit flow's sequent depth differs from the normal depth, expect the jump to move above or below the point of slope change. If the normal depth is below the critical depth, no jump occurs. Is there a jump? If so, where is it?

4 A flow of 50 cfs joins a flow of 100 cfs. Design the junction. Assume smooth concrete is the material. Cross sections are $n = 0.02$, $z = 3$, and $b = 2$ ft. The main channel slope is $S = 0.01$, and the slope of the joining channel is 0.02. Use some drag-inducing piers if needed.

5 A flow of 100 cfs joins a flow of 200 cfs. The channels are made of rough concrete ($n = 0.025$, $z = 3$, and $b = 4$ ft). The main channel slope is 0.005, and the joining channel slope is 0.05. Design the junction, using drag-inducing piers of needed.

6 A 10 cfs flow in a trapezoidal concrete channel ($z = 2$, $b = 2$) on a 1% slope intersects (90-degrees) a channel carrying 20 cfs in a trapezoidal channel ($z = 2$, $b = 5$) on a 0.005 ft/ft slope. Design the intersection.

7 Compute the momentum coefficient β of the channel in Example 2.1.

8 A rectangular drainage channel on a slope of 0.01, roughness $n = 0.02$, bottom width $= 1.5$ ft, has a flowrate of 1.5 ft^3/s. The channel has a drainage rack ($C_D = .4$ and $\varepsilon = .35$) the same width as the channel. How long must be the rack to capture all the flow. Solving Equation 6.22 with $y = 0$ yields $x = -1.1$ ft. The rack must be 1.1×1.5 ft. Build and present a spreadsheet to solve this problem.

9 Project: Use HEC-RAS to solve problem 1.

References

Brunner, G.W. (2016). *HEC-RAS River Analysis System: User's Manual*. Davis, CA: US Army Corps of Engineers, Hydrologic Engineering Center.

Chow, V.T. (1959). *Open-Channel Hydraulics*. New York, NY: McGraw-Hill.

French, R.H. (2007). *Open Channel Hydraulics*. Highlands Ranch, CO: Water Resources Publications.

Graf, W.H. and Altinakar, M.S. (1998). *Fluvial Hydraulics: Flow and Transport Processes in Channels of Simple Geometry*. Chichester, UK: Wiley.

Haan, C.T., Barfield, B.J., and Hayes, J.C. (1994). *Hydrology and Sedimentology of Small Catchments*. New York, NY: Academic Press.

Henderson, F.M. (1966). *Open Channel Flow*. New York, NY: Macmillan Publishing Co.

Jain, S. (2001). *Open-Channel Flow*. New York, NY: Wiley.

Kindsvater, C.E. (1944). The hydraulic jump in sloping channels. *Transactions of the American Society of Civil Engineers* 109: 1107–1154.

Streeter, V.L. (1971). *Fluid Mechanics*, 5e. New York, NY: McGraw-Hill.

Sturm, T.W. (2010). *Open Channel Hydraulics*, 2e. New York, NY: McGraw-Hill.

Unites States Bureau of Reclamation (1987). *Design of Small Dams*, 3e. Washington, DC: Water Resources Technical Publications.

Yen, B.C. and Wenzel, H.G. (1970). Dynamic equations for steady, spatially varied flow. *American Society of Civil Engineers. Hydraulics Division* 96 (HY3): 801–814.

7

Hydraulics of Water Management Structures*

Water management structures help stabilize actively eroding areas, control and monitor flows, and convey stormwater runoff. Various hydraulic structures facilitate research to measure runoff coefficients in Chapter 5. Structural systems often form the backbone for many Best Management Practices (BMPs) for on-farm or on-job sediment control in the United States and developed countries. Biswas (1970) reports on ancient civilizations' use of hydraulic structures before 600 BCE for flow measurements and controls. Flow controlling mechanisms include orifices, weirs, pipes, and open channels. Pipe slope, length, and degree of inlet/outlet submergence determine the flow category.

Structures covered in this chapter find applications in waste treatment, irrigation, ponds and basins, gully stabilization, sediment control, and food processing. Worldwide, aquaculture increases create a continuing demand for new structures associated with fishponds (Gopalakrishnan and Coche 1994). Figures 7.1 through 7.5 illustrate structural approaches for gully stabilization, conveyance under roadways, drop structure for a vegetated waterway, runoff control, and irrigation management.

Weirs, orifices, and pipes describe the pond or basin spillway. Integration of spillway hydraulics with the water supply impoundments or sediment basin remains for discussion in Tollner's texts (2016).

Goals

- to identify structures concerning various inlets, conveyances, and outlets;
- to determine the structure hydraulics (weir, orifice, flume, channel, and culvert) based on inlet and conveyance hydraulics and appreciate the need to dissipate energy at the outlet;
  - to touch on potential ecological impacts of placing structures in natural streams.

---

*This chapter draws heavily upon the following: Tollner, E.W. (2016). *Introduction to Engineering Hydrology for Natural Resources Engineers*, 2e. Chapter 11.

Open Channel Design: Fundamentals and Applications, First Edition. Ernest W. Tollner.
© 2022 John Wiley & Sons Ltd. Published 2022 by John Wiley & Sons Ltd.
Companion website: www.wiley.com/go/tollner/openchanneldesign

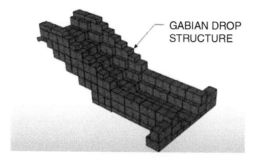

GABIAN DROP
STRUCTURE

Figure 7.1 Gully stabilization structure schematic showing packages of rock typically in wire-mesh packages. Broad crested weir hydraulics controls the flow. *Source:* Courtesy of Maccaferi Gabions, Inc., Williamsport, MD.

Figure 7.2 A culvert outlet with riprap for flow energy dissipation.

Structure Types

Figures 7.1 through 7.5 show a variety of structure types. Figure 7.6 shows schematics of additional structure types. Standard designs are available from sources such as Brakensiek et al. (1979) and USBR (1987). Hydraulic design is the primary interest in this presentation.

The attributes characterizing a structure are the *inlet, conduit, or conveyance, and the outlet with energy dissipation* features. One classifies structures according to the following characteristics. Inlet (e.g., weir) conditions and sometimes conduit (pipe or channel hydraulics), or outlet hydraulics (e.g., the effect of depth of flow in the outlet channel) limits flow through weirs, flumes, chutes, and drop inlets. Hydraulics of the inlet and conduit can restrict flow in pipe type spillways. Erosion control is often the primary design consideration for outlet specifications. Erosion or scour[1] is a critical problem when the free outfall condition exists. Typically used models consist of various sills and block arrangements.

---

1 The program HY-8, discussed below, contains provisions for calculating scour resulting from culvert outfalls.

Figure 7.3 Concrete drop structure for vegetated waterway under construction. Drop structures enable lower channel slopes for erodible channels. *Source:* Courtesy of NRCS.

Figure 7.4 Flow control structure in an urban area to control and measure runoff flow rates. Sharp crested weir hydraulics applies to this structure.

Additional components include wing walls, sidewalls, headwall extensions, cut off walls, and toe walls. Figure 7.7 shows some structural examples.

Adequate foundations are essential to the success of permanent structures. One should remove surface soils. Pipes should have seepage collars to prevent seepage-induced erosion from causing eventual failure. Much of the design information for structures comes from the testing of models in the laboratory. Beasley et al. (1984) give approximate guidance on outlet selection considering head drop and flow, shown in Table 7.1.

Figure 7.5 Parshall flume for measuring the flowrate in an irrigation channel.

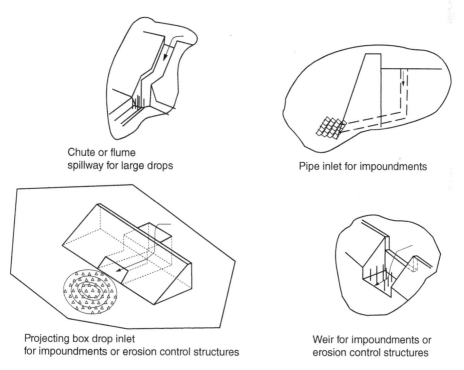

Chute or flume
spillway for large drops

Pipe inlet for impoundments

Projecting box drop inlet
for impoundments or erosion control structures

Weir for impoundments or
erosion control structures

Figure 7.6 Schematic showing how four common spillway types are applicable for applications ranging from gully stabilization to pond principal spillways.

Hydraulic Concepts

Specific energy: The specific energy of flow, depth y plus velocity energy ($v^2/2g$), was defined in Chapter 5 as flow energy. Figures 7.8 and 7.9 contain typical specific energy plots. A flow may possess the same specific energy at two different depths and corresponding velocities.

Figure 7.7 Selected spillway inlets, conduits or conveyances, and outlets with energy dissipation structures. *Source:* Courtesy of Wikimedia with permission to reuse per the Creative Commons Attribution-Share Alike Unported License provisions, unless otherwise indicated. (a) Installation of a drop inlet spillway. *Source:* Courtesy of the NRCS. (b) Pipe conveyance and plunge pool, Nebraska. *Source:* Courtesy of National Register of Historic Places, Culvert on 522 Ave (Antelope County, NE). (c) Chute conveyance, Mountain Chute Dam, Quebec. *Source:* Courtesy of Qui1che. (d) Small bell-mouth pipe-drop spillway, Great Meadows National Wildlife Refuge, Concord, MA. *Source:* liz west - Flickr,https://commons.wikimedia.org/w/index.php?curid=18098251. Licensed under CC BY 2.0.

Table 7.1 General guidance for selecting outlet structure based on the head drop and discharge.

Controlled head (ft)	Discharge (ft³/s)					Controlled head (m)
4–8	10–25	50–100	100–150	150–800	800–1500	1–2.5
	Drop inlet or hooded inlet	Drop inlet or hooded inlet	Drop	Drop	Drop	
12–30	Hooded or pipe drop	Hooded or pipe drop	Pipe Drop	Pipe drop or another custom drop	Chute	3.5–9.3
40–80	Pipe drop	Pipe drop	Pipe drop	Custom drop	Chute	12–24
	0.25–0.75	1.4–2.8	2.8–4.3	4.3–22.6	22.6–43	
	Discharge (m³/s)					

Source: Adapted from Beasley et al. (1984) and NRCS (1964).

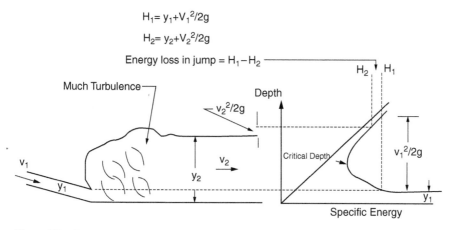

Figure 7.8 Hydraulic jump schematic showing energy relationships (not to scale).

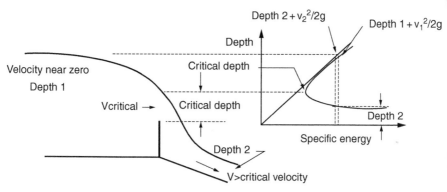

Figure 7.9 Energy relationships for a subcritical–supercritical flow transition used in quantifying discharge (not to scale).

Critical depth is the depth at which water flowing in a channel has the lowest energy head. Flow with a Froude number = 1 enables unique stage–discharge relationships, which is critical to flow measurement.

Hydraulic jump positioning: The hydraulic jump, shown in Figure 7.8, was discussed in detail in Chapter 6. The jump may move up or downstream as flow varies unless measures are taken to fix the jump location. A moving jump is undesirable due to safety concerns. Blocks and sills in the outlet portion of the structure control the hydraulic jump location and increase the energy loss. The focus here is to consider ways of controlling the jump location and energy dissipation. Friction and turbulence result in lost energy in the jump zone. Example 6.2 discussed energy loss in hydraulic jumps. Because the hydraulic jump position depends on depth and flowrate above and below the jump, stilling basins hold the hydraulic jump at a fixed location and enhance energy dissipation. These basins and structures take momentum away from the flow under a range of flow conditions and thus stabilize the jump location.

Sub-supercritical flow transitions: Important hydraulic concepts related to hydraulic structures' function are the critical depth and specific energy. Figure 7.9 shows a specific energy curve for a sub-super critical flow transition. These transitions are chokes, as discussed in Chapter 5. They derive from the unique flow- depth relationship at the critical point. Flow measurement requires a three-step transition from subcritical (i) to critical (ii) to supercritical (iii) flow, harking back to flow choking discussed in Chapter 5. Thus, in the flow structure throat, the Froude number equals one. Coupling with the continuity equation and setting the Froude number (F) as unity enables a unique relationship between stage and discharge.

Stage–Discharge Relationships of Weir Inlets and Flumes

The drop spillway, box-inlet drop spillway, and formless flume (Blaisdell 1948; Blaisdell and Donnelly 1951; Wooley et al. 1941) behave according to weir hydraulics with a free discharge. A simplified derivation of the weir formula follows. At critical depth the Froude number, defined as $F = v/(gD_h)^2$ is equal to 1, therefore $v = (gD_h)^{1/2}$. For a rectangular (Topwidth T_w = bottom width L) weir flow $D_h = A/T_w = LH/L =$ depth $= H$. One may insert the appropriate definition for D_h for other cross-sections. Coupling the Froude number relationship with the continuity ($q = Av = L * H * v$ for a rectangular weir) creates the basis for a heuristic stage–discharge relationship for many structures. One can write $q =$ Const $* g^{1/2} * A * H^{1/2} =$ Const $* g^{1/2} * L * H * H^{1/2}$, where Const is a discharge coefficient approximately equal to 0.6. The previous sentence leads directly to the weir equation given in Equation 7.1:

$$q = \phi_{dim} C_w L H^{3/2} \tag{7.1}$$

where φ_{dim} is a dimensional constant equal to 1 in English units and 0.55 in SI units; C_w represents the weir constant:

The constant C_w corrects for converging flow effects and energy loss in the weir cross-section and includes the square root of the gravitational constant g. Other more formal derivations of Equation 7.1 proceed directly from the energy equation and do not invoke the Froude number. Sturm (2010) provides the derivations for the rectangular, triangular, and trapezoidal weirs. Using the energy equation-based derivation enables us to measure the flow depth upstream where the velocity is nearly zero.

Be careful to check the consistency of units before proceeding, as many references use imperial units. For stage measurement devices, follow established procedures for the measurement location, allowable upstream flow conditions, and acceptable exit conditions (e.g., outlet submerged or free discharge). Haan et al. (1994) summarize hydraulic relations for several types of weirs in Figure 7.10. Grant and Dawson (2001) provide additional stage-flow equations and various weir inlet devices and flumes.

Drop spillways enable lower channel slopes with discrete drops in elevation, resulting in potential erosion control and more stable channels. Flow passes through a weir, falls onto an apron or stilling basin, and passes into a downstream channel.

Selected weir applications: For free flow, one calculates capacity for a drop spillway with the weir formula. The length L is the length of the three sides of a box inlet, the arc length

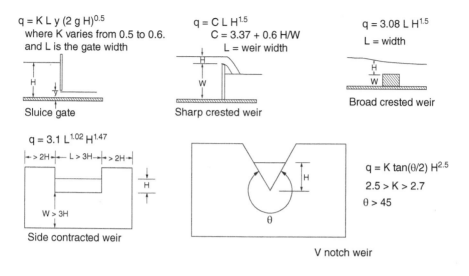

$q = K L y (2 g H)^{0.5}$
where K varies from 0.5 to 0.6.
and L is the gate width

Sluice gate

$q = C L H^{1.5}$
$C = 3.37 + 0.6 \, H/W$
L = weir width

Sharp crested weir

$q = 3.08 \, L \, H^{1.5}$
L = width

Broad crested weir

$q = 3.1 \, L^{1.02} \, H^{1.47}$

W > 3H

Side contracted weir

$q = K \tan(\theta/2) \, H^{2.5}$

2.5 > K > 2.7

θ > 45

V notch weir

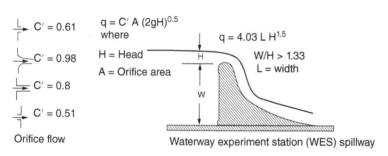

C' = 0.61

C' = 0.98

C' = 0.8

C' = 0.51

Orifice flow

$q = C' A (2gH)^{0.5}$
where

H = Head

A = Orifice area

$q = 4.03 \, L \, H^{1.5}$
W/H > 1.33
L = width

Waterway experiment station (WES) spillway

Figure 7.10 Hydraulics of common flow control and inlet devices. Notational notes: English (FPS) units assumed; flowrate Q here is identical to q; weir coefficient C is similar to C_w; and orifice coefficient C' is similar to C_{orf}. *Source:* Redrawn following Haan et al. (1994).

of the arch inlet, and the straight inlet's crest length. We measure the head from the weir or flume bottom near the point of critical flow.

For many situations, a value of $C_w = 3.2$ produces satisfactory results for design purposes. One can describe other structures using the following C_w values: box inlets, $C_w = 3.2$ drop spillways, $C_w = 3.2$; formless flume, $C_w = 3.9$. For the condition with the ratio of head to box width is <0.2, refer to Blaisdell and Donnelly (1951). Figure 7.10 also gives accepted values of the weir coefficient. For weirs designed for flow measurement, use these relationships and associated C_w values.

Brater and King (1976), Bos et al. (1984), and French (2007) give additional information concerning weir constants. The design relation allows a direct calculation of flowrate from depth. One can determine the expected depth at a given flowrate. Size the structure to keep the minimum flow between 10% and 100% of full depth. The approach velocity must be near zero upstream of the structure for the fundamental weir equation's validity. A spreadsheet that covers computations for several types of weirs is given in the Excel "Weirs" folder. French (2007) and Stephens (1991) provides additional information relating to weir selection, placement, and sizing.

Example 7.1

What is the required length of a sharp-crested weir ($C_w = 3.2$) for moving $0.5\,m^3/s$ at an available head of $0.85\,m$? Solve for the case of a broad crested weir ($C_w = 3.9$).

Begin with Equation 7.1. $0.5\,m^3/s = 0.55(3.2)\,(L)\,((0.85\,m)^{1.5})$ leading to $L = 0.36\,m$. For the broad crested weir we have $0.5\,m^3/s = 0.55(3.9)\,(L)\,((0.85\,m)^{1.5}) = 0.30 = L$. If this were a projecting box inlet spillway then the length of each of the three flowing sides of an equal sided box would be $L/3 = 0.10\,m$.

The Parshall flume, shown in operation in Figure 7.5, provides a choke that enables measurement of flows with minimal head loss. Figure 7.11 shows a schematic of the Parshall flume. The Parshall flume is relevant in surface irrigation applications. Many processing applications involving open channel flows employ Parshall flumes. Equation 7.2 gives the stage–discharge:

$$q = AH_a^b \tag{7.2}$$

where A ranges from 0.99 (3 in. throat), 2.06 (6 in. throat), 3.07 (9 in. throat), $4W$ (12 in. to 8 ft throat), and $3.6875W + 2.5$ (10–50 ft throat);

b ranges from 1.54 (3 in. throat), 1.58 (6 in. throat), 1.53 (9 in. throat), $1.522W^{0.026}$ (12 in. to 8 ft throat), or 1.6 for throat widths from 10 to 50 ft;

H_a is the inlet head (ft).
W is the throat width (ft).

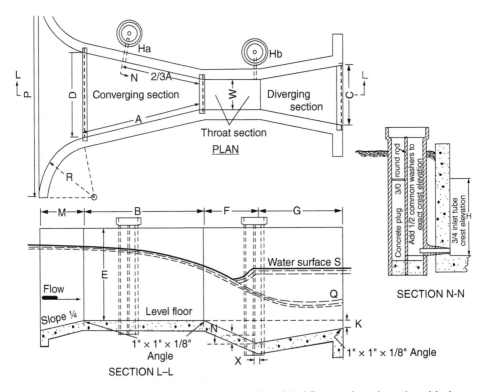

Figure 7.11 Schematic of Parshall flume, showing the critical flow section where the critical section may be submerged. *Source:* Courtesy of NRCS.

Note the similarity with the weir equation. The critical flow section's unusual shape explains the (empirically determined) deviation from the weir coefficient of 3.2 and the exponent deviation from 1.5.

The critical flow section may be submerged in a Parshall flume. Submergence affects the flow when the water level is from 60 to 100% (or greater) of the inlet head. Refer to works such as Chow (1959) and Brakensiek et al. (1979) for factors and diagrams providing submergence correction factors. They also provide additional construction details. Parshall flumes perform best with low sediment concentrations.

Example 7.2

Given a Parshall flume, 50% submerged, throat width of 1 ft, and inlet head of 0.5 ft. Compute the flow.

Since submergence is less than 50%, no correction is needed. From Equation 7.2 and coefficients for the flume with a 1 ft throat,

$$Q = 4(1\text{ft})(0.5\text{ft})^{1.522(1\text{ft})^\wedge(0.026)} = 1.39\,\text{cfs}.$$

Refer to Chow (1959) or Brakensiek et al. (1979) if submergence is higher than 50%.

The H_x flume, where subscript $x =$ S, M, L, (e.g., <u>S</u>mall, <u>M</u>edium, <u>L</u>arge) is a critical flow device shaped not to trap sediment with sediment-laden flows. Figure 7.12 shows a field view of an H_M flume. Figure 7.13 shows the relative proportions of the H_s flume (see Brakensiek et al. (1979) for similar drawings for the H_m and H_l flumes).

The H_x flume is a critical flow device based on the trapezoid projecting to a surface normal to the flow direction (see the photograph in Figure 7.12).

Figure 7.12 Flume and sampling equipment for measuring runoff hydrographs, sedimentation concentration vs. time, and pesticide concentration vs. time. A mechanical stage recorder is partially visible under the raised cover on the right. The location is the University of Georgia Phil Campbell research farm (formerly USDA-ARS station) at Watkinsville, GA.

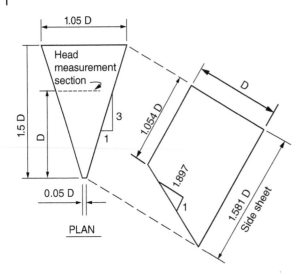

1.05 D

Head measurement section

0.05 D

1.5 D

D

3

1

PLAN

1.054 D

1.897

1

D

1.581 D

Side sheet

Figure 7.13 Three views of an H_s flume. The H_m and H_l flumes are similar in style with modified proportions to accommodate better larger flows. *Source:* From Brakensiek et al. (1979).

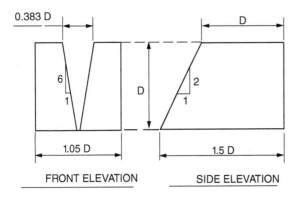

0.383 D

6

1

1.05 D

FRONT ELEVATION

D

D

2

1

1.5 D

SIDE ELEVATION

Using the heuristic approach of Equation 7.1 and substituting the hydraulic elements of the projected trapezoidal section, one can then expand the weir equation, as shown in Equation 7.3:

$$q = 10^{-6}\sqrt{g}\left[c_o + c_1\frac{y}{2B_o} + c_2\log_{10}\left(\frac{y+6.092}{0.5\text{Des}}\right)\right]\left[\frac{2B_o +\left(B_1 + b_1\right)\left(y + h\right)^2}{\sqrt{2B_o + [2\left(B_1 + b_1\right)\left(y + h\right)}}\right]^{1.5} \quad (7.3)$$

where B_o is one-half the bottom width of the flume at its outlet (in Figure 7.12, Plan view, B_o is 0.025D, mm);

Des is the design depth of the H flume (D in Figure 7.12, mm);
y is the depth of flow in the flume (mm);
g is gravity (9810 mm/s^2);
q is flowrate (l/s);

B_1 is the projected side slope value for the flume as seen from the front elevation (e.g., in Figure 7.12, front view, $B_1 = 1/6 = 0.167$);

b_1 is a projected slope offset found using nonlinear regression

h is a depth offset found to be nonzero using nonlinear regression. The virtual projection explains the offset.

c_0, c_1, c_2 are components of the weir C_w coefficient.

The lowercase coefficients are coefficients estimated using nonlinear regression approaches. The author wrote the coefficients following an expansion of a formula for weir losses from Brater and King (1976).

Setting the Froude number equal to one and applying the continuity equation with the projected trapezoidal geometry leads to Equation 7.2. We included offsets for fitting purposes, as shown. The first large bracket in the above equation is the C_w value. Table 7.2 gives selected geometric elements for the various H flumes. Table 7.3 gives values for the coefficients of Equation 7.3. Using Equation 7.2 presumes a determined H_x flume geometry. One may first solve Equation 7.3 using coefficients for all flow values for the approximate flowrate calculation. Refinements to the calculated flow are then possible from using appropriate coefficients corresponding to the approximate flow. Equation 7.3 appears useful for modeling stage–discharge of H_x flumes with geometric characteristics bounded by the standard models given by Brakensiek et al. (1979).

Equation 7.3 fails when the depth y is less than 0.1 Des because the subcritical–supercritical transition does not happen consistently with shallow flow depths. The flow depth in the flume is, of course, less than the flume height. Brakensiek et al. (1979) contain detailed rating tables and construction drawings for H_x flumes.

Table 7.2 Selected geometric elements for H_x flumes.

Type	2^*B_0 (mm)	D (mm)	B_1 (–)	H_{max} (mm)	q_{max} (L/s)
H_S	6.096	121.92	0.1667	118.87	2.274
H_S	9.14	182.88	0.1667	179.83	6.258
H_S	12.19	243.84	0.1667	240.79	12.941
H_S	15.24	304.8	0.1667	301.75	22.738
H_M	15.24	152.4	0.5	149.35	9.373
H_M	22.86	228.6	0.5	228.6	27.102
H_M	30.48	304.8	0.5	301.75	54.368
H_M	45.72	457.2	0.5	454.15	150.928
H_M	60.96	609.6	0.5	606.55	310.919
H_M	76.2	762.0	0.5	758.95	543.683
H_M	91.44	914.4	0.5	911.35	860.832
H_M	137.16	1371.6	0.5	1368.55	2378.6
H_L	243.84	1219.2	1.0	1219.2	3313.1

Source: From Brakensiek et al. (1979).

Table 7.3 Summary of estimated parameters for the generalized H_x flume critical flow equation at selected flow conditions.

Term	Coefficients applicable to all flumes in Brakensiek et al. (1979), where $H \sqsupset 0.1D$, $H < D$, and flowrate q is less than the indicated flow rate.				
	q = all	q#115 L/s	q#30 L/s	q# 15 L/s	q# 8 L/s
h(mm)	−0.57	−0.39	−0.21	0	0
b_1	0.059	0.083	0.075	0.078	0.0793
c_0	0.673	0.684	0.689	0.685	0.684
c_1	0.003	−0.0023	−0.0022	−0.0025	−0.0025
c_2	0.027	0.0875	0.0924	0.0895	0.0901
Residual range (L/s)	−22 to +14	−2.6 to 0.6	−0.26 to 0.15	−0.05 to 0.07	−0.03 to 0.05

For a given H_x flume and maximum flow for that flume, select the appropriate flow equation parameters. The minimum flow is that which meets the $H \geq 0.1D$. One may also choose the size based on the maximum anticipated flow.
Source: Tollner (2016) unpublished data.

Example 7.3

A sediment-laden flow not expected to exceed $0.1\,\text{m}^3/\text{s}$ (100 L/s) is anticipated from a watershed. Select an H_x flume and compute the maximum expected depth.

Select an H_m flume with $D = 457.2\,\text{mm}$ from Table 7.2. Use coefficients with q #115 L/s from Table 7.3 in Equation 7.3 to determine the maximum flow depth.

Discharge Relations of Orifices and Sluice Gates Inlet Devices

Consider the energy, shown as Equation 7.4, assuming negligible losses.

$$\frac{p_1}{\gamma} + \alpha_1 \frac{v_1^2}{2g} = \frac{p_2}{\gamma} + \alpha_2 \frac{v_2^2}{2g} \tag{7.4}$$

The α_i represents the respective energy correction factors (taken as 1 throughout this chapter); the datum is the channel bottom. One can assume velocity v_1 and p_2 to be near zero, resulting in Equation 7.5a:

$$\frac{p_1}{\gamma} = \frac{v_2^2}{2g} = y \Rightarrow v = \sqrt{2gy} \tag{7.5a}$$

Substituting $q = Av$ into Equation 7.5 corrected with an orifice coefficient C_{orf} leads directly to the orifice equation in Equation 7.5b.

$$q = C_{orf} A_{orf} \sqrt{2gy} \tag{7.5b}$$

where A_{orf} is the orifice cross-sectional area (L^2); C_{orf} is the orifice coefficient (−); and, Y is the distance from the orifice center (L).

The inclusion of the C_{orf} coefficient accounts for energy losses and momentum effects due to convergence. Sluice gate analysis proceeds along similar lines. These equations are valid in any consistent unit system.

Orifice and sluice gate applications: This computation is a straightforward application of the orifice equation. An orifice constant of 0.6 is adequate in most common design situations. Refer to Figure 7.10 for accepted orifice coefficients. Brater and King (1976) also give orifice constants for many orifice configurations, including submerged discharge cases. Orifice and sluice gates may be submerged. Chin (2013) shows a combined energy and momentum equation analysis for submerged structures. Flows through sluice gates based on energy and continuity equation solutions were discussed in Chapters 5 and 6.

Example 7.4

What is the orifice area required to move $0.5\,m^3/s$ with a head of 2.85 m, measured from the center of the orifice to the free water level behind the orifice?

One may begin with Equation 7.5b, substituting known quantities. $0.5\,m^3/s = 0.6\,A$ Sqrt($2(9.8\,m/s^2)\,(2.85\,m)$) leading to $A = 0.1114\,m^2$. The diameter is Sqrt($4(.1114)/\pi$) = 0.376 m.

Flow Hydraulics of Closed Conduits

Either the inlet section or the conduit may control the flow through closed conduits such as culverts and other pipe structures. Inlet flow control is either weir-limited (unsubmerged) or orifice-limited (submerged). Conduit flow control is either open channel (unsubmerged inlet and outlet) or pipe limited (inlet and outlet submerged or unsubmerged). Figure 7.14 shows culvert nomenclature associated with culvert analyses, and Figure 7.15 shows possible flow scenarios.

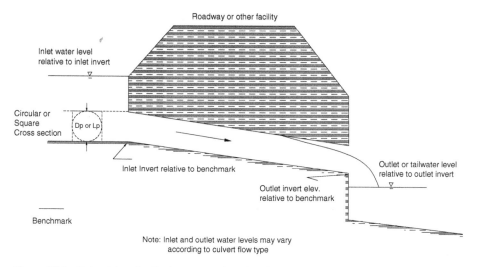

Figure 7.14 Culvert longitudinal cross-section showing nomenclature.

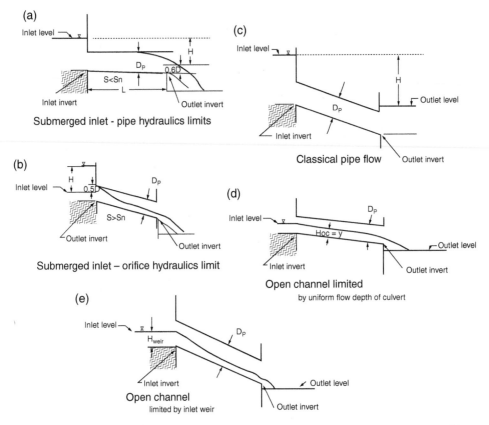

Figure 7.15 Selected possible scenarios of flow through a closed conduit. Consider the definition of the head in Categories A through C. Categories A and B are required when designing a free discharge culvert to pass a peak discharge subject to head limitations. These problems require a test (e.g., calculation of flow for each type and selecting minimum flow) to determine which of A or B controls. Category C is unique to communicating water bodies, using the Category A equation with a modified definition for head. Categories D and E are solved utilizing the weir equation at the entrance and the Manning equation plus energy equation as in the transition problem. The controlling Category is the one providing minimum discharge. Categories D and E are helpful in stage-discharge relationships for spillways and related structures. Culverts whose outlets are submerged to an intermediate extent require advanced treatment beyond the scope of this text.

Depending on the water stage at the inlet and outlet, weir flow, orifice flow, open channel flow, or pipe flow may control the flowrate. This text discusses only the case where the exit water level does not control (except full pipe flow, not requiring the simplification). We defer a more detailed analysis of exit water conditions to advanced courses. Weir flow nearly always occurs at the onset of flow through the structure. Orifice flow is most likely the controlling mechanism when pipes are large and short, with steep slopes. Pipe flow generally controls when pipes are relatively long with low slopes.

Submerged inlets-free discharge: If the inlet is submerged, orifice or pipe flow dictates the flowrate. Equation 7.6 shows the orifice flow (category B):

$$q = C_{orf} A_{orf} \sqrt{2gH} \quad \text{with} \quad \begin{aligned} H &= H_{inlet} - 0.5D_{orf} \quad \begin{pmatrix} \text{pipe horizontal} \end{pmatrix} \\ H &= H_{inlet} \quad \begin{pmatrix} \text{pipe vertical} \end{pmatrix} \end{aligned} \tag{7.6}$$

where, q is flow capacity (L^3/T); A_{orf} ($=A_p$) is conduit cross-sectional area (L^2); H is the head causing full pipe flow (L); see Figure 7.15; D_{orf} ($=D_p$) is the diameter of the orifice or conduit; H_{inlet} is the (submerged) inlet water surface elevation from the inlet invert; g is gravity (L/T^2); C_{orf} is the orifice coefficient, typically 0.6.

Definition: invert elevation refers to the pipe's bottom elevation with respect to some benchmark elevation datum.

As shown in Equation 7.5, the orifice equation was derived with the datum in the center of the orifice. The pipe flow (Category A) condition is described by writing an energy balance, as shown in Equation 7.7:

$$H = y_i - y_o = \frac{v_{pip}^2}{2g} + K_e\frac{v_{pip}^2}{2g} + K_b\frac{v_{pip}^2}{2g} + S_f L \tag{7.7}$$

where, v_{pip} is the velocity in the pipe (m/s); y_o is the pressure head at the pipe exit (m); y_i is the pressure head at the pipe inlet (m); S_f is the unit friction loss per unit length; and K_e, K_b are loss coefficients for the entrance and bend, respectively.

Figure 7.16 shows the components of the energy balance along with the energy grade line.

One may derive the friction slope S_f from the Manning equation (Equation 8.14), taking φ_{dim} as one, giving Equation 7.8a:

$$S_f = v_{pip}^2\frac{n^2 4^{4/3} 1000^{4/3}}{D_{pip}^{4/3}} \tag{7.8a}$$

where, D_{pip} is the pipe diameter (mm). This approach works well with rough piping because large ε/D ratios and high Reynolds numbers lead to constant friction factors. One can see from a Darcy–Weisbach friction factor diagram (see fluid mechanics texts such as Chin 2013) that the friction factor is a function only of roughness in the fully developed turbulence regime. Multiplying both sides by $2g$ gives Equation 7.8b:

$$2gS_f = v_{pip}^2\frac{n^2 4^{4/3} 1000^{4/3}}{D_{pip}^{4/3}} 2g = K_c v_{pip}^2 \tag{7.8b}$$

where, K_c is introduced for simplification. Substituting the right-hand side of Equation 7.8b into Equation 7.7, solving for v_{pip}, and multiplying by pipe cross-sectional area (A_p) leads directly to the discharge relationships summarized below in Equation 7.9a:

$$q = \frac{A_{pip}\sqrt{2gH}}{\sqrt{1 + K_e + K_b + K_c L}} \tag{7.9a}$$

$$H = \left(H_{inlet} + H_{inlet\ invert}\right) - H_{outlet\ invert} - 0.6D_{pip} \tag{7.9b}$$

where L is conduit length (m); K_b is the bend coefficient (Figure 7.17); K_e is the entrance coefficient (Figure 7.17); K_c is the conduit loss coefficient (Equation 7.9b or c); H_{inlet} is the (submerged) inlet water surface elevation above the inlet invert; $H_{inlet\ invert}$ is the pipe bottom elevation at the inlet relative to a surveying benchmark; $H_{outlet\ invert}$ is the pipe bottom elevation at the outlet relative to a surveying benchmark.

Note: other minor losses are usually neglected in culvert pipe flow analyses.

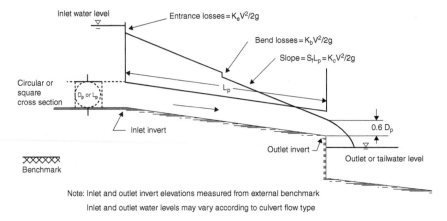

Figure 7.16 Culvert situation of Figure 7.14 showing the energy grade line and respective inverts and crown nomenclature.

$\dfrac{R}{D} = \dfrac{\text{Bend radius to pipe center line}}{\text{Pipe diameter}}$	Bend coefficient, K_b	
	45° Bend	90° Bend
0.5	0.7	1.0
1	0.4	0.6
2	0.3	0.4
5	0.2	0.3

Source: U.S. soil conservation service (1951). *Engineering Handbook*, Section 5: "Hydraulics." SCS, Washington, DC.

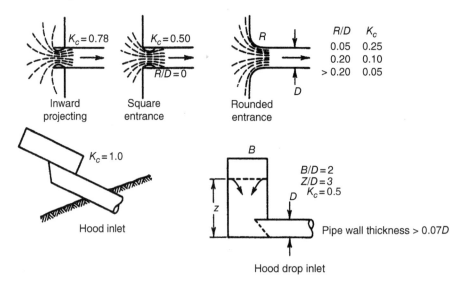

Figure 7.17 Entrance and bend loss coefficients. *Source:* Courtesy of NRCS.

Charts (see Figure 7.17) or formulas below provide coefficient estimates. The culvert pipe flow equation is simply an extension of the orifice equation, both of which follow from the Bernoulli equation. The distance of $0.6\,D_p$ accounts for the reality of critical flow at the pipe outlet, as can be shown in more detailed analyses. Critical flow for a range of diameters, slopes, and discharges is "approximately" just over half of the pipe diameter.

The coefficient K_c, the conduit loss factor for circular pipes, is (Equation 7.9c):

$$K_c = \frac{\mathrm{Const}_{\mathrm{dim\,C}}\, n^2}{D_{\mathrm{pip}}^{4/3}}$$
(7.9c)

where K_c = the conveyance coefficient (m^{-1} or ft^{-1}); D_p is the pipe diameter in mm or inches; $\mathrm{Const}_{\mathrm{dimC}} = 1\,244\,522$ (D_p in mm; K_c in m^{-1}) or 5087 (D_p in inches; K_c in ft^{-1}); circular pipe; and n is the Manning roughness factor for the pipe or conduit material.

For square conduits, Equation 7.9d gives the loss factor.

$$K_c = \frac{\mathrm{Const}_{\mathrm{dim\,S}}\, n^2}{R^{4/3}}$$
(7.9d)

where K_c = the conveyance coefficient (m^{-1} or ft^{-1}) as above; R is the hydraulic radius in meters or feet; $\mathrm{Const}_{\mathrm{dimS}} = 19.6$ (R in m; K_c in m^{-1}) or 29.16 (R in ft; K_c in ft^{-1}). One may derive the dimension factors in Equation 7.9, starting with Equation 7.8 using the appropriate φ_{dim} in the Manning equation.

One may differentiate pipe flow from orifice flow (test of Category A or Category B) with a submerged inlet–unsubmerged outlet using Equation 7.10:

$$S_{\mathrm{neut}} = K_c \frac{v_{\mathrm{pip}}^2}{2g}$$
(7.10)

where s_n is the neutral slope.

The rationale for Equation 7.10 is apparent in Figure 7.16. The neutral slope condition occurs when the energy grade line slope is the slope of the pipe. For actual pipe slopes higher than the computed neutral slope, orifice flow prevails. For pipe slopes less than the neutral slope, pipe flow hydraulics prevails. The neutral slope calculation requires a proper estimate of velocity. If pipe flow controls, the calculated neutral slope always exceeds the design slope. If the neutral slope is near the design slope, expect orifice flow to control.

Category B and Category A design are the most used design scenarios for culverts. The usual design goals are as follows. Size the culvert pipe such that a specified flow does not exceed a given head. Determine the head required for a particular flow and pipe size and compute the flow for a given pipe and head. Finding the *pipe size for a known discharge and head* requires iteration; Finding the *needed head for a given discharge-known pipe solution* is a one-pass solution. The *flow with a given pipe-known head* is a one-pass solution. Compute orifice flow for the final design to check the flow controlling mechanism.

Write Equation 7.9 and substitute known values. Write the equation for head H and substitute known values and solve using iteration or immediate solution as needed.

Note: if H_{inlet} is unknown, and diameter and flow are known, assume initial inlet submergence. Find the head H in Equation 7.9. Put the found H in the head equation to solve for H_{inlet}. If H_{inlet} is negative, then the inlet is not submerged. A negative H_{inlet} suggests weir control or open channel within the pipe control. *One would miss this outcome if one immediately solved for the neutral slope using continuity.* The procedures given above provide reliable results in typical design situations.

Example 7.5   Solve for flow with known head-known pipe size

Given a 762 mm (30 in.) pipe size corrugated culvert 12 m (39.36 ft) long. Inlet invert elevation is 127.9 m; outlet invert elevation is 127.7 m. Headwater elevation is 129.5 m (e.g., H_{inlet} = 1.6 m = 129.5–127.9 m) and water elevation at the pipe exit is 126.8 m. What is the discharge?

The discharge is a free discharge since the water level at the pipe exit is below the outlet invert. The actual culvert slope is (127.9–127.7 m)/12 m = 0.0167. Assume pipe flow holds. Use Equation 7.9. The Head H = 129.5–127.7 m–0.6 * (2.5 ft/3.28 ft/m) = 1.342 m; K_e = 0.5, K_c = 0.112 m^{-1} (n = 0.025 and Equation 7.9b), K_b = 0, A_p = 0.456 m^2. Then, from the pipe flow equation, one has:

q = (0.456 m^2 Sqrt(2*9.8 m/s^2 *1.342 m))/Sqrt(1 + 0.5 + .112 m^{-1} *12 m) = 1.386 m^3/s.

Now we must check the neutral flow condition.

v_{pip} = q/A_p = (1.386 m^3/s)/0.456 m^2 = 3.03 m/s; S_n = 0.112 m^{-1} *(3.03 m/s)^2/ (2 * 9.8 m/s^2) = 0.053. Since the actual slope is less than the neutral slope (0.0167 < 0.053), pipe flow is valid. One could also compute orifice flow to find that it was larger than pipe flow.

Example 7.6   Solve for the head with a given discharge-known pipe size

Given a culvert 12 m (39.36 ft) long, no bends. The discharge rate is 5 m^3/s. Inlet invert elevation is 127.9 m; outlet invert elevation is 127.7 m. The water level at the exit is 126.8 m (e.g., free discharge). What inlet water elevation is required to convey the flow through a 1 m diameter corrugated pipe?

Assume free outfall. Starting with Equation 7.9,

Head H = [(H_{inlet} + 127.9 m)–127.7 m–0.6(1 m)] = H_{inlet}–0.4 m where H_{inlet} is the distance from the inlet invert to the water surface at the inlet. a = π (1 m)^2/4 = 0.785 m^2. K_c = 0.0796 m^{-1} (n = 0.025), K_e = 0.5, K_b = 0. Using Equation 7.7 gives

5 m^3/s = [0.785 m^2 Sqrt[2(9.81 m/s^2)(H_{inlet}–0.4 m)]]/Sqrt(1 + 0.5 + 0.0796 m^{-1} (12 m))

Note that Sqrt(1 + 0.5 + 0.0796 m^{-1} (12 m)) = 1.57

[5 m^3/s (1.57)/0.785 m^2]^2 = 2(9.81 m/s^2) (H_{inlet}–0.4 m)

H_{inlet} = 4.43 m. v_{pip} = 5 m^3/s/0.785 m^2 = 6.37 m/s; S_n = 0.0796 ((6.37 m/s)^2/2(9.81 m/s^2)) = 0.16 Pipe flow OK. HY-8 gave similar numbers.

Example 7.7   Solve for pipe size with known discharge and known head

Find the straight corrugated pipe size to convey 5 m^3/s with an inlet depth of 3 m. The length of the culvert is 12 m. Inlet and outlet elevations as above, as are the roughness n and inlet conditions.

Starting with Equation 7.9, assuming K_e = 0.5 and K_b = 0 we can write 5 m^3/s = [(3.14 D_p^2/4)Sqrt(2(9.81 m/s^2)[(3 m + 127.9 m)–127.7 m–0.6(D_p)])]/Sqrt(1 + 0.5 + (12) (1 244 522(0.025^2)/(1000 mm/m D)$^{1.33}$). Using an iterative solver results in D_p = 1.03 m pipe diameter. To check neutral slope K_c = 0.076; A_p = 0.838 m^2; v_{pip} = 5 m^3/s/0.838 m^2 = 5.96 m/s. S_n = 0.076 ((5.96 m/s)^2/(2(9.81 m/s^2))) = 0.137. Pipe flow OK. HY-8 leads to nearly the same pipe diameter.

Example 7.8 Orifice flow prevails

Find the flow with a straight 1 m diameter corrugated pipe 4 m long on a 20% slope. The inlet invert is at 100 m; the inlet depth is 10 m free-fall conditions exist. Use the inlet entrance conditions given above.

A_p = 0.785 m^2, K_e = 0.5, K_c = 1 244 522 (0.025^2)/1000$^{1.33}$ = 0.079; H = ((100 +10 m)–(100 m–0.2(4 m))–0.6(1 m)) = 10.2 m. Equation 7.9 appears as

Q = 0.785 m^2 Sqrt(2(9.81 m/s^2(10.2)))/Sqrt(1 + 0.5 + 0.079(4)) = 8.24 m^3/s; v_{pip} = 8.24/0.785 = 10.5 m/s.

Assume orifice flow. Using Equation 7.6, $H_{orifice}$ = 10–0.5(1) = 9.5 m; $C_{orifice}$ = 0.6.

Q = 0.785 m^2(0.6) Sqrt(2(9.81 m/s^2(2.5 m))) = 6.43 m^3/s. Orifice flow prevails. HY-8 puts the inlet at the same elevation as the stated road elevation at the Orifice flow value. Assuming full flow at the orifice flow condition yields a velocity of 8.1 m/s. The neutral slope is 0.27. These conditions suggest that pipe flow is controlling; however, the value is close.

Submerged inlet–submerged outlet (Category C): Treat this situation as a regular pipe flow problem. The Category C equation is identical to the Category A as above, except for the head computation. H is now the difference between the upper and lower water surfaces. The design relationship is the same as Equation 7.9a with the head computed, as shown in Equation 7.11.

$$H = H_{inlet} + H_{inlet\ invert} - \left(H_{outlet\ invert} + H_{outlet\ subm} \right) \quad (7.11a)$$

where $H_{outlet\ subm}$ is the water level elevation at the pipe outlet. The remaining nomenclature is defined as for Equation 7.9a and b.

If one knows the distance from the inlet and outlet culvert inverts to a head-limiting reference above the culvert (e.g., a roadway), one may compute the head H as shown in Equation 7.11b:

$$H = \left(H_{road} - H_{upper\ water\ surf} \right) - \left(H_{road} - H_{road\ water\ surf} \right) \quad (7.11b)$$

where H_{road} is the road surface (or other head limiting condition) elevation relative to a surveying benchmark; $H_{upper\ water\ surf}$ is the distance from H_{road} down to the upper water surface; and $H_{lower\ water\ surf}$ is the distance from H_{road} down to the lower water surface level.

Example 7.9 Solve for flow with submerged inlet and outlet

Suppose a horizontal 24″ culvert crosses a 5 ft wide trail 30 ft below the water surface on the upper side and 2 ft below a water surface on the lower side. The culvert pipe length is 30 ft. The submerged corrugated steel pipe connects the two sides. Will the culvert transfer 50 cfs without flooding? The driving head H is the difference in water elevations relative to the trail surface, which is 30 – 2 = 28 ft. Using the pipe flow equation, loss factors, and by converting to metric units, write Equation 7.11a (using Equation 7.11b for head computations) as

50 cfs/(35.28 cfs/m^3) = (0.292 * Sqrt(2 * 9.8 m/s^2 * H))/(Sqrt(1 + 0.5 + 0.15(30 ft/ (3.28 ft/m))))

Solve for H = 3.45 m = 11.32 ft. Thus, a water surface differential of 11.32 ft transfers 50 cfs. Under these stated conditions, the culvert would transfer:

$q = 0.292 \, \text{m}^2 * \text{Sqrt}(2 * 9.8 \, \text{m/s}^2 * 28 \, \text{ft}/(3.28 \, \text{ft/m}))/(\text{Sqrt}(1 + 0.5 + 0.15 \, \text{m}^{-1} \, (30 \, \text{ft}/ (3.28 \, \text{ft/m}))) = 2.22 \, \text{m}^3/\text{s} = 79 \, \text{cfs}$. HY-8 requires one to find a flow that brings the inlet to the top of the road while having a tailwater channel that maintains a 2 ft submergence at the exit.

Example Note: If the problem is to determine pipe diameter passing specified flow at the specified head, propose pipe diameters, solve for discharge at the set head. Continue until the selected diameters give a range of flows bracketing the needed flow. Select the next largest available diameter.

Unsubmerged inlets–unsubmerged outlets (Categories D and E): When the inlet and outlet are unsubmerged, then either open channel flow (Category D) or weir flow (Category E) controls the flow.

The FHWA generalized on the dimensionless relationships as shown in Equation 7.12a and b:

$$\frac{H_{inlet}}{D_p} = K\left[\frac{q}{AD_p^{.5}}\right]^M \quad \text{(Weir)} \tag{7.12a}$$

$$\frac{H_{inlet}}{D_p} = c\left[\frac{q}{AD_p^{.5}}\right]^2 - 0.NS + O \quad \text{(Orifice)} \tag{7.12b}$$

Parameters K, M, c, N, and O are constants related to inlet style, pipe material, and cross-section geometry. Figure 7.18 provides a typical plot for the weir and orifice cases with a circular conduit. Consult FHWA (2012) and works such as Walski et al. (2015) for parameter values describing various conduit shapes and entrance conditions. FHWA (2012) and Walski et al. (2015) provide more generalized relationships based on nondimensional analyses and more advanced hydraulic studies (except for category D, open channel). The parameters of Equation 7.12a and b should not be extrapolated beyond the ranges shown in Figure 7.19.

The open channel flow (Category D) is identical to the flow transition problem of Chapter 5. With open channel flow, the Manning equation, coupled with the energy equation, describes flow transition hydraulics. Refer to the hydraulic elements (Table 2.1), the Manning equation, and flow transitions discussion in Chapter 5. Taking the minimum of weir or channel flow for a given stage is adequate for standard design purposes.

Example 7.10 Inlet and outlet both unsubmerged

Suppose a 30″ corrugated pipe at a 1% slope was flowing at a depth of 15″. Compute the flow by weir flow and open channel flow.

Using Figure 7.19 for weir flow, for $H_{inlet}/D_p = 0.5$, read $q/D^{5/2} = 0.8$ (English, ft). Since $D_p = 2.5 \, \text{ft}$, $q = 0.8(2.5^{5/2}) = 7.9 \, \text{cfs}$.

For open channel flow (refer to Table 2.1 or Figure 2.3 for channel elements for a circle) and write the Manning equation. The energy equation and the Manning equations must be satisfied. Assume zero velocity above the inlet. We will first solve the Manning equation, then calculate the acceleration and loss term, adjust head downward, recalculate the Manning equation, and adjust until the losses equal the initial adjustment. Consider that $(1 + K_{tran}) v_{pip}^2/2g$ is the acceleration and transition head loss with $K_{tran} = 1$. Select a guess

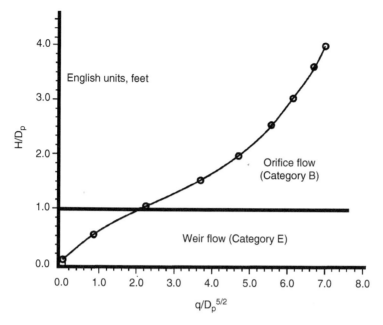

Figure 7.18 Flow relations for unsubmerged and submerged flow through a pipe entrance, assuming no pipe effects as an open channel or closed pipe, English units of cfs for discharge, and feet for head and diameter. *Source:* Crossing data in excel.

depth of 1.25 ft. $R/R_0 = 1$ and $A/A_0 = 0.5$ from Figure 2.3. Computing Manning, first pass where $R = 1(2.5\,\text{ft}/4) = 0.625\,\text{ft}$ and $A = 0.5 * 3.14((2.5\,\text{ft})^2)/4 = 2.45\,\text{ft}^2$. Then $v = (1.49/0.025)$ $(0.625\,\text{ft})^{67}$ Sqrt(0.01) = 4.35 ft/s and $q = 4.35\,\text{ft/s}\,(2.45\,\text{ft}) = 10.67\,\text{cfs}$. The acceleration and losses = $2((4.35\,\text{ft/s})^2/(2*32.2\,\text{ft/s}^2)) = 0.58\,\text{ft}$. Then the depth of flow is about 0.67 ft instead of 1.25 ft. Recalculate the hydraulic radius to be $0.6 * 2.5\,\text{ft}/4 = 0.375\,\text{ft}$ and area to be $0.15(3.14 * (2.5\,\text{ft})^2)/4 = 0.74\,\text{ft}^2$. (Verify these estimates using Figure 2.3). Then $v = (1.49/0.025)\,(0.375)^{67}$ Sqrt(0.01) = 3.08 ft/s. $q = 3.08\,\text{ft/s}\,(0.74\,\text{ft}^2) = 2.27\,\text{cfs}$. Losses and acceleration are now $2((3.08\,\text{ft/s})^2)/(2*32.2\,\text{ft/s}^2) = 0.3$. Let depth = 0.95 ft. Then $R = 0.7(2.5\,\text{ft}/4) = 0.44\,\text{ft}$ and $A = 0.35 * 3.14 * (2.5\,\text{ft/s})^2/4 = 1.72\,\text{ft}^2$. $V = (1.49/0.025)\,(0.44^{0.67})$ Sqrt(0.01) = 3.43 ft/s and $q = 3.43\,\text{ft/s}\,(1.72\,\text{ft}^2) = 5.91\,\text{cfs}$. Losses and acceleration are then 0.36 ft, acceptably close to the assumed 0.3 ft. Continue the iteration until solutions converge to within 15%. Since open channel flow results in the minimum flow, open channel flow is limiting. Had the pipe been shorter and/or steeper, weir flow would be the controlling mechanism. In viewing the HY-8 summary table, one will observe near equality in depth for inlet and outlet in the culvert, thus uniform open channel flow is controlling.

Culvert design software: The Federal Highway Administration (FHWA 2012) provides the most used approach based on coefficients determined using similitude analysis and experimentation first attributed to Mavis (1943). Highway culverts are designed to pass the 50–100-year storm. The 10- or 25-year storm is frequently the criteria for rural road culverts. The HY-8 software (FHWA 2014) is an excellent culvert design tool for a range of runoff values. Roads are treated as broad crested weirs when the flow exceeds culvert pipe capacity. Examples are available in the downloadables. The HY-8 program efficiently solves realistic design situations and does not require explicit dealing with the flow

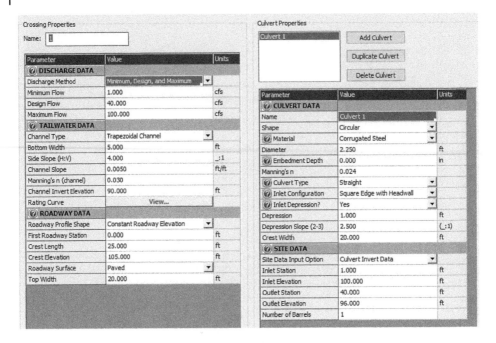

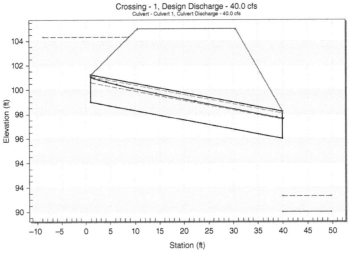

Figure 7.19 Annotated input page (partial) and associated graphic output with HY-8 for a typical culvert design situation.

controlling mechanism details. In the following problems, we manipulate conditions by setting various elevations at the inlet and outlet, and manually iterate to satisfy the specified requirements. HY-8 computes design values for energy dissipation structures often required for culverts. Figure 7.19 shows an annotated HY-8 input sheet. The downloadables contain the model file. One is strongly advised to consult the quick start guide and other documentation available at the download site for this public domain software. The latest version of HY-8 was recently released with a module for simulating and evaluating impacts of the structure of aquatic organism passage. The downloadables contain example HY-8 projects.

Personnel at the NRCS[2] have developed a spreadsheet for culvert analysis and design, presented in the excel "Culverts" folder. This spreadsheet contains useful culvert data, entrance condition descriptions, and other help. It is convenient for computing flows with known pipe specifications and inlet/outlet heads. The spreadsheet iterates the flow classes and determines which mechanism governs and computes the resulting flow. The FHWA design relationships mentioned above, used with HY-8, form the core of the analysis. The NRCS package does not include roadway overflow modeling.

HEC-RAS (Brunner 2016) evaluates culvert design effects on water levels above and below the structure. HEC-RAS is not well suited for modeling stage–discharge relations through complex pond outlet structures. We discuss HEC-RAS for structures, including culverts, in more detail in a subsequent section.

Stage–Discharge Curves for Culverts and Spillways

How could one describe the flow-depth relation of a culvert or spillway structure? An inlet and a closed conduit (e.g., pipe) comprise a compound structure. Table 7.1 provides suggestions for spillway type selection. The following approach enables the generation of stage–discharge curves for the selected culvert or spillway.

One simple spillway structure is the siphon, shown in Figure 7.20. Treat the siphon as a variation of full flow culvert flow, Category A or C. The most distinguishing feature of the siphon is the elevated middle pipe section. Pressures less than atmospheric may exist within the middle pipe section. The inlet is usually a projecting style inlet. A siphon spillway in a pond may appear, as shown in Figure 7.20. The siphon pipe must have adequate strength not to collapse under negative pressures.

An oft-used pond spillway is related to the culvert, having a vertical riser ahead of the culvert. Weir, orifice, and pipe hydraulics may be relevant depending on the stage. Other outlet structures may also involve weir flow, followed by orifice flow, and finally, pipe flow. Considering the problem before and after inlet submergence is useful. Before submergence, weir flow or open channel flow mechanisms control the flow. After inlet submergence, orifice flow or pipe flow could prevail. The condition giving the minimum flow is in control. In situations involving additional spillways carrying runoff, add each spillway's contributions at each value of depth. Culvert and pond spillway analyses and modeling build upon the hydraulic structure's attributes discussed in this chapter.

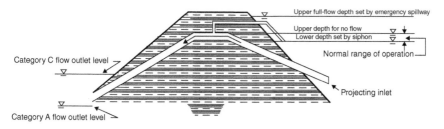

Figure 7.20 Schematic of a siphon spillway with the projecting inlet. The schematic drawing indicates the tailwater elevations defining Category A and Category C flow types. Inlet flow control levels are noted. The sketch is not to scale.

2 Provided courtesy of Annette A. Humpal, P.E., Hydraulic Engineer, NRCS-Appleton, Wisconsin, Area Office.

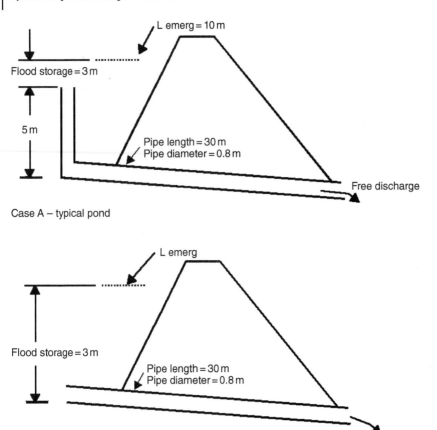

L emerg = 10 m

Flood storage = 3 m

5 m

Pipe length = 30 m
Pipe diameter = 0.8 m

Free discharge

Case A – typical pond

L emerg

Flood storage = 3 m

Pipe length = 30 m
Pipe diameter = 0.8 m

Case B – large culvert

Free discharge

Figure 7.21 Spillway cases for typical spillways.

Example 7.11 Stage–discharge computation for spillways – vertical riser
Compute the stage–discharge curve for the principal spillway vertical riser shown in Figure 7.21 (Case A, vertical riser), which uses a corrugated, 0.8 m diameter pipe.

The horizontal pipe method is identical to that of the vertical pipe except for the possible flow mechanisms. Category D and E culvert flows are possible with the horizontal pipe, no vertical riser. The spreadsheet, "SimpleResRoute" shows the details of the structure model. We refer again to this spreadsheet in Chapter 8 to focus on the effect on the flow hydrograph.

If the pond level is below the vertical riser elevation, there is no flow. The pipe always functions as an orifice due to the steep slope, as one may see from the table of vertical riser results. The pipe edge first acts as a weir and then as an orifice. Finally, pipe hydraulics becomes applicable if the physical slope is less than the neutral slope. The mechanism resulting in the minimum flow is the controlling mechanism. Thus, one may calculate the flow by each flow mechanism and select the minimum flow. One measures the head for the weir and orifice from the top of the riser. One measures the head for pipe flow following Figure 7.15, case A assuming free discharge. Culvert flows of categories D and E are possible with the horizontal pipe.

With a rising hydrograph limb, the horizontal pipe experiences open channel and weir flow before inlet submergence. Open channel flow is modeled by applying the energy equation coupled with the Manning equation (see Example 7.10). One may model the problem using circular channel elements (Figure 2.3 or Table 2.1). Empirical regression functions based on Figure 2.4 enable modeling the hydraulic elements of circular and elliptical cross-sections before submerging the open channel mechanism. Because of the need to iterate for the open channel flow class, one must set the channel's depth of flow (y_{chan}) as a guess.

Weir flow is modeled measuring head from the inlet bottom. Use circular channel elements to estimate weir length (wetted perimeter). The FHWA (2012) approach, which generalizes on the Mavis (1943) charts shown in Figure 7.19, could have also been used.

Pipe flow and orifice flow are possible once submergence occurs. One measures the orifice head from the center of the conduit. Measure the pipe flow head, as shown in Figure 7.15 (case A assuming free discharge). Enter the head variable (H_{inlet} in Example 7.11) values ranging over the desired range of stage and solve the flow (Q_{tot}) at each input stage value.

Spillway design is inherently trial and error. The common goal is to meet prescribed peak flow targets for return period storms ranging from as low as the one year storm to the emergency spillway required to pass the 100-year storm. We enter the stage variable (called H_{inlet}) as an input list. One should include several points before submergence occurs. Solve for the flow (Q_{tot}) at each input stage. Then, develop a function with a range of Q_{tot} and a domain of H_{inlet}.

We evaluated the suite of flow equations over a range of stages using a spreadsheet in "ResReachRoute" folder, "SimpleResRouting." This spreadsheet evaluates a pond spillway with a vertical riser coupled to a culvert. Flow into the riser may be controlled by weir, orifice, or pipe hydraulics, depending on the stage. The outlet system includes an emergency spillway above the principal spillway, which is modeled as a broad crested weir. The total outflow is evaluated as a function of the stage. The selected flow is the minimum of weir, orifice, or pipe flow at the indicated stage (except for the emergency spillway). The total flow is the sum of the minimum pipe flow plus the emergency spillway flow once the emergency spillway flows.

The focus presently is on the description of the flow into the outlet structure. We revisit the "SimpleResRouting" spreadsheet in Chapter 8 regarding the pond and its outlet structure's impact on the output hydrograph.

Proprietary computer tools are available for erosion and sediment detention ponds and post-development flow control. Several proprietary software approaches for culvert design are available. For example, SedCad (Warner 1998), HydroCAD (2020), and PondPack (Walski et al. 2015) discuss computer software (Bentley) for culvert design. CIVIL3d, an add-on to AutoCAD, also handles complex retention pond design. These packages are popular choices for detention pond design.

We discuss reach and reservoir routing in Chapter 8.

The culvert/spillway pipe in either the vertical or horizontal case must stand up to the loads in question, including negative pressures found in siphons. Culverts should be at least 0.6 m below the surface in a farm field and require a more sophisticated Geotech analysis in most civil engineering settings. Tollner (2016) discusses the strength requirements of pipes in various soils and loading configurations.

Closed Conduit Systems for Urban Stormwater Collection

Stormwater collection: The category E (closed conduit open channel flow) situation discussed with culvert flow applies to stormwater collection systems flow. These structures are commonly known as storm sewers. Storm sewer networks are designed by first identifying the outlet invert elevation. Move up the slope, and identify grade change points and junctions. At the first junction, proceed upgrade along with one of the branches, identifying grade changes and additional connections. Maintain the grade change and junction identification until the complete network has been tentatively identified. Following the initial layout, one computes the design storm (usually 10-year, TC) using the rational method.

Curb gutters or other channels provide initial collection. Curb inlets then divert the flow of sewers. Variations of the weir formula or other empirical formulas are used to design curb and sewer inlets (see Walski et al. 2015). In small drainage networks, one may ignore travel time. Sum the peak flows at flow path junctions.

Note the flows at each inlet and junction. Design the component pipes to accommodate the flows as closed conduit open channels based on the known flow, slope, and pipe material using the category E methodology. Excellent computer software is available for designing storm sewer collection systems, including curb inlets (e.g., see Walski et al. 2015).

Texts on water resources engineering such as Linsley et al. (1992) provide detailed design information for stormwater collection, as does FHWA (2012). The rational formula forms the basis for peak flow prediction. Local codes must be followed concerning pipe and inlet placement, depth, diameter, materials, and clean-outs. Fifield (2004) provides an excellent discussion of how one may incorporate silt fences, bale barriers, rock check dams, and other structures into erosion and stormwater management plans. He provides simple design relations for these structures. Computer software (see Appendix A) is helpful in urban stormwater and sewerage collection system design.

As urbanization continues to encroach on agricultural or forested lands, the natural resources engineer frequently becomes more involved in developing stormwater drainage systems using closed conduits. Discharge from storm sewers is becoming regulated. One should anticipate increased regulation regarding water drainage in the rural or urban environment.

Water measurement structures for irrigation and aquaculture: Flow measurement is customary in arid areas, where growers often purchase water. The techniques also apply to humid regions. In humid regions, one may wish to quantify flows for general information or regulatory purposes. One may find examples of each technology discussed above in irrigation and aquaculture applications. These devices accomplish a sub- to supercritical flow transition. The unique stage–discharge at critical flow is the core attribute for these flow measurement devices. The selection of parameters and measuring position is essential for accurate measurements. Bos et al. (1984), Brater and King (1976), and Walker and Skogerboe (1987) provide additional information regarding device selection, placement, and field use.

The need to control irrigation water requires a variety of permanent structures. Adjustable gates enable control of flows from irrigation canals into farm ditches. These are like sluice gates. Distribution boxes with appropriate baffling are employed to pass water from pump outlets into multiple channels. Weirs and flumes are widely used to measure flowrates. Figure 7.22 shows a flow division box used when delivering irrigation water via open channels. Diversion boxes incorporate weirs and sluice gate structures.

Yoo and Boyd (1994) and Coche and Muir (1992) discuss the construction of weirs to measure and control water flows in ponds for aquaculture. They also discuss inlet and outlet devices for fishponds.

Modeling structures with HEC-RAS: We conclude this discussion with an HEC-RAS (see Brunner 2016) analysis of a channel with a culvert and bridge, with a free discharge and normal depth entrance.

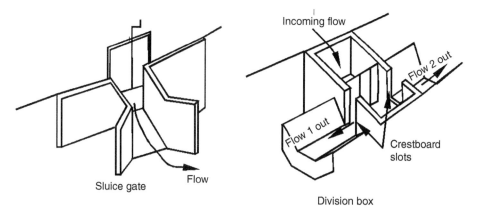

Incoming flow

Flow 2 out

Flow 1 out

Crestboard slots

Sluice gate

Flow

Division box

Figure 7.22 A schematic drawing showing an irrigation canal distribution box.

Example 7.12

Construct a 2000 ft straight concrete trapezoidal channel with a 4:1 side slope and a 10 ft bottom width. Half of the side slope and the bottom have a Manning $n = 0.015$, while the top of the side slope has a Manning $n = 0.03$. The channel entrance is at river station 2000, and the exit is at river station zero. The surface elevation is 110 ft at the entrance. The slope is 0.0005.

At river station 1500, a 9 ft clean circular corrugated pipe ($n = 0.024$) that is 20 ft long serves as the culvert. The headwalls on both ends are vertical; thus, the road width is also 20 ft. There is a 20 ft wide bridge at river station 1000. The bridge has a 2:1 embankment on each side and a 2 ft diameter pier in the center of the span. Set boundary cross-sections above and below the structures putting the 20 ft wide structures in the middle of a 100 ft channel span. The flow is 500 cfs, which enters as a normal depth at river station 2000 and exits as a free discharge at river station zero. Also, evaluate a scenario of known water surfaces at the entrance and exit.

The step-by-step analysis is presented in Table 7.4. Figure 7.23e–f presents flow profiles for the free discharge/normal depth and the known water surface scenarios. The example shown suggests the necessity of redesigning the bridge and culvert to enable passage of the flow. HEC-RAS excels at flow analysis involving structures, although data entry time is extensive, and errors can be challenging to identify. Problems involving structures often become significant projects.

HEC-RAS treats other structures such as gates, weirs, side discharge outlets, and the like. There is an extensive suite of solved examples of all these structures in the HEC-RAS documentation. One should examine examples closely for gaining insights into how these structures function. One can extend this advice to many topics in unsteady flow and sediment transport also. Space does not allow for a more detailed discussion of these HEC-RAS cases.

Ecologic Suitability

Structures dealt with in this chapter purposed for stormwater management after construction can have some harmful ecological effects. Streams with a seasonal or permanent flow may support migrating fish species. The change in design criteria is as follows: from "avoid

Table 7.4 Procedure for constructing a channel with a culvert and bridge structure in HEC-RAS.[a]

1) Construct the reach as a whole as in Chapter 2
2) Define the upper and lower reach cross-sections with the entire reach length as in Chapter 2
3) Go to the upper end, copy with the downstream cross-section within 100 ft of the first structure
4) Go to this reach, and copy with the downstream cross-section of 100 ft
5) Go to the new cross-section and again copy to within 100 ft of the next structure (if any)
6) Repeat for additional structures as needed.
7) Adjust elevations of all cross-sections to achieve the desired channel slope
8) Begin to specify the first structure by pressing Bridge/Culv
9) On the popup, go to options and add a new structure, defining the station to be halfway between the 100 ft cross-sections. Assume a culvert to be added.
10) On Bridge/Culvert, press the Culvert tool
11) Set the shape, span (if needed), diameter, length, various coefficients, and elevations. Then set the centerline station (usually at having the total width).
12) Press the sloping abutment button.
13) Starting at station zero, enter the elevation; repeat at the last station. Notice that the area is filled except for the pipe.
14) Now press deck/roadway
15) Enter the distance to the upper cross-section, width (equal pipe length if square headwall), and use the default weir coefficient.
16) Set the high cord at the roadway elevation and the low chord a little below that at stations 0 and 90 (the right extent). Press OK.
17) NOTE: as one sets the stations, the geometric editor may not reflect true scale and conditions. Be sure the stationing and structures are consistent. When you save the project and restart, it appears in correct proportions.
18) Add another structure (if applicable). Set the stationing to occur halfway between the next 100 ft cross-section interval. Assume a bridge, for example.
19) Press Deck/Roadway. Fill the popup with data as with the culvert. Assume the lower chord to be 1 ft below the upper cord.[b]
20) Press the sloping barrier and configure to give a slope of 2:1, for example.[b]
21) Define the surface of the barrier, starting with station 0. Note that the barrier height is zero at the channel bottom. Continue the profile to the extent of the channel on the right side. Usually copy the upstream data to downstream
22) One may want to place a support pier in the middle. Fill the pier width and bottom elevation, and top elevation. Copy upstream to downstream. Ignore debris data. This completes the bridge.
23) Add other structures as needed.
24) Now interpolate between the cross-sections not containing a structure. Use a short distance for interpolation at the lower end, assuming a free discharge[b] at the end. Use a longer interpolation distance in the upstream reaches.
25) Save the project.
26) On the main menu, go to the steady flow editor. Enter the flow rate.
27) Press Reach Boundary Conditions and set the boundaries. Apply data. Save the project.
28) Go to the main editor and press Run. Notice any errors that may tell where data fixes are required.
29) View results.
30) Collect plots and tables as needed

[a] See BridgeCulv1 Folder in the HEC-RAS directory.
[b] In practice, use actual river conditions, measured, or specified data.

(a)

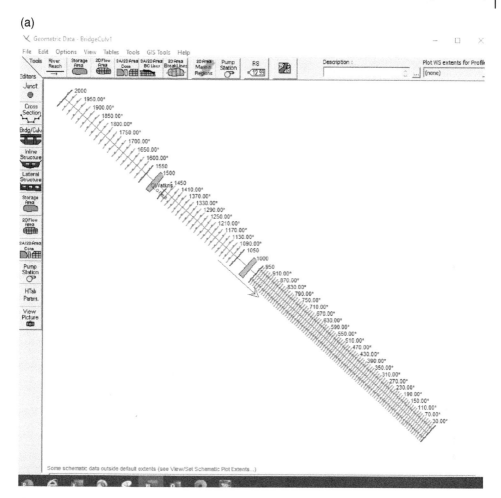

Figure 7.23 (a) HEC-RAS geometric editor showing the channels, culvert, bridge, and interpolated cross-sections. (b) Bridge culvert editor showing the culvert at river station 1500. The embankment is shown, and the channel profile is shown. The channel portions with *n* = 0.015 are contained within the red dots away from the culvert (although not relevant at the culvert site). (c) Bridge with the upper chord at the road surface and lower chord 1 ft below shown with 2 : 1 embankment and support pier. (d) Flow profile inputs are shown. Reach boundary conditions for the case of known water surface depth at both the entrance and exit. The popup showing the known water surface for the channel entrance is also shown. (e) Flow profile for the channel with a 500 CFS flow with a free discharge at the exit and normal depth at the entrance. Both the culvert and the bridge restrict the flow, causing flooding at the channel entrance. (f) Flow profile for the channel except for a known water surface of 108 ft at the exit and known water surface of 110 ft at the entrance. The entrance is flooded by the culvert flow restriction.

(b)

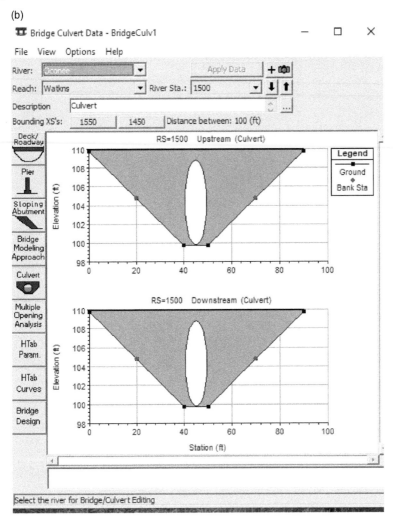

Figure 7.23 (Continued)

flooding of the roadway with a specified return period storm" to "avoid high velocities, shallow water depths and jump heights hindering migrating species." The impact of the change is larger culverts with diameters partially submerged in the channel. Figure 7.24 shows a partially embedded culvert conduit. Downloadables contain Mathematica software for sizing a circular pipe with specified embeddedness for flow capacity. Low flow conditions frequently determine suitability. One should consult an ecologist for key species and their limiting conditions. Ecological engineers are developing new design methodologies and design storm specifications that minimize velocity changes and plunge pool depths. These have reduced harmful effects on migrating species. HY-8 provides tools for analyzing aquatic organism passage, which requires threshold values provided by stream ecologists. More contemporary design approaches build on existing hydraulic methods for culvert and structure design covered in this chapter.

(c)

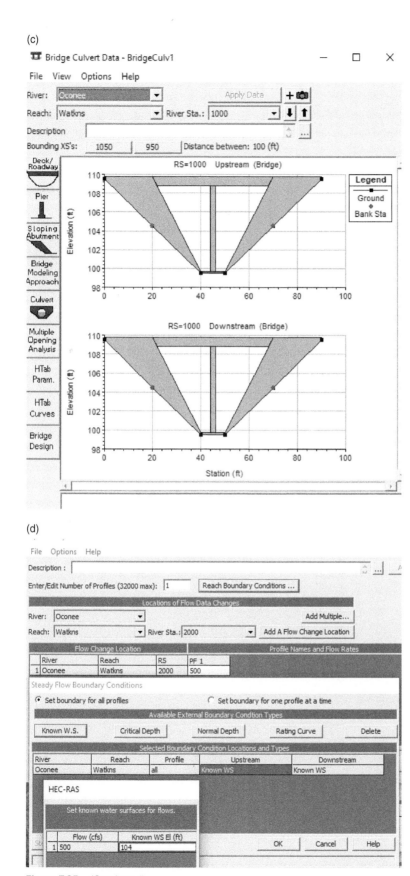

(d)

Figure 7.23 (Continued)

(e)

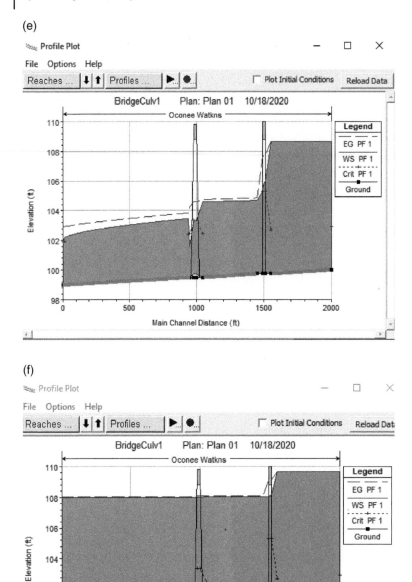

(f)

Figure 7.23 (Continued)

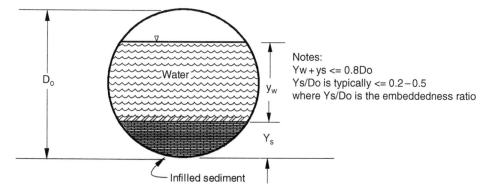

Figure 7.24 Sketch of a partially embedded culvert, which can serve as a fish passage.

Summary and Conclusions

The methods summarized in this chapter define the hydraulic relations for various outlet spillways, drop structures, and culverts, which are used on-farm and on construction job sites. More advanced treatments of each spillway type are available (e.g., see Walski et al. 2015; FHWA 2012). The inlet and conduit usually define the hydraulics, with some indirect impacts on flow hydraulics contributed by conditions at the outlet. NRCS (1964), in an old section of the National Engineering Handbook, provides background on classical but still used hydraulic structures.

Avoiding overflow failures requires designed structures. Similarly, obtaining accurate flow measurements requires properly designed installations to avoid freeze-thaw, seepage, and rodent-induced failures. The primary reasons for the failure of conservation structures are the following:

- insufficient capacity;
- insufficient energy dissipation, including unanticipated supercritical flows, and;
- improper installation leading to seepage failures due to structure undermining.

Structure design and specification begin with an assessment of customer and technical requirements. Tables 7.5 and 7.6 provide prompt sheets that guide the evaluations. The process for structure selection and design is as follows:

1) Compute the design peak flow using 50-, 25-, 10-year, 30 minutes storm consistent with state and local codes and agencies such as the NRCS.
2) If the structure is a principal spillway or gully stabilization structure, see Table 7.1 to select structural designs based on peak flow.
3) If the structure is a weir, inlet, orifice, or sluice gate, see Figure 7.10 for a stage–discharge relationship. If the situation warrants a weir, know how to get the length of a rectangular weir, box inlet, and other structures in Figures 7.6 and 7.7. See Examples 7.1 and 7.2. If Parshall flume or H_x flume is present, see Equations 7.2 and 7.3.
4) With a culvert structure, identify the flow class from Figure 7.15. Draw a picture. Most design problems involve selecting road height (e.g., H_{inlet}) with a free discharge or

Table 7.5 Prompt sheet for assessing customer requirements.

Customer requirements
Purpose?
Peak flow reduction?
Sediment content and other nutrients?
Handle flows over the flow range adequately
Flow measurements and head regulation?
Energy dissipation within the structure?
Stability of flows over time
Headloss in line with available Head and flow
Discharge regulation or erosion stabilization/prevention?
Pump storage?
Consequences of short/long-term flooding near structure
Discharge from a location along the depth profile?
Land use/value near structure and upstream
Is leakage a problem?
Maximum depth?
Storage/surface area needed if an impoundment
Life?
Available funding
Environmental/Regulatory Issues
Impact of structure on native species

submerged discharge. If we are solving for H_{inlet} knowing q, and there is a free discharge, solve for the weir condition (see Figures 7.14 and 7.15), the orifice condition (Equation 7.6), and the pipe condition (Equation 7.9). The situation giving the minimum H_{inlet} controls.

See Figure 7.17 for K_b and K_e; Equation 7.9b and c for K_c.

See Examples 7.3 and 7.4.

If one is solving for H_{inlet} and the outlet is submerged, one need only analyze the pipe flow case (see Equation 7.11).

If one knows distances from a head limiting reference above the culvert (e.g., a roadway), use Equation 7.11b. If one is solving for the diameter needed to pass a given flow and given H_{inlet} (free discharge), use Equation 7.9, iterating for D. If the discharge is submerged, use Equation 7.11, and iterate for D.

5) The principal spillway may act as a weir, orifice, and pipe. Examples 7.6 and 7.7 exemplify the development of a stage–discharge relationship for the spillway.
6) Provide for energy dissipation at the structure outlet, if applicable.

Proprietary and public domain software discussed or mentioned previously provides considerable detail on stage–discharge modeling. These include software such as SedCad (Warner 1998), TR-20 (NRCS 1983), HydroCAD (2020), Civil3D (Autodesk

Table 7.6 Prompt sheet for technical requirements.

Technical requirements
Withstand flow now and through time given water quality
Energy dissipation at the discharge of structure?
Capacity/volume
Accurately measure flows over the required flow range?
Foundation requirements?
Freeze/thaw issues?
Materials specifications
Energy dissipation approaches
Soil stability
Sediment issues?
Hydraulic conductivity of surrounding soils
Surrounding soil physical/chemical characteristics
Sand/silt/clay/rock content (e.g., texture) of surrounding soils
Stability of soil to piping failure
Is plastic/bentonite feasible if sealing is needed?
Anticipated pressure heads (positive or negative)
Costs of pipe, excavation, liners, concrete work, labor.
Nature of discharge (free or submerged) based on a need for species passage

2020), HEC-RAS (Brunner 2016), SWMN (USEPA 1994), and CulvertMaster and PondPak (Walski et al. 2015). BMPs involving channelization of flows such as basin drains and roadway slope drains (see Fifield 2004) are combinations of weirs, orifices, and pipes. One or more of the above software packages address the above scenarios.

Problems and Questions

Note: Understand how to measure the effective weir length for all the structures in Figures 7.6 and 7.7. Draw sketches of all completed designs and provide supporting reasons for discussion questions. Show all work and offer a rationale for assumptions made, remembering to use appropriate significant digits. Feel free to use downloadable software where applicable.

1   Given 1 m^3/s flow moving through a rectangular weir with a bottom width of 1 m, what is the depth of flow in the weir throat? Assume a weir coefficient of 3.1. Sketch the weir.

2   Determine the design dimensions of a box inlet (1.5×1.5 m, see Figure 7.25); we need the depth here) to carry 7 m^3/s. Assume broad crested weir control prevails.

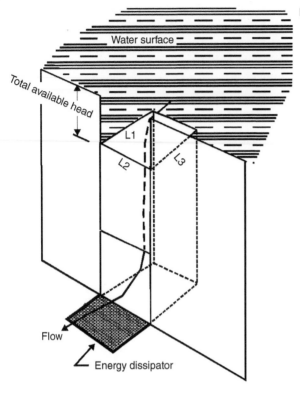

Figure 7.25 Projection box inlet.

3 What is the maximum capacity of a straight, broad crested drop spillway having a crest length of 3 m and a depth of flow of 0.9 m?

4 Determine the crest length of a straight inlet drop spillway to carry 7.5 m³/s is the depth of flow is not to exceed 0.9 m. What are the dimensions of a square box under the same conditions?

5 A spillway has the Waterways Experiment Station (WES) spillway shape. What is the length required to accommodate a 3 m³/s flowrate if one constrains the head not to exceed 0.4 m?

6 Given the following situation: A concrete pipe culvert 2 ft in diameter and 50 ft long containing a 90-degree bend ($R/D = 2$). The entrance is flush with a retainer wall, with an inlet invert at 101 ft. The outlet invert is at 98 ft, exit water level at 90 ft, and the inlet height maximum at 5 ft above the inlet invert. Determine the expected maximum culvert flow.

7 (i) A 0.25 ft diameter corrugated pipe trickle tube spillway (designed as a culvert) projects into a pond with a water level 2 ft above the inlet invert. The discharge is free. If the spillway lies on a 5% slope and is 5 ft long, what is the discharge flowrate? (ii) If the spillway lies on a 0.05% slope and is 100 ft long, what is the flowrate? (iii) If the water level is 20 ft above the inlet invert, and the pipe is 5 ft long at a 20% slope, what is the discharge?

8 A culvert conveys $3\,m^3/s$ under a roadway. The vertical distance from the road surface to the culvert inlet invert is $4\,m$. The concrete pipe is $15\,m$ long and has no slope. Assume a free discharge. Assume a square-edged inlet. What pipe diameter is required to pass the flow without flooding?

9 A road at $20\,m$ elevation has a straight culvert $100\,m$ in length to be placed at an inlet invert elevation of $15\,m$. The system must carry an estimated peak runoff of $0.4\,m^3/s$. The outlet invert is $14.0\,m$. The maximum exit water level elevation is $10.09\,m$. The entrance is square with a concrete headwall flush with the end of the pipe. A corrugated metal pipe with $n = 0.025$ is specified. Compute the pipe diameter required to avoid flooding.

10 A $0.3\,m$ diameter smooth concrete pipe connects two reservoirs, one with a $3.5\,m$ level above the inlet invert, the other $2.5\,m$ above the outlet invert. The pipe lies on a 0.01% slope and is $50\,m$ long. Assume a square-edged inlet. What is the flow through the pipe?

11 A culvert is $1\,m$ in diameter, $25\,m$ long with one bend, made of corrugated pipe, and laying on a 0.5% slope. The inlet pool is ten times the culvert diameter for $10\,m$ upstream. Assume a square-edged flush inlet. Estimate the flowrate through the culvert when the inlet depth is $0.1, 0.3, 0.5, 0.7, 1.5$, and 3 times the pipe diameter, assuming the discharge is free?

12 Compute the stage–discharge curve for a $0.6\,m$ diameter corrugated metal pipe riser connected to a $25.24\,m$ pipe of the same material and diameter. The pipe contains a 90-degree bend ($R/d = 1$) and has no slope. Assume a square-edged inlet. The riser top is $5.05\,m$ above the outlet bottom.

13 Develop the stage–discharge curve for the structure in Example 7.7, Case B.

14 A flow of $10\,m^3/s$ must pass down a steep slope with a total head drop of $11\,m$. Suggest a structure type and size the structure.

15 Compute the diameter of a $20\,m$ straight corrugated pipe culvert needed to conduct $10\,m^3/s$ under a road. The road surface is $8\,m$ above the upper inlet invert, the entrance is projecting, and the exit is a free discharge.

16 The purpose of this problem is to explore the effect of culvert situations on controlling flow mechanisms. Be sure to evaluate orifice/pipe and weir/channel possibilities when appropriate. Suppose one fixed the pipe diameter because stock pipe happens to be available. In each situation below, (i) compute the effective head and, (ii) identify the controlling flow category, and (iii) show the energy grade line in a sketch.
   A A $457\,mm$ diameter straight culvert $100\,m$ in length conveys an estimated peak runoff of $0.28\,m^3/s$. The invert elevation of the pipe inlet is $15.24\,m$, and the outlet invert is $14.0\,m$. The maximum exit water level elevation is $10.09\,m$. The entrance is square with a concrete headwall flush with the end of the pipe. Corrugated metal pipe with $n = 0.025$ is available. What pipe diameter is required?

B Rework case A when an exit water level elevation is 18 m. All other variables are the same as in Case A.

C Rework Case A with a 10 m long culvert, all other variables as in Case A.

D Rework Case B with the 10 m long culvert. Given 4 m elevation between an inlet invert and the road surface, is one pipe adequate, or are multiple pipes needed?

17 Analyze the flow through the flowing siphon spillway shown in Figure 7.20, with the tailwater in the Category A state. The total pipe (corrugated) length of 40 m, the diameter is 500 mm. An upper level of 2 m is maintained above the lower invert of the projecting inlet (100 m). The lower invert elevation is at 98 m. Bends are approximately 30°, and $R/D = 2$. (Interpolate between the tabulated value for 90° and the value of $K_b = 0$ for zero-degree bend). Compute the flow, neglecting the effects of the siphon control tube.

18 From what point does one measure the upper invert elevation in a culvert pipe? The Lower invert elevation?

19 What categories of flow are possible with culvert pipe, submerged, or non-submerged inlet with a free discharge?

20 Derive the culvert equation for Category B flow conditions for the square conduit case. HINT: Follow the steps outlined in the derivation of the circular conduit case.

21 What flow control mechanism operates for each structure shown in Figures 7.6 and 7.7? How does one calculate the head for each flow mechanism on the various structural systems?

22 Redesign the structures of Example 7.12 using HEC-RAS to enable passage of the design storm without upstream flooding. One may change bridges to culverts and vice versa as needed.

References

Autodesk. (2020). CIVIL3D reference manual. 111 McInnis Parkway, San Rafael, CA

Beasley, R.P., Gregory, J.M., and McCarty, T.R. (1984). *Erosion and Sediment Pollution Control*, 2e. Ames, IO: Iowa State University Press.

Biswas, A.K. (1970). *History of Hydrology*. Amsterdam: North-Holland Publishing Company.

Blaisdell, F.W. (1948). Development and hydraulic design - Saint Anthony falls stilling basin. *Transactions of the American Society of Civil Engineers* 113: 483–520.

Blaisdell, F.W. and Donnelly, C.A. (1951). Hydraulic design of the box inlet drop spillway. USDA-SCS-TP-106. USDA Soil Conservation Service, Washington, DC.

Bos, M.G., Replogle, J.A., and Clemmens, A.J. (1984). *Flow Measuring Flumes for Open Channel Systems*. New York, NY: Wiley.

Brakensiek, D.L., Osborne, H.B., and Rawls, W.J. (1979). Field manual for research in agricultural hydrology. Agr. Handbook 224. USDA-SEA, Washington, DC.

Brater, E.F. and King, H.W. (1976). *Handbook of Hydraulics*, 6e. New York, NY: McGraw-Hill.

Brunner, G.W. (2016). HEC-RAS, river analyses system hydraulic reference manual, Version 4.1. US Army Corps of Engineers, Institute for Water Resources, Hydrologic Engineering Center, Davis, CA.

Chin, D.A. (2013). *Water Resources Engineering*, 3e. Upper Saddle River, NJ: Prentice-Hall.

Chow, V.T. (1959). *Open Channel Hydraulics*. New York, NY: McGraw-Hill.

Coche, A.G. and Muir, J.F. (1992). Pond construction for freshwater fish culture: Pond-farm structures and layouts. FAO training series 20/2. Food and Agriculture Organization of the United Nations, Rome.

Federal Highway Administration (2012). *Hydraulic Design of Highway Culverts, Hydraulic Design Series No. 5*, 3e. Washington, DC: Federal Highway Administration.

Federal Highway Administration. (2014). HY-8 Version 7.3. http://www.fhwa.dot.gov/engineering/hydraulics/software/hy80/ (accessed May 2019).

Fifield, J.S. (2004). *Designing for Effective Sediment and Erosion Control on Construction Sites*. Santa Barbara, CA: Forester Press.

French, R.H. (2007). *Open Channel Hydraulics*. Highlands Ranch, CO: Water Resources Publications.

Gopalakrishnan, V. and Coche, A.G. (1994). Handbook on small-scale freshwater fish. FAO Training Series No. 24. Food and Agriculture Organization of the United Nations, Rome, Italy.

Grant, D.M. and Dawson, B.D. (2001). *ISCO Open Channel Flow Measurement Handbook*, 5e. Lincoln, NE: ISCO.

Haan, C.T., Hayes, J.C., and Barfield, B.J. (1994). *Hydrology and Sedimentology of Small Catchments*. New York, NY: Academic Press.

HydroCAD. (2020). HydroCAD reference manual, Version 10-4b. HydroCAD Software Solutions LLC, Chocorua, NH.

Linsley, R.K., Franzini, J.B., Freyberg, D.L., and Tchobanoglous, G. (1992). *Water Resources Engineering*, 4e. New York, NY: Irwin McGraw-Hill.

Mavis, F.T. (1943). The hydraulics of culverts. Penn. Engineering Experiment Station Bull. 56. Pennsylvania State Univ., College Station, PA.

NRCS. (1964). Structures, Chapter 6, Engineering Field Manual, Part 650, Natural Resources Conservation Service (formerly SCS), US Dept. of Agriculture-NRCS, Washington, DC.

NRCS. (1983). Computer program for project formulation-hydrology. SCS Tech. Release 20. Soil Conservation Service (Natural Resource Conservation Service), Washington, DC.

Stephens, T. (1991). *Handbook on Small Earth Dams & Weirs: A Guide for Siting, Design & Construction*. Concord, NH: Paul & Company Publishers Consortium.

Sturm, T.W. (2010). *Open Channel Hydraulics*, 2e. New York, NY: McGraw-Hill.

Tollner, E.W. (2016). *Engineering Hydrology for Natural Resources Engineers*, 2e. Cambridge, UK: Wiley.

United States EPA. (1994). SWMM-Stormwater management model. EPA 823-C-94-001. USEPA, Washington, DC.

US Bureau of Reclamation (USBR) (1987). *Design of Small Dams*, 3e. Washington, DC: Government Printing Office.

Walker, W.R. and Skogerboe, G.V. (1987). *Surface Irrigation: Theory and Practice*. New York, NY: Prentice-Hall.

Walski, T.M., Barnard, T., Durrans, R. et al. (2015). *Computer Applications in Hydraulic Engineering*, 8e. Exton, PA: Bentley Institute Press.

Warner, R.C. (1998). SEDCAD7 4 for Windows95/NT. Notes for short course August 5–7, 1998. Biosystems and Agricultural Engineering Department, University of Kentucky, Lexington, KY.

Wooley, J.C., Clark, M.W., and Beasley, R.P. (1941). The Missouri soil saving dam. Missouri Agr. Expt. Station Bull. 434. Univ. of Missouri, Colombia, MO.

Yoo, K.H. and Boyd, C.E. (1994). *Hydrology and Water Supply for Pond Aquaculture*. New York, NY: Chapman and Hall.

8

Gradually Varied Unsteady Flow

When a variation in flowrate occurs with time, how long does it take for the flow variation, or hydrograph, to move downstream? How does the shape of the hydrograph change as the flow moves downstream? Open channel flow phenomena of great importance involve unsteady flows, which implies that either the depth of flow and/or velocity varies with time. Practically, unsteady flow analysis begins with an input hydrograph and examines the hydrograph's progress through a reservoir or river. Do we use the energy balance or the momentum balance? The focus is on momentum, given that the acceleration or deacceleration of mass is involved. We start with an overview of the continuity and momentum equations and then examine methods beginning with simple continuity-based approaches and progress toward incorporating momentum equation components. In this chapter, we pursue the following goals.

Goals

- Have a grasp of the continuum of routing techniques from reservoir-reach (continuity equation only) routing to complete dynamic routing, including both the continuity and momentum equations.
- Understand and perform reservoir routing
- Understand, parameterize, and perform Muskingum river reach routing
- Understand aspects of the relation between kinematic routing and runoff hydrograph analysis.

Following French (1985), consistent with our emphasis on one-dimensional flow, we have Equation 8.1 (continuity) and Equation 8.2 (momentum) to describe these flows.

$$\frac{\partial y}{\partial t} + y\frac{\partial v}{\partial x} + v\frac{\partial y}{\partial x} = 0 \tag{8.1}$$

$$\frac{\partial v}{\partial t} + v\frac{\partial v}{\partial x} + g\frac{\partial y}{\partial x} - g\left(S_x - S_f\right) = 0 \tag{8.2}$$

Open Channel Design: Fundamentals and Applications, First Edition. Ernest W. Tollner.
© 2022 John Wiley & Sons Ltd. Published 2022 by John Wiley & Sons Ltd.
Companion website: www.wiley.com/go/tollner/openchanneldesign

where y is the depth (L), v is the velocity (L/T) at point x, g is the gravity (L/T^2), S_x is the channel slope at point x (–), S_f is the slope of the water surface at point x (–), and t is the time (T).Equations 8.1 and 8.2 are commonly known as the S_t. Venant equations. After beginning with the most straightforward approaches, we explore these equations and work up to the full S_t. Venant solution.

Alternate but equally valid forms of Equation 8.1 are shown in Equations 8.3a and b. Equation 8.2 may be written as shown in Equation 8.4.

$$T\frac{\partial y}{\partial t} + \frac{\partial(Av)}{\partial x} = 0 \tag{8.3a}$$

or,

$$\frac{\partial A}{\partial t} + \frac{\partial Q}{\partial x} = 0 \tag{8.3b}$$

and

$$\frac{1}{g}\frac{\partial v}{\partial t} + \frac{v}{g}\frac{\partial u}{\partial x} + \frac{\partial y}{\partial x} + S_f - S_x = 0 \tag{8.4}$$

where T is the top width (L) and A is the flow area (L^2).Equations 8.1 and 8.2 or 8.3 and 8.4, are termed the complete dynamic model.

These equation sets provide accurate results for unsteady flow situations; however, they can be very demanding of computer resources.

Two groups of simplified models stem from various assumptions regarding the momentum equation conservation terms' relative importance. One can write a generalized routing equation for unsteady flow, as in Equation 8.5.

$$Q = Q_n = \Gamma A R^m \sqrt{S_f} \tag{8.5}$$

In unsteady flow, S_f varies with both the wave slope and the flow depth. The parameter Γ is the unsteady flow variable. If the flow is steady, one may write Equation 8.5 as follows.

$$Q = Q_n = \Gamma A R^m \sqrt{S_x} \tag{8.6}$$

or

$$\Gamma A R^m = \frac{Q_n}{\sqrt{S_x}}$$

Substituting Equation 8.6 into 8.5 leads to Equation 8.7.

$$Q = Q_n \sqrt{\frac{S_f}{S_x}} \tag{8.7}$$

Writing Equation 8.4 in terms of S_f yields Equation 8.8.

$$S_f = S_x - \frac{1}{g}\frac{\partial v}{\partial t} + \frac{v}{g}\frac{\partial v}{\partial x} + \frac{\partial y}{\partial x} \tag{8.8}$$

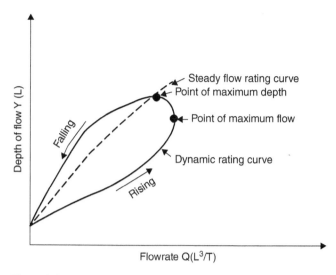

Figure 8.1 Looped rating curve emphasizing dynamic effects.

Substituting Equation 8.8 into Equation 8.7 allows us to write Q as a function of Q_n in the momentum equation, as shown in Equation 8.9.

$$Q = Q_n \left(1 - \frac{1}{S_x} \frac{\partial y}{\partial x} - \frac{v}{S_o g} \frac{\partial u}{\partial x} - \frac{1}{S_o g} \frac{\partial u}{\partial t} \right)^{1/2}$$

(8.9)

Hydrologic routing-| includes only Equation 8.3 and no terms of Equation 8.4

Kinematic wave ------| includes Equation 8.3 and where $S_o = S_f$ in Equation 8.4

Diffusion analogy-------------| includes Equation 8.3 and $S_f = S_o - \dfrac{\partial y}{\partial x}$ in Equation 8.4

Complete dynamic solution----------------------| all terms included in Equations 8.3 and 8.4

Equation 8.9 enables a convenient partitioning of flow solutions ranging from simple to complete solutions to the dynamic equations. Before considering Equation 8.9, we examine approaches based only on the continuity equation (Equations 8.3) and then move toward including increments of Equation 8.9. This text considers the kinematic wave, the diffusion analogy, and then proceeds to some full dynamic solutions. The looped rating curve shown in Figure 8.1 captures the range of variations shown in Equation 8.9.

Hydrologic Routing Approaches

Many engineering problems involving unsteady flow do not require a complete dynamic solution to the momentum and continuity equations. One can focus on the continuity equation, which we write as Equation 8.10.

$$\frac{dS}{dt} = I - O$$

(8.10)

where S is the storage (L^3), I is the inflow rate (L^3/T), and O is the outflow rate (L^3/T)
One can integrate Equation 8.3b, as shown in Equation 8.11a.

$$\int_{x_i}^{x_o} \frac{\partial Q}{\partial x} dx + \int_{x_i}^{x_o} \frac{\partial A}{\partial t} dx = 0 \tag{8.11a}$$

The first integral is the difference between the discharge evaluated at x_i and x_o for the control volume. One can apply the Leibniz rule to Equation 8.11a to find Equation 8.11b.

$$Q_o - Q_1 + \frac{d}{dt}\int_{x_i}^{x_o} A \, dx \tag{8.11b}$$

One can see the relationship between Equations 8.11b and 8.10. One can write the storage term of Equation 8.10, as shown in Equation 8.12.

$$S = \theta \left[XI + (1 - X)O \right] \tag{8.12}$$

where θ is a time constant, and other nomenclature is defined above. A physical basis for Equation 8.12 stems from the argument that channel storage consists of prism storage and wedge storage, as shown in Figure 8.2. Prism storage is associated with steady uniform flow, and wedge storage is the time-dependent storage component. The value of the Muskingum X generally falls between 0 and 0.5 or 0.6. These limits on the range of X values establish two different behaviors concerning the propagation of a flood wave. For $X \approx 0$, storage is related to outflow only, maximizing the wedge effect. The case of $X \approx 0$ is frequently called reservoir routing, which is further discussed below. In the case of $X \approx 0.5$, there is a significant wedge contribution, resulting in a translation of the hydrograph in time, known as river routing. The time constant θ in Equation 8.12 determines the extent of the translation. Methods for estimating X and θ are further discussed below.

Figure 8.3 shows extremes of the value of X on hydrographs. Reservoir routing implies that the body of water acts uniformly with the position in the reservoir. In contrast, river routing implies that the hydrograph translates down the reach as a wave. Reservoir routing

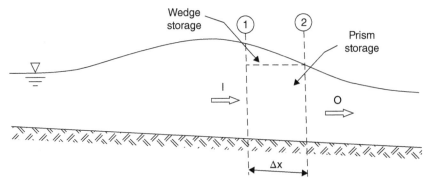

Figure 8.2 Hydrologic routing prism and wedge storage. *Source:* From Jain (2001), used with permission of John Wiley & Sons.

(a)

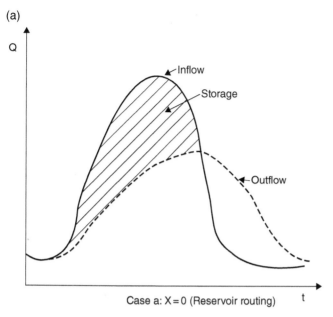

Case a: X = 0 (Reservoir routing)

(b)

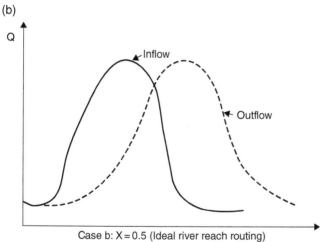

Case b: X = 0.5 (Ideal river reach routing)

Figure 8.3 Routing of hydrographs with pure storage (a) and pure translation (b).

is a particular case of reach routing. In reality, based on experience, most river routing applications have X values that fall between 0 and 0.5. We explore details of reservoir routing and river routing below.

Reservoir routing: Reservoir routing is sufficient for spillway design in small dams, stormwater management structures, and farm ponds. One may discretize Equation 8.10, as shown in Equation 8.13a, called the storage indication method.

$$\frac{S_2 - S_1}{\Delta t} = \frac{I_1 + I_2}{2} - \frac{O_1 + O_2}{2} \tag{8.13a}$$

Equation 8.13a may be rearranged to place the S_2 and O_2 terms on the left-hand side to arrive at Equation 8.13b.

$$\frac{2S_2}{\Delta t} + O_2 = I_1 + I_2 + \frac{2S_1}{\Delta t} - O_1 \qquad (8.13b)$$

One can develop a table of the stage, storage, outflow, and $2S/\Delta t + O$ vs. O for the structure. The Excel folder "ResReachRoute" contains "SimpleResRouting," which examines a hydrograph routing through a pond with a circular vertical spillway riser plus an emergency spillway. As discussed in Chapter 7, when the water level approaches the top of the riser, weir flow, orifice flow, and pipe flow may control the flow. We compute each possibility and take the maximum to determine the riser outflow. The emergency spillway (a broad crested weir) contributes when the flow reaches the weir's bottom. One must carefully characterize each outlet to capture the hydraulics of proposed outlet(s) structures. The flow control exerted by the outlet structure causes storage changes in the reservoir, affecting the outlet hydrograph. The "SimpleResRouting" spreadsheet demonstrates the effect on the outlet hydrograph.

Using reservoir routing techniques requires much pre-calculation work for each structure, as the example above demonstrates. The stage–storage relationship must be determined, as does the stage–outflow relationship. Software such as HydroCad[1] and SedCad,[2] provide help in computing these relationships but do not obviate the need to gather data for each situation.

River reach routing (Muskingum): For river reach routing, Equation 8.12 enables the evaluation of S_2 and S_1 with results substituted into Equation 8.13a to yield Equation 8.14.

$$\theta \left[X \left(I_2 - I_1 \right) + \left(1 - X \right) \left(O_2 - O_1 \right) \right] = \frac{\Delta t}{2} \left[\left(I_1 + I_2 \right) - \left(O_1 + O_2 \right) \right] \qquad (8.14)$$

Collecting terms in Equation 8.14 and solving for O_2 yields Equations 8.15a–d.

$$O_2 = C_0 I_2 + C_1 I_1 + C_2 O_1 \qquad (8.15a)$$

where

$$C_0 = \frac{-\theta X + 0.5 \Delta t}{\theta - \theta X + 0.5 \Delta t} \qquad (8.15b)$$

$$C_1 = \frac{\theta X + 0.5 \Delta t}{\theta - \theta X + 0.5 \Delta t} \qquad (8.15c)$$

$$C_2 = \frac{\theta - \theta X - 0.5 \Delta t}{\theta - \theta X + 0.5 \Delta t} \qquad (8.15d)$$

The routing coefficients C_0, C_1, and C_2 sum to one. Thus, Equations 8.15a–d can be viewed as weighting factors. Equations 8.15a–d are applied to the inflows at the beginning

---

1 HydroCad, P.O. Box 477, Chocorua, NH 03817.
2 SedCad, Civil Design Software, LLC. Box 706, Ames, IO 50010.

and end of the time step and the beginning outflow for the flow solution at the time interval end (Sturm 2010).

Sturm (2010) reported that $\Delta t \geq 2\theta X$, where θ is the wave travel time within the reach length. Also, $\Delta t \leq t_p/5$ where t_p is the time to peak of the input hydrograph. Further, $\Delta x \leq \dfrac{V_w \Delta t}{2X}$ where V_w is the wave velocity through the reach length Δx. This version of river routing is known as the Muskingum–Cunge method of stream routing.

Suppose one has inflow and outflow hydrographs for a given river reach. In that case, cumulative storage can be computed from a rearrangement of Equation 8.13, as shown in Equation 8.16.

$$S_2 = S_1 + \frac{\Delta t}{2}\left(I_2 + I_1 - O_2 - O_1\right) \tag{8.16}$$

A repeated application of Equation 8.16 enables the determination of cumulative storage. If one takes the initial value of storage as zero, then one can relate storage to the weighted flow value $[X I + (1 - X) O]$. One can do this for various X values and find X values' choice to provide the tightest fit. The slope of the best fit line is the time constant θ.

One can configure a least-squares approach for estimating X and θ (Gill 1977; Singh and McCann 1980; Aldama 1990). With an error function written as Equation 8.17, we have

$$E = \sum_j \left(AI_j + BO_j + S_1 - S_j\right)^2 \tag{8.17}$$

where $A = \theta X$, $B = \theta(1 - X)$, S_1 is the initial storage, S_j is the observed storage at time step j, and j is the the the running step.

The error E of Equation 8.17 is minimized by differentiating A and B with respect to error in Equation 8.17 then setting the resulting equations to zero. The resulting equations are solved for A and B to give Equation 8.18a and b.

$$A = \frac{1}{C}\left[\left(\sum I_j O_j \sum O_j - \sum I_j \sum O_j^2\right)\sum S_j + \left(N\sum O_j^2 - \left(\sum O_j\right)^2\right)\sum I_j S_j \right.$$
$$\left. + \left(\sum I_j \sum O_j - N\sum I_j O_j\right)\sum O_j S_j\right] \tag{8.18a}$$

$$B = \frac{1}{C}\left[\left(\sum I_j \sum I_j O_j - \sum I_j^2 \sum O_j\right)\sum S_j + \left(\sum I_j \sum O_j - N\sum I_j O_j\right)\sum I_j S_j \right.$$
$$\left. + \left(N\sum I_j^2 - \left(\sum I_j\right)^2\right)\sum O_j S_j\right] \tag{8.18b}$$

With C being defined as Equation 8.19c

$$C = N\left[\sum I_j^2 \sum O_j^2 - \left(\sum I_j O_j\right)^2\right] + 2\sum I_j \sum O_j \sum I_j O_j$$
$$- \left(\sum I_j\right)^2 \sum O_j^2 - \sum I_j^2 \left(\sum O_j\right)^2 \tag{8.18c}$$

With A and B, one can compute θ and X from Equations 8.19.

$$\theta = A + B \tag{8.19a}$$

$$X = \frac{A}{A+B} \tag{8.19b}$$

Equations 8.18 and 8.19 provide a way to compute the parameters for Equation 8.15a–d. with the provision that $\Delta t \geq 2\theta X$, (θ is the wave travel time within the reach length) and $\Delta x \leq \dfrac{V_w \Delta t}{2X}$ (V_w is the wave velocity through the reach length, Δx. Further, $\Delta t \leq t_p/5$ where t_p is the time to peak of the input hydrograph. Sturm (2010) provides an example calculation, which we reproduce as Example 8.1.

Example 8.1 Utilize the observed inflows and outflows for a river reach given in Table 8.1 (from Hjelmfelt and Cassidy 1975) to obtain values of θ and X using the graphical and the least-squares approach. Then determine the routing coefficients and route the inflow hydrograph through the river reach.

The storage is calculated from the average inflow and outflow rates over a single time step using Equation 8.16 and accumulated beginning with zero initial storage, as shown in Table 8.1. Then various values of X are substituted to obtain the weighted inflow and

Table 8.1 Computation of storage and Muskingum parameters for Example 8.1.

T_1 (days)	I, (m³/s)	Q, (m³/s)	I_{avg}	O_{avg}	S (m³)	$XI + (1-X)\, O$		
						$X = 0.1$	$X = 0.25$	$X = 0.4$
0.0	2.2	2.0			0.0	2.0	2.1	2.1
0.5	14.5	7.0	8.4	4.5	1.66E+05	7.8	8.9	10.0
1	28.4	11.7	21.5	9.4	6.89E+05	13.4	15.9	18.4
1.5	31.8	16.5	30.1	14.3	1.38E+06	18.0	20.3	22.6
2	29.7	24.0	30.8	20.3	1.83E+06	24.6	25.4	26.3
2.5	25.3	29.1	27.5	26.6	1.87E+06	28.7	28.2	27.6
3	20.4	28.4	22.9	28.8	1.62E+06	27.6	26.4	25.2
3.5	16.3	23.8	18.4	26.1	1.29E+06	23.1	21.9	20.8
4	12.6	19.4	14.5	17.4	9.76E+05	18.7	17.7	16.7
4.5	9.3	15.3	11.0	13.3	7.00E+05	14.7	13.8	12.9
5	6.7	11.2	8.0	9.7	4.73E+05	10.8	10.1	9.4
5.5	5.0	8.2	5.9	7.3	3.07E+05	7.9	7.4	6.9
6	4.1	6.4	4.6	5.8	1.88E+05	6.2	5.8	5.5
6.5	3.6	5.2	3.9	5.8	1.04E+05	5.0	4.8	4.6
7	2.4	4.6	3.0	4.9	2.16E+04	4.4	4.1	3.7

Source: Adapted from Sturm (2010).

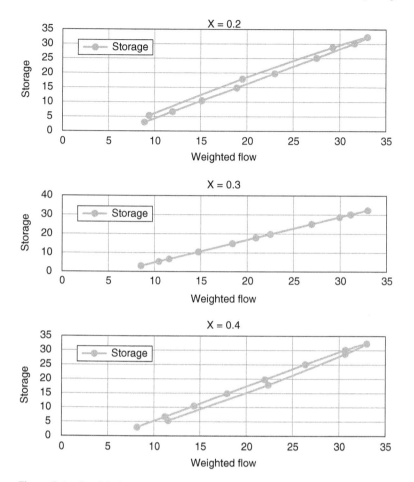

Figure 8.4 Graphical method for determining Muskingum *K* and *X* using data of Example 8.1.

outflow quantity, $XI + (1 - X) O$, at the end of each time step. The storage is related to this quantity by Equation 8.12. Figure 8.4 shows the results for $X = 0.10, 0.25$, and 0.4. By trial and error, the narrowest loop occurs for $X \approx 0.25$. The best line fit of the data for $X = 0.25$ gives a value of $\theta = 0.92$. Alternatively, Equations 8.18 and 8.19 give $X = 0.243$ and $\theta = 0.897$ days. Using the later values, we compute the Muskingum routing coefficients to be $C_0 = 0.034$, $C_1 = 0.504$, and $C_2 = 0.462$. For this case, $\Delta t > 2\theta X$. The solution proceeds, as shown in Table 8.2, and the results are plotted in Figure 8.5.

As with reservoir routing, Muskingum river routing is used in many stormwater management programs. The luxury of having data to compute the Muskingum parameters is usually not present. Programs often suggest routing parameters for approximate use. We revisit the Muskingum parameter determination in our discussion of Diffusion routing below.

Table 8.2 Muskingum routing ($C_0 = 0.034$, $C_1 = 0.504$, and $C_2 = 0.462$) for Example 8.1.

T_1 (days)	I, (m³/s)	$C_0 I_2$	$C_1 I_1$	$C_2 O_1$	O_2 (m³/s)
0.0	2.2				2.2
0.5	14.5	0.50	1.11	1.02	2.62
1	28.4	0.97	7.31	1.21	9.49
1.5	31.8	1.09	14.32	4.38	19.79
2	29.7	1.02	16.03	9.13	26.18
2.5	25.3	0.86	14.98	12.09	27.93
3	20.4	0.70	12.76	12.89	26.34
3.5	16.3	0.56	10.29	12.16	23.00
4	12.6	0.43	8.22	10.62	19.27
4.5	9.3	0.32	6.35	8.89	15.56
5	6.7	0.23	4.69	7.18	12.10
5.5	5.0	0.17	3.38	5.59	9.14
6	4.1	0.14	2.52	4.22	6.88
6.5	3.6	0.12	2.07	3.17	5.37
7	2.4	0.08	1.82	2.48	4.37

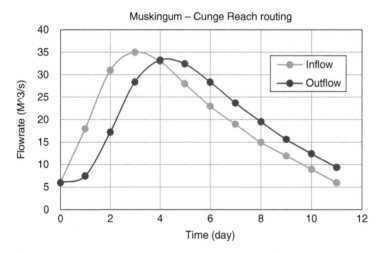

Figure 8.5 Observed inflow and computed outflow hydrograph for Example 8.1.

Kinematic Wave Method

Referencing Equation 8.9, kinematic routing is based on variations of velocity and depth with respect to distance x and variation of velocity with respect to time t. In other words, $Q = Q_n$, or $S_f = S_o$, and $Q = K \sqrt{S}$, which is the Manning equation. One can write Equation 8.20 based on the chain rule.

$$\frac{dQ}{dt} = \frac{dQ}{dA}\frac{\partial A}{\partial t} \tag{8.20}$$

The continuity equation for unsteady flow may be written as Equation 8.21.

$$\frac{\partial Q}{\partial x} + \frac{\partial A}{\partial t} = 0 \tag{8.21}$$

Eliminating $\dfrac{\partial A}{\partial t}$ allows one to write Equations 8.20 and 8.21 as Equation 8.22.

$$\frac{\partial Q}{\partial t} + c_k\frac{\partial Q}{\partial x} = 0 \tag{8.22}$$

where $c_k = \dfrac{\partial Q}{\partial A}$.

Equation 8.22 may be represented, as shown in Equation 8.23.

$$\frac{dQ}{dt} = \frac{\partial Q}{\partial t} + \frac{dx}{dt}\frac{\partial Q}{\partial x}\cdots\text{if }\frac{dx}{dt} = c_k \tag{8.23}$$

Parameter c_k is the speed of a uniformly progressive wave traveling down the channel at a constant speed from an upstream region of uniform flow to a downstream area of uniform flow. This uniform progressing wave is referred to as a monoclinal wave by Jain (2001).

The integration of Equation 8.23 yields $Q =$ constant along the path of Equation 8.23, which defines the characteristics of the first-order partial differential equation (Equation 8.22). Such a wave is termed a kinematic wave (Lighthill and Whitham 1955). Jain (2001) shows that the ratio β of the kinematic wave velocity to the flow velocity V is given as Equation 8.24.

$$\beta = \frac{c_k}{V} = \frac{A}{TK}\frac{dK}{dy} \tag{8.24}$$

In Equation 8.24, K is the conveyance (note that $Q = K\sqrt{S_0}$), and other terminologies are as defined above. Table 8.3 gives values for β for some channel types. Ponce et al. (1978) claim the kinematic wave approach to be 95% as accurate as the complete dynamic solution if Equation 8.25 is satisfied.

Table 8.3 Values of $\beta = c_k/V$ for indicated channel types based on the Chezy and Manning Equations.

Channel type	$\beta = c_k/V$	
	Manning	Chezy
Wide rectangular	1.67	1.50
Wide parabolic	1.44	1.33
Triangular	1.33	1.25

Source: From Jain (2001).

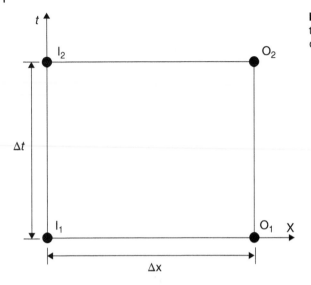

Figure 8.6 Rectangular mesh for the kinematic wave space–time discretization.

$$T_p \geq \frac{171 y_n}{S_o V_n} \tag{8.25}$$

In Equation 8.25, y_n is the normal depth, Vn is the normal flow velocity, and T_p is the wave propagation period.

Cunge (1969) showed a discretization of the temporal derivative in Equation 8.22, utilizing a weighted central difference temporal derivative and a non-weighted central difference spatial derivative (see Figure 8.6). These operations resulted in Equation 8.26.

$$\frac{X(I_2 - I_1) + (1 - X)(O_2 - O_1)}{\Delta t} + c_k \frac{(O_1 - I_1) + (O_2 - I_2)}{2\Delta x} \tag{8.26a}$$

where X is a weighting coefficient between 0 and 0.5.

It is interesting to note that Equation 8.26a may cast as similar in form to Equations 8.15a–d, which are the Muskingum routing equations. Kinematic wave motion analysis neglects the local and convective acceleration along with the water surface slope terms. It also assumes a simplified stage–discharge relationship. Thus far, we have mainly relied on various forms of the continuity equation to develop routing equations. The momentum equation's uniform flow term was tapped in the kinematic wave method, leaving the dynamic components to be further explored below. Ponce et al. (1978) claim that kinematic routing is a satisfactory solution to the dynamic routing program when Equation 8.26 is satisfied.

$$T_p \geq \frac{171 y_n}{S_o V_n} \tag{8.26b}$$

For a sinusoidal disturbance in a steady uniform flow, Equation 8.26b, T_p is the disturbance wave period. S_o is the bottom slope, and y_n is the normal depth.

Perhaps the most applied version of the kinematic wave method is for overland flow analysis. Figure 8.7 shows the overland flow scenario.

Figure 8.7 Rainfall excess over a small catchment. *Source:* From Jain (2001); used with permission of John Wiley & Sons.

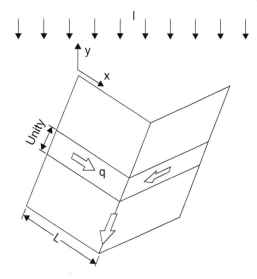

Following Jain (2001), we begin with modifying Equation 8.21, as shown in Equation 8.27.

$$\frac{\partial Q}{\partial x} + \frac{\partial A}{\partial t} = i \tag{8.27}$$

In Equation 8.27, i is the intensity of rainfall excess. One can write the discharge equation as $q = \alpha y^m$. Substituting this into Equation 8.27 and simplifying leads to Equation 8.28.

$$\frac{\partial y}{\partial t} + m\alpha y^{m-1}\frac{\partial y}{\partial x} = i \tag{8.28}$$

Equation 8.28 can be expanded, as shown in Equations 8.29.

$$\frac{dy}{dt} = i \tag{8.29a}$$

$$\frac{dy}{dt} = \frac{\partial y}{\partial t} + \frac{dx}{dt}\frac{\partial y}{\partial x} \tag{8.29b}$$

$$\frac{dx}{dt} = c = m\alpha y^{m-1} \tag{8.29c}$$

Integration of Equations 8.29a and 8.29c leads to overland flow characteristic equations, which appear as Equations 8.30.

$$y - y_0 = i(t - t_0) \text{ where } t \le t_r \tag{8.30a}$$

$$x - x_0 = m\alpha \int_{t_0}^{t}\left(i(t - t_0) + y_0\right)^{m-1} dt \tag{8.30b}$$

In Equations 8.30, t_r is the time of rainfall excess, x_o is the initial position, and y_o is the initial depth of flow. Equations 8.30 provide a basis for three characteristics, as shown in Figure 8.7. These characteristics are termed the J-characteristic, the limiting characteristic, and the K-characteristic. The J-characteristic CD initiates from point $C(x_j,0)$ on the x-axis where $t_o = 0$ and $y_o = 0$. The K-characteristic EF initiates from point E $(0, t_k)$ on the t-axis where $x_o = 0$ and $y_o = 0$. The J and K characteristics are extreme cases above and below the primary governing case. The labeled limiting (primary governing case) characteristics in Figure 8.8 represents a major concern. Jain (2001) shows that one can write the following equations for the characteristics as limited by time and distance.

$$y = \left(\frac{xi}{\alpha}\right)^{\frac{1}{m}}, \quad x_w \geq x \geq 0, \ t \leq t_c \tag{8.30c}$$

(a)

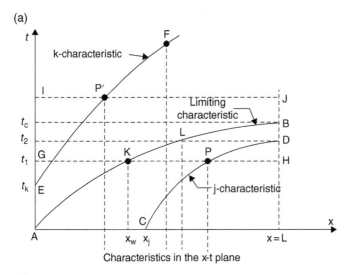

Characteristics in the x-t plane

(b)

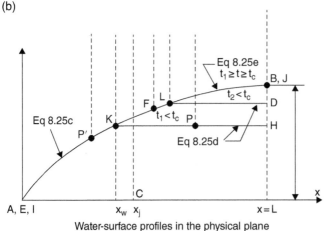

Water-surface profiles in the physical plane

Figure 8.8 Characteristics diagram and water surface profile for an overland flow situation. *Source:* From Jain (2001); used with permission of John Wiley & Sons.

$$y = it, \quad x \geq x_s \tag{8.30d}$$

$$y = \left(\frac{xi}{\alpha}\right)^{1/m}, \quad L \geq x \geq 0, \ t \geq t_c \tag{8.30e}$$

$$x = \alpha i^{m-1}\left(t - t_i\right)^m \tag{8.30f}$$

Equation 8.30g gives the x value of the limiting characteristic.

$$x = \alpha i^{m-1} t^m \tag{8.30g}$$

By substituting $x = L$ into Equation 8.30f, one can get the time of concentration t_c. The t_c is the time taken by a small disturbance to travel along the limiting characteristic from $x = 0$ to $x = L$.

$$t_c = \left(\frac{L i^{1-m}}{\alpha}\right)^{1/m} \tag{8.30h}$$

See Jain (2001) for the cases when t_r approaches infinity flow profiles after the excess rainfall ceases.

Kinematic routing for overland flow stems from developing the appropriate parameters m and α for the sheet flow equation $q = \alpha y^m$. The kinematic wave approach is suitable for paved areas. It is more challenging to apply on earthen surfaces where flows tend to channelize.

Diffusion Wave Method

The diffusion wave brings in an additional term of Equation 8.9 such that the momentum equation becomes as shown in Equation 8.31.

$$S_f = S_o - \frac{\partial y}{\partial x} \tag{8.31}$$

Using the conveyance form of the Manning equation to substitute for S_f gives Equation 8.32.

$$\frac{Q^2}{K^2} = S_o - \frac{\partial y}{\partial x} \tag{8.32}$$

Going back to the continuity equation (Equation 8.3b) and differentiating with respect to x and Equation 8.32 with respect to t (using $dA = T\,dy$) yields Equations 8.33a, 8.33b.

$$\frac{1}{T}\frac{\partial^2 Q}{\partial x^2} + \frac{\partial^2 y}{\partial x \partial t} = 0 \tag{8.33a}$$

$$\frac{2Q}{K^2}\frac{\partial Q}{\partial t} - \frac{2Q^2}{K^3}\frac{dK}{dA}\frac{\partial A}{\partial t} = -\frac{\partial^2 y}{\partial x \partial t}$$ (8.33b)

Elimination of the mixed partial derivative in Equations 8.33b, using the continuity equation to set $\frac{\partial A}{\partial t} = -\frac{\partial Q}{\partial x}$, and rearrangement of the resulting equation yields Equation 8.34.

$$\frac{\partial Q}{\partial t} + \left(\frac{Q}{K}\frac{dK}{dA}\right)\frac{\partial Q}{\partial x} = \left(\frac{K^2}{2TQ}\right)\frac{\partial^2 Q}{\partial x^2}$$ (8.34)

The term $\left(\dfrac{Q}{K}\dfrac{dK}{dA}\right) = c_f$ is the speed of the diffusion wave, and the term $\left(\dfrac{K^2}{2TQ}\right) = \mu$ is the diffusion coefficient. Equation 8.34 was the source of the diffusion routing approach because the partial differential equation is similar to a well-known mass diffusion and heat transfer solution. Sturm (2010) noted that Carslaw and Jaeger (1959) gave a symbolic solution to a slightly rearranged version of Equation 8.34a and b. The rearranged Equation 8.34a and solution 8.34b appear as follows.

$$\frac{\partial \varphi}{\partial t} + c_k \frac{\partial \varphi}{\partial x} = D\frac{\partial^2 \varphi}{\partial x^2}$$ (8.34a)

In Equation 8.34a, $\phi = y'$, $c_k = 1.5\, V_i$, and $D = V_i y_i/(2\,S_o)$. The boundary condition upstream is an abrupt increase in stage from the initial value of $\phi_d = \dfrac{y-y_i}{y_f-y_i}$. The solution appears as Equation 8.34b.

$$\varphi_d = 0.5\left[\operatorname{erfc}\left(\frac{x-c_k t}{\left(4Dt\right)^{1/2}}\right) + \exp\left(\frac{c_k t}{D}\right)\operatorname{erfc}\left(\frac{x+c_k t}{\left(4Dt\right)^{1/2}}\right)\right]$$ (8.34b)

In Equation 8.34b, the erfc is the complementary error function.

A Mathematica numerical solution with slightly different boundary conditions is provided in the "Diffusion Equation" folder. The numerical solution offers more flexibility in assigning initial and boundary conditions.

Extending an analogy from the c_f term, one can write a ratio of the diffusion wave velocity to the flow velocity to arrive at Equation 8.35.

$$\beta' = \frac{c_f}{V} = \frac{A}{TK}\frac{dK}{dy}$$ (8.35)

Note that Equation 8.35 is identical to Equation 8.24. The flow velocity comes from two different forms of the momentum equation (compare Equation 8.26 with and without the differential). Still, the differences are frequently assumed to be negligible, leaving $\beta' \approx \beta$.

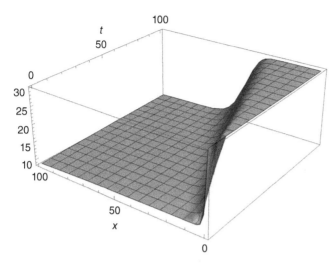

Figure 8.9 Diffusion equation solution for a simple wave from Mathematica with cf = 2.3 and μ = 3. The initial and boundary conditions are u(x,0)=5, and u(0,t) = 10*Sin[0.5*t/tmax].

The diffusion method designation arises from the similarity of Equation 8.34 with the classical diffusion equation from mass transport literature. Equation 8.34 is numerically solved in the "Mathematica Diffusion" equation Figure 8.9 shows a system experiencing a severe flow drop along with a sinusoidal variation. Note that one can see the propagation of the flow reduction down the channel with time. Also, the sharpness of the decline lessens with time, which is due to the diffusion term. Figure 8.9 is consistent with the closed-form solution of Carslaw and Jaeger (1959) in terms of the solution form, based on graphical comparisons. If diffusion were negligible, the result would revert to a kinematic wave solution.

Note that Equation 8.35 is identical in form to Equation 8.24. The flow velocity comes from two different forms of the momentum equation (compare Equation 8.26 with and without the differential). Still, the differences are frequently assumed to be negligible, leaving $\beta' \approx \beta$.

Similar to Equation 8.26, which provides a criterion for the adequacy of kinematic routing, Ponce et al. (1978) claim that diffusion routing is a satisfactory solution to the dynamic routing problem when Equation 8.36 is satisfied.

$$T_p \geq \frac{30}{S_z}\left(\frac{y_n}{g}\right)^{1/2} \tag{8.36}$$

The nomenclature in Equation 8.36 is the same as that for Equation 8.26, except that S_x is the water surface slope.

Tie back to Muskingum to arrive at the Muskingum–Cunge method: Cunge (1969) showed that Equations 8.15a–d is also a finite difference representation of the diffusion wave equation (Equation 8.34) when

$$\mu = (0.5 - X)c_f\Delta x \tag{8.37a}$$

$$c_f = \frac{\Delta x}{\hat{K}}$$
(8.37b)

Instead of past flood data, Equations 8.37a, b and 8.38 enable the determination of routing parameters $\hat{K}$ and X. In general, both $\hat{K}$ and X are functions of Q (or y). Note that one can easily relate the θ of Equations 8.15a–d to $\hat{K}$ Practically, $\hat{K}$ and X are assumed constant at a reference discharge Q_o using the average or peak discharge. After manipulating Equations 8.37, using the definitions of μ and β ($\approx \beta'$), Jain (2001) arrives at Equations 8.38 for Muskingum–Cunge routing parameters.

$$X = \frac{1}{2}\left(1 - \frac{Q_o}{\beta V_o S_o T_o \Delta x}\right)$$
(8.38a)

$$\hat{K} = \frac{\Delta x}{c_f} = \frac{\Delta x}{\beta V_o}$$
(8.38b)

The routing equations (Equations 8.15a–d) then appear as Equations 8.39a–c.

$$C_0 = \frac{\left(\frac{\Delta t}{\hat{K}}\right) - 2X}{2(1-X) + \left(\frac{\Delta t}{\hat{K}}\right)}$$
(8.39a)

$$C_1 = \frac{\left(\frac{\Delta t}{\hat{K}}\right) + 2X}{2(1-X) + \left(\frac{\Delta t}{\hat{K}}\right)}$$
(8.39b)

$$C_2 = \frac{2(1-X) - \left(\frac{\Delta t}{\hat{K}}\right)}{2(1-X) + \left(\frac{\Delta t}{\hat{K}}\right)}$$
(8.39c)

Example 8.2   (Adapted from Sturm 2010)
Determine the routing parameters of the Muskingum–Cunge method for the following flood and channel characteristics: The reference discharge is 48 m^3/s; the channel is rectangular with $T = 12$ m. The channel slope is 0.0002, and the roughness $n = 0.012$. Take the reach length to be 10000 m.

Using the Excel sheet "Trapezoid channel design Example8.2 Muskingum" the knowns, one can find that $V = 1.68$ m/s, depth $= 2.38$ m, $K = 3394$ m^3/s. Note that $z = 0$. Next, we develop a Mathematica notebook (see "Muskingum Parameters") to solve Equations 8.33 for X and $\hat{K}$, which requires a solution of Equation 8.38. Parameter $X = 0.098$ and $\hat{K} = 1.11$ hours.

A spreadsheet is available in the Excel "Res ReachRoute" folder for executing the Muskingum–Cunge reach routing method using Equations 8.38 and 8.39. The spreadsheet is configured to solve the Jain (2001) Example 8.2. This spreadsheet contains a page for estimating the routing parameters based on known inflow and outflow data. Alternatively, one may use the parameters generated by the above Mathematica notebook to accomplish the routing.

The diffusion analogy's practical contribution is to provide a way to parameterize the Muskingum–Cunge river routing model without the requirement to collect observed input–output data. The long, narrow reach of Example 8.2 is typical of where river reach routing is impactful. Short stream reaches respond more like reservoirs than rivers. Muskingum–Cunge is relevant in artificial prismatic channels. One may idealize the natural channel by a prismatic section that best approximates it, such as the parabola. The channel should have a consistent cross section over the reach length. The compound channel requires a numerical analysis of the derivative in Equation 8.36, with careful selection of the reference flow. The Muskingum–Cunge method is not relevant with backwater flows. The complete dynamic solution must be invoked for natural, variable cross sections with in-stream structures.

Dynamic Routing

Thus far, we have focused on decreasing degrees of simplicity for solving the unsteady flow problem. Perhaps the most common unsteady flow problem is hydrograph routing in a stream but maybe a gate opening or closure. Now we take a direct view of solving Equations 8.1 and 8.2, building on the numerical approaches already presented in simplified form.

Graf and Altinakar (1998) present three methods to solve the St. Venant equations: characteristics, explicit finite difference, and implicit finite-difference. The method of characteristics is presented schematically herein. The method of characteristics is the classical method and is presented by nearly all texts on open channel flow. We offer a thumbnail overview of how to arrive at the characteristic equations for a wide rectangular channel, where hydraulic depth is the same as actual depth. The reader is advised to consult Henderson (1966) or Graf and Altinakar (1998) and reference cited therein to fill in calculation gaps in the presentation below. For a rectangular channel, the St. Venant equations appear as Equations 8.40a and b.

$$h\frac{\partial V}{\partial x} + V\frac{\partial y}{\partial x} + \frac{\partial y}{\partial t} = 0 \tag{8.40a}$$

$$\frac{1}{g}\frac{\partial V}{\partial t} + \frac{V}{g}\frac{\partial V}{\partial x} + \frac{\partial y}{\partial x} = S_f - S_o \tag{8.40b}$$

Now, $c^2 = gh$, where c is the wave velocity or celerity. One can write $d(c^2) = 2c\,dc = d\,(gh)$, which implies that celerity c becomes a measure of flow depth. Equations 8.41 arise from substituting the differentials into Equations 8.40 (Graf and Altinakar 1998).

$$c\frac{\partial V}{\partial x} + 2V\frac{\partial c}{\partial x} + 2\frac{\partial c}{\partial t} = 0 \tag{8.41a}$$

$$\frac{\partial V}{\partial x} + U\frac{\partial V}{\partial x} + 2c\frac{\partial c}{\partial x} = g\left(S_f - S_o\right) \tag{8.41b}$$

Taking the sum and differences of Equations 8.41a and b allows one to write the equations as total derivatives of the following form.

$$\left(\frac{\partial}{\partial t} + (V+c)\frac{\partial}{\partial x}\right)(V+2c) = g\left(S_f - S_o\right) \tag{8.42a}$$

$$\left(\frac{\partial}{\partial t} + (V-c)\frac{\partial}{\partial x}\right)(V-2c) = g\left(S_f - S_o\right) \tag{8.42b}$$

Equations 8.42a and b, because they contain total derivatives, can be written as follows.

$$\frac{d}{dt}(V+2c) = g\left(S_f - S_o\right) \tag{8.43a}$$

$$\frac{d}{dt}(V-2c) = g\left(S_f - S_o\right) \tag{8.43b}$$

$$(V+c) = \frac{dx}{dt} \tag{8.43c}$$

$$(V-c) = \frac{dx}{dt} \tag{8.43d}$$

Equations 8.43a–d are the characteristic equations for the wide rectangular channel. The term, $V \pm 2c$, is approximately constant if S_f-S_o is near zero, which is commonly assumed for simple approximations. Jain (2001) presents an alternative discussion leading to the same result, as shown in Equations 8.43. Chow (1959) attributes these developments to Massau (original work in French not seen). Chow (1959) solves the dynamic equations for $\frac{\partial y}{\partial x}$, then lets the denominator go to zero, and then forces the numerator to zero. He then arrives at the characteristic equations as given above. The characteristics' development's serendipitous nature is fortuitous and occurred with excellent working knowledge of multivariate calculus and differential equations. Substituting the definition of celerity into Equations 8.43 yields equations for two characteristics, as shown in Equations 8.44.

$$dV - \sqrt{\frac{g}{y}}\,dy + dt\left[g\left(S_f - S_o\right)\right] = 0 \quad \text{and} \quad dx = \left(V - \sqrt{gy}\right)dt \;\; \rightarrow C- \tag{8.44}$$

The curves, given by $dx = V \pm \sqrt{gy}\,dt$, are plotted in Figure 8.10, emanating from common points $L(x_L, t_L)$ and $R(x_R, t_R)$ where the velocities V_L and V_R and flow depths y_L and y_R are known. Figure 8.10 portrays subcritical flow. Thus, by solving the equations for V and y in different points $P, P_1, P_2 ...$, which are common to the two characteristics curves throughout the x-t plane, one can describe the unsteady flow.

After substituting first-order finite differences for differentials and solving four unknowns (x_P, t_P, V_P, and y_P) one finds Equations 8.45 (see Graf and Altinakar 1998).

$$t_p = \frac{\left(x_L - x_R\right) + t_R\left(V_R - \sqrt{gy_R}\right) - t_L\left(V_L + \sqrt{gy_L}\right)}{\left(V_R - V_L\right) - \left(\sqrt{gy_L} + \sqrt{gy_R}\right)} \tag{8.45a}$$

$$x_P = x_L + \left(V_L + \sqrt{gy_L}\right)\left(t_P - t_L\right) \tag{8.45b}$$

$$y_P = \frac{\left(V_L - V_R\right) + \left(\sqrt{gy_L} + \sqrt{gy_R}\right) - \left(t_P - t_L\right)\left[g\left(S_{oL} - S_f\right)\right] + \left(t_P - t_R\right)\left[g\left(S_{oR} - S_f\right)\right]}{\sqrt{\dfrac{g}{y_L}} - \sqrt{\dfrac{g}{y_R}}} \tag{8.45c}$$

$$V_P = V_L - \sqrt{\frac{g}{y_L}}\left(y_P - y_L\right) - \left(t_P - t_L\right)\left[g\left(S_{oL} - S_f\right)\right] \tag{8.45d}$$

See Figure 8.10b for putting Equations 8.45a–d in perspective. It is possible to calculate another series of points (e.g. 11, 12, and 13) at an advanced time level. Another series of points (101, 104, ... also L and R) are subsequently calculated using the previously known series of points (11, 12, 13, etc.). Thus, the calculation repeats itself.

Before starting, one must specify the initial and boundary conditions (see Figure 8.10c). The initial state refers to the depth and average velocity at all points in the channel at $t = 0$. The boundary conditions refer to the depth and velocity (or discharge) at the upper and lower ends of the reach at all times after calculations begin. Equations 8.45 may be used for supercritical flows also. Figure 8.11 shows a comparison of the characteristics of supercritical and subcritical flow.

The method of characteristics does not apply to problems containing discontinuities such as a hydraulic jump or sudden gate opening or closure. The S_t. Venant equations break down, and pressure distribution fails to be hydrostatic at these locations.

The method of characteristics is a method having an explicit scheme. Thus, if at a given time, $t = t_1$, all values of y and V are known, then at another time level $t = t_1 + \Delta t$, all values of y and V can be determined. The numerical stability of this scheme requires satisfaction of the Courant condition (see Figure 8.10d). Equation 8.45e gives the Courant condition.

$$\Delta t \le \frac{\Delta x}{V + c} \tag{8.45e}$$

The Courant condition can be very restrictive as it imposes an upper limit on Δx. Overcoming this restriction and removing the frustration of picking a Δx and Δt to follow characteristics has motivated the explicit and implicit methods.

The method of characteristics has a variable Δx and Δt, which causes some frustration. The explicit method fixes Δx and Δt at the outset, which results in a grid of points, as shown in Figure 8.12. The explicit method uses a central difference scheme for estimating the derivatives instead of a forward difference scheme. The Courant condition must be

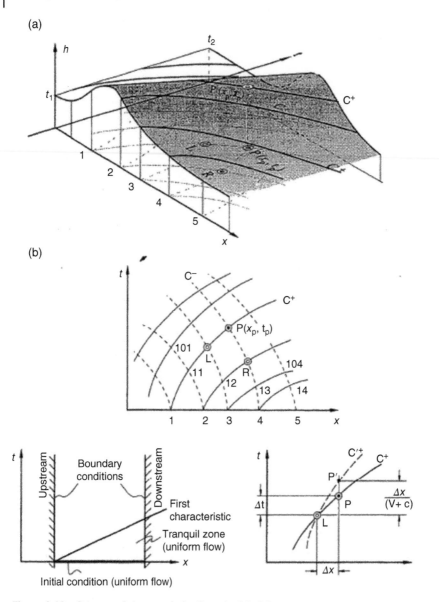

Figure 8.10 Scheme of characteristics for subcritical flow:
a) Profile of an unsteady flow at t_1 and t_2;
b) Surface around $P(x_P, t_p)$ projected into the x–t plane;
c) Boundary conditions;
d) Courant condition;
Source: From Graf and Altinakar (1998); used with permission of John Wiley & Sons.

satisfied for numerical stability. Graf and Altinakar (1998) and references cited therein provide details.

The implicit method uses numerical derivatives similar to the explicit method but applies a general weighting factor. The implicit method simultaneously computes all the distance nodes at a given time step. The weighting factor can move the scheme from being explicit

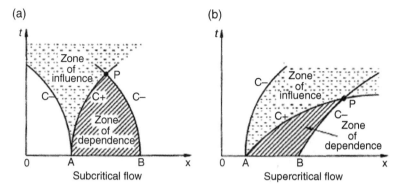

Figure 8.11 Zone of influence of disturbance at point A and zone of dependence on point P in subcritical flow and (b) supercritical flow. The zone of influence refers to the space-time zone that can be affected by conditions at A at time zero. The zone of dependence denotes the region under point P reflecting the influence on P by initial conditions at A and B. *Source:* From Jain (2001), used with permission of John Wiley & Sons.

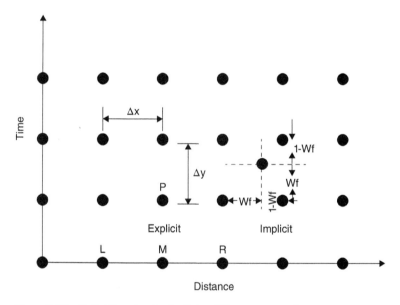

Figure 8.12 Definition sketch of a finite difference network.

(Wf = 0), completely implicit (Wf = 1), and four-point center-implicit. Graf and Altinakar (1998) summarize several investigators' recommendations suggesting that Wf ≥ 0.65, which bodes well for numerical stability. Figure 8.12 provides a schematic view of how the weighting factor interacts with Δt and Δx. Graf and Altinakar (1998) give a thumbnail sketch of the implicit method. The implicit method is the starting point for 2-D analysis, which Chaudhry (1993) extensively explores. In addition to the references cited above, Sturm (2010) provides a concise summary of the explicit and implicit methods for solving the

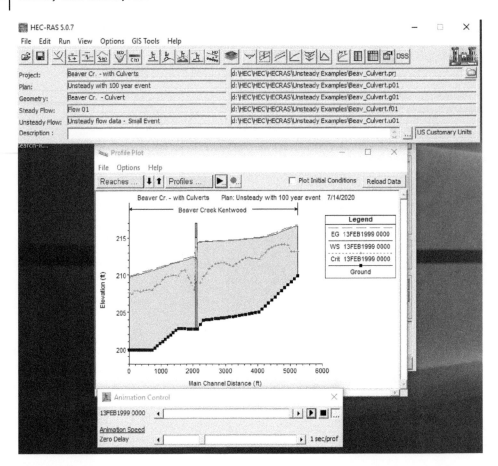

Figure 8.13 Selected output of the HEC-RAS Beaver Creek model (with culverts) showing the main menu, a profile plot at the initial condition, and the animation control. *Source:* HEC-RAS Computer program. © United States Army Corps of Engineers

dynamic equations. This presentation only touches the surface of a vast body of literature examining the dynamic equations' full solutions.

The HEC-RAS (Brunner 2016) software embodies the full one-dimensional dynamic equation set solution using state-of-the-art numerical methods. An example HEC-RAS unsteady flow problem, "Beaver Cr. with culverts," is included in the HEC-RAS software folder. This project routes a 100-year hydrograph down a stream that contains culverts midway. Figures 8.13 and 8.14 present profile plots for this example. Many other unsteady flow examples are available with the sample data provided by HEC-RAS. The unsteady flow analysis requires careful attention to entries of the simulation time window. The start and end times must be synchronized with the available data. Be prepared to experience a learning curve when learning to use HEC-RAS, unsteady flow capabilities. Solving unsteady flow problems with HEC-RAS or any other software requires considerable time.

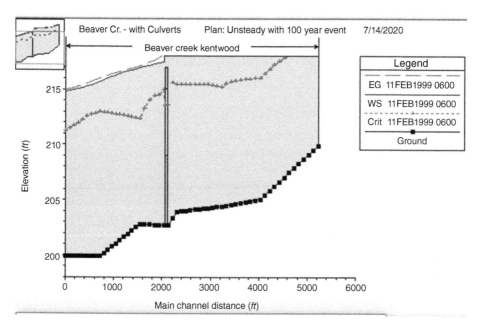

Figure 8.14 Profile plot of the Beaver Creek model at maximum flow. Note that the culvert fails to handle the 100-year peak flow.

Summary and Conclusions

We looked first at hydrologic routing, which did not consider momentum effects. The presentation began with hydrologic routing (reservoir and reach) that used only the continuity equation. Equation 8.9, based on the momentum equation, shows the continuum between simple kinematic routing and full dynamic routing. The analysis herein focused on the wide rectangular geometry where hydraulic depth is essentially the actual depth. Chow (1959), French (1985), Chaudhry (1993), and Sturm (2010) provide details for writing and solving the dynamic equation set for other geometries. Figure 8.1 shows a relation between flow depth and flow rate. The rising limb typically has a steeper water surface slope than the bottom slope.

Conversely, the falling limb may have a lesser water surface slope than the bottom slope. The dashed line is the steady flow rating curve, which one would typically find with kinematic routing. Muskingum reach routing gave similar results to those of hydrologic routing (continuity equation only) and kinematic routing. Kinematic routing assumes a simple relationship between depth and flow. A collateral benefit of the diffusion analysis was that it provided a way to compute the Muskingum routing parameters from channel attributes. The diffusion routing and full dynamic routing capture the wave transmission effect of Figure 8.1.

Reservoir and reach routing suffice for simple retention pond design and stream routing. Simple approaches do not suffice for wave transmission and subsequent effects due to wave interaction with structures. Full dynamic routing using numerical methods rise to this challenge and provide the basis for extending to multidimensional analysis. Rapidly varying unsteady flows represent a challenge for complete dynamic analyses, especially when moving beyond the rectangular channel to other geometries. We explore rapidly varying unsteady flows in the next chapter.

Problems and Questions

1 This problem aims to route a hydrograph through a levee pond: First, find the stage–discharge relation for a spillway and reservoir with the following specifications. The riser height is 5 m. An emergency spillway, $L = 30$ m, is located 6 m above the riser height. Assume the straight outlet pipe to be corrugated steel, 35 m long, the inlet invert is at 100 m, and the outlet invert is 90 m. The diameter is 0.8 m. The riser diameter is twice the pipe diameter. The levee pond reservoir has the following dimensions: $L = 100$ m, $W = 100$ m, and $z = 1$ from a depth of 0 to a maximum depth of 15 m. The volume-depth of the reservoir is given by the following:

$$V = \left[LW + \left(L - 2zd \right)\left(W - 2zd \right) + 4\left(L - zd \right)\left(W - zd \right) \right]\left(\frac{d}{6} \right)$$

W is the width at water surface, L is the length at water surface, z is the side slope ratio, and d is the depth. Note that L and W must exceed the $2 \times z \times d$ product. The pond has an emergency spillway ($L = 10$ m) 1.5 m above the principal spillway. Now use this stage–storage relationship to route the storm in the excel pond routing worksheet (or another storm of your instructor's choosing) through the levee.

2 Route the following inflow hydrograph through a reservoir created by a dam and determine the maximum reservoir elevation reached during the passage of the hydrograph. The flow from the reservoir passes over a spillway with a crest elevation of 122 m. The outflow O (m³/s) is given by $O = 15H^3$, where H is the head at the spillway crest in meters. The reservoir has near-vertical banks. The surface area is 65 ha, and the base flow is 10 m³/s. The inflow is in m³/s.

Time (hr)	0	2	4	6	8	10	12	14	16	18	20
Inflow	10	100	180	210	160	90	60	46	30	16	10

3 Using the Mathematica notebook for Muskingum parameter analysis along with spreadsheets for lined channel design, determine the Muskingum routing parameters for the case of Example 8.2, where the channel is a trapezoid with $z = 1$.

4 Repeat problem 3 for the case of a triangular channel where $z = 10$.

5 Repeat problem 3 for the case of a parabolic channel where $To = 24$ m.

6 Evaluate problem 3 for cases of reference flow $Q = 10, 20, 40,$ and 60 m³/s.

7 Given the inflow and outflow hydrographs on the Iowa River, determine K and X for the river reach. (inflow and outflow in m^3/s)

Time (h)	0	1	2	3	4	5	6	7	8
Inflow	70	110	130	194	256	266	222	186	153
Outflow	70	76	102	129	182	237	254	228	195
Time	9	10	11	12	13	14	15	16	17
Inflow	117	92	72	70	70	70	70	70	70
Outflow	161	127	99	79	74	73	72	71	70

8 Modify the reach routing spreadsheet to compute the least-squares routing parameters using Equations 8.18 and 8.19.

9 Using the reach routing spreadsheet, route the hydrograph of problem 7 using the routing parameters determined in problem 7. Develop a plot comparing the inflow, predicted outflow, and observed outflow hydrographs.

10 Find the relationship between θ in Equations 8.15a–d to $\hat{K}$ in Equation 8.33b.

11 Assuming the Iowa River is approximated by a parabolic cross-section of $To = 200$ m, calculate the Muskingum parameters using the Mathematica Muskingum routing parameters notebook.

12 Set up a spreadsheet to route the hydrograph of problem 7 through the Iowa River's reach. The reach length is 5000 m. Use the Muskingum method.

13 Modify (e.g. increase the diameter) the Beaver Creek HEC-RAS project's culverts to pass the 1.25 times the maximum flow in the example.

References

Aldama, A.A. (1990). Least-squares parameter estimation for Muskingum flood routing. *ASCE Journal of Hydraulic Engineering* 116 (4): 580–586.

Brunner, G.W. (2016). *HEC-RAS river analysis system: User's manual*. US Army Corps of Engineers, Hydrologic Engineering Center, Davis, CA.

Carslaw, H.S. and Jaeger, J.C. (1959). *Conduction of Heat in Solids*, 2e. Oxford, UK: Oxford Science Publications.

Chaudhry, M.H. (1993). *Open-Channel Flow*. Englewood Cliffs, NJ: Prentice-Hall.

Chow, V.T. (1959). *Open-Channel Hydraulics*. New York, NY: McGraw-Hill Publishers.

Cunge, J.A. (1969). On the subject of a flood propagation computation method (Muskingum method). *Journal of Hydraulic Research* 7 (2): 205–230.

French, R.H. (1985). *Open Channel Hydraulics*. New York, NY: McGraw-Hill.

Gill, M. (1977). Routing of floods in river channels. *Nordic Hydrology* 8: 163–170.

Graf, W.H. and Altinakar, M.S. (1998). *Fluvial Hydraulics: Flow and Transport Processes in Channels of Simple Geometry*. Chichester, UK: Wiley.

Henderson, F.M. (1966). *Open Channel Flow*. New York, NY: Macmillan Publishing Co.

Hjelmfelt, A.T. Jr. and Cassidy, J.J. (1975). *Hydrology for Engineers and Planners*. Ames, IO: Iowa State University Press.

Jain, S.C. (2001). *Open-Channel Flow*. New York, NY: Wiley.

Lighthill, M.J. and Whitham, G.B. (1955). On kinematic waves: I. – Flood movement in long rivers. *Proceedings of the Royal Society of London* A229: 281–316.

Ponce, V.M., Li, R.M., and Simons, D.B. (1978). Applicability of kinematic and diffusion models. *Journal of Hydraulics Division, ASCE* 100 (hy3): 353–360.

Singh, V.P. and McCann, R. (1980). Some notes on Muskingum method of flood routing. *Journal of Hydrology* 48: 343–361.

Sturm, T.W. (2010). *Open Channel Hydraulics*, 2e. New York, NY: McGraw-Hill.

9

Rapidly Varying Unsteady Flow Applications – Waves

Placing a structure such as a weir or gate in a channel brings about deviations from the uniform flow condition. Gradually varied flow analysis quantifies the extent of those deviations. If the structure is placed or set at a specified time, how long is required for the flow profile upstream to develop? This effect propagation is but one of the types of problems addressed by rapidly varying unsteady flow analyses. Any rapidly varying flow may be described as a wave. The chapter briefly considers three problem types and organizes them from field scale to regional scale: surface irrigation, opening/closing of sluice gates, and the dam-break. Practically, surface irrigation initiates flow in a furrow, infiltrates water into the soil as a function of wetting time, then turns off water at the furrow inlet. Likewise, the sluice gate can mimic flows in canals with regulated flows such as power generation flows. The rapidly opening sluice gate segues into the failure of a dam. The chapter concludes with a mention of oscillatory waves found on water bodies.

As with gradually varying unsteady flows, the continuity and momentum equations provide the starting point for these analyses. However, the Saint Venant equations per se are not strictly valid due to the streamlines' curvature near locations where sharp flow changes occur. The presentation's thrust is to provide an overview while referring the reader to more advanced references for additional details.

Goals

- Examine cases where $dQ/dt \neq 0$ and $dQ/dx \neq 0$
- Provide an overview of surface irrigation hydraulics
- Examine sluice gate openings and closures and related phenomenon
- Examine dam-break hydraulics in smooth, wide rectangular channels

Surface Irrigation

Surface furrow irrigation is economical when water is plentiful and available without pumping. The design of a surface irrigation system is another example of how hydrologic,

Open Channel Design: Fundamentals and Applications, First Edition. Ernest W. Tollner.
© 2022 John Wiley & Sons Ltd. Published 2022 by John Wiley & Sons Ltd.
Companion website: www.wiley.com/go/tollner/openchanneldesign

kinematic, and dynamic routing have been applied to predict the rate of advance of water down a furrow.

Figure 9.1 shows the phases of furrow irrigation occurring as water flows down a furrow. Figure 9.2 shows applications of the processes in Figure 9.1b. Infiltration at a point varies with the time of wetting. We show later that infiltration at a point varies with time. Thus, infiltration varies with time and distance.

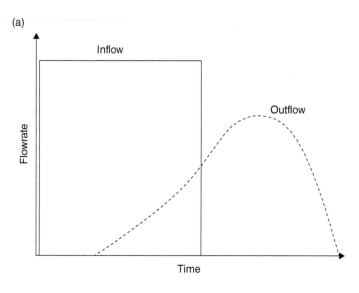

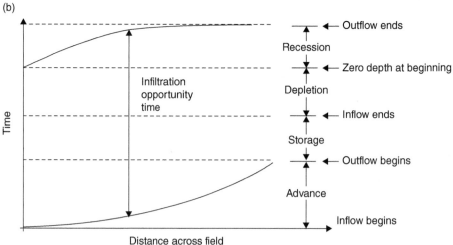

Figure 9.1 The complete surface irrigation furrow inflow and outflow hydrographs in part (a) shows the effect of stream routing along with the infiltration that leads to less area under the outflow hydrograph compared to the inflow pulse. Part (b) shows the advance of the wetting front with time, and the recession front from the time the depth was zero at the inflow point to the time of zero depth at the outflow. The interval between the curves is the infiltration opportunity time.

(a)

(b)

Figure 9.2 Furrow irrigation systems: (a) Furrow irrigation system using siphon tubes to supply individual rows from a lined supply channel. Note that the channel has drop structures that take flat sections down the field slope via spaced drop structures. (b) Furrow system supplied by a pressurized pipe with spaced orifices. *Source:* Photos courtesy of Wikipedia creative commons.

The continuity and momentum equations for surface irrigation applications appear as Equations 9.1 (James 1988).

$$\frac{\partial Q}{\partial x} + \frac{\partial A}{\partial t} + I = 0 \tag{9.1a}$$

$$\frac{1}{Ag}\left(\frac{\partial Q}{\partial t}\right)+\left(\frac{2Q}{A^2g}\right)\left(\frac{\partial Q}{\partial x}\right)+\left(1-F^2\right)\left(\frac{\partial y}{\partial x}\right)=S_o-S_f \tag{9.1b}$$

In Equations 9.1,

A is the cross-sectional area of flow (L^2), Q is the stream size (L^3/T), I is the infiltration rate per unit furrow length (L^2/T), Y is the flow depth (L), g is the gravitational acceleration (L/T^2), t is the time (T), x is the distance along the furrow from the furrow entrance (L), T is the top width of the flow cross-section (L), F is the Froude number (-), S_o is the slope of the furrow (-), and S_f is the slope of the energy grade line or water surface (-).

As was the case in Chapter 8, various approaches have been directed toward solving Equations 9.1. Walker and Skogerboe (1987) summarize hydrologic routing, kinematic wave, and full dynamic modeling approaches with input hydrographs to solve the rate of advance problem of water down a furrow. Figure 9.3 shows conceptual cumulative irrigation during the profile advance along a furrow (space between crop rows). We give a sketch of the hydrologic modeling process for the advance phase shown in Figure 9.1

Figure 9.3 shows how $z(x, t)$ is related to $z(0, t)$, where z refers to infiltration depth in a hydrologic routing. This relation builds on the continuity equation and ignoring momentum effects. The profiles portray successive instances of time, a constant interval δt apart. Once $z(t)$ is known, the horizontal lines labeled z_{01}, z_{02}, ..., etc., separated by the constant time interval, are drawn. The verticals x_{A1}, x_{A2}, etc., represent successive locations of the stream front (unknown), at $t = \delta t$, $t = 2\delta t$, $t = 3\delta t$, etc. Hall (1956), as summarized by Walker and Skogerboe (1987), showed that the successive subsurface profiles could be constructed by joining opposite corners of the resulting rectangles, as shown. It is assumed that the wetted perimeter is constant along the furrow. Equation 9.2a defines the distance the profile travels during δt_1.

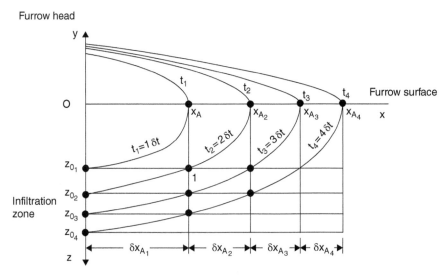

Figure 9.3 Cumulative furrow infiltration and profile advance with a constant rate of input at the furrow head. *Source:* Adapted from Strelkoff and Clemmens 2007.

$$\delta x_1 = \frac{Q_o \delta t}{\sigma_y A_0 + \sigma_z Z_1 + p_f}$$ (9.2a)

In Equation 9.2a,

A_o is the cross-sectional flow area at the field inlet defined by the normal depth at that point, p_f is the puddle factor that accounts for dead storage on the soil surface, σ_z is a subsurface shape factor for the tip cell (to x_{At1} at this point) defined by a ratio of areas $\{O\text{-}Z_{o1}\text{-}x_{A1}\}/\{O\text{-}Z_{o1}\text{-}1\text{-}x_{A1}\}$, Z_1 is the depth of infiltration during δt_1, calculated using $Z = kt^a + f_o t$, t is time, f_o is the asymptotic intake rate in volume units per unit time per unit length, and a, k is the soil property constants measured or assessed from soil texture data.

Equation 9.2a generalizes to Equation 9.2b for predicting δx_i during the advance phase (Walker and Skogerboe 1987).

$$\delta x_i = \frac{Q_o \delta t}{A + \sigma_z} - \left[\sum_{k=1}^{i-1} \frac{Z_{i-k+1} - Z_{i-k-1}}{2(A + \sigma_z Z_1)} \delta x_k \right]$$ (9.2b)

Walker and Skogerboe (1987) provide additional analysis for the storage phase (time of constant water input after the profile advances to $x =$ furrow end) and the recession phase (time of turning the water off at the furrow entrance to time water ceases to flow from the furrow end). In addition to the hydrologic model, Walker and Skogerboe (1987) provide details on a kinematic wave model, zero inertia model, and characteristics solutions.

Strelkoff and Clemmens (2007) have devoted effort to solving the dynamic equation set for surface irrigation purposes. Bautista et al. (2012) present a software package containing the Strelkoff and Clemmens (2007) work that predicts surface and subsurface flow of water for a known system geometry, infiltration, and roughness conditions. Finite difference methods (see Chaudhry 1993) were employed. A complete dynamic solution is available from the Strelkoff and Clemmens (2007) model. Their software,[1] "WinSRFR 4.1," is a public domain package. Surface irrigation is moving out of favor as water becomes limited, pushing the industry toward more efficient irrigation approaches. Even though the above work was directed toward furrow irrigation, it is easily adapted to solving similar problems for any channel.

Sluice Gate and Related Operations

Interesting hydraulic phenomena resulting in rapidly varying unsteady flow arise from natural and man-induced water management. A realistic example is a hydraulic or tidal bore. Chow (1959) gives a photograph of a bore in China. Tidal bores result from high tidal ranges, usually found in the high latitudes, converging to narrow streams of low slope

---

1 WinSRFR 4.1, USDA-ARS U.S. Arid Land Agricultural Research Center, Maricopa, AZ USA.

when the tide is incoming. On a smaller scale, the bore is similar to a moving hydraulic jump. We neglect the effects of infiltration (or exfiltration) in this analysis. The sluice gate problems addressed here assume a steady reservoir of flow above and below the surge wave.

Figure 9.4 shows a catalog of flow types resulting from sluice gate operations and external controls. The continuity and momentum equations provide a basis for developing a relationship between the upstream velocity, downstream velocity, and surge wave velocity for the six situations depicted in Figure 9.4.

Insert Figure 9.4.

Following Jain (2001), we write the continuity and momentum equations, as shown in Equations 9.3, for the control volume between points 1 and 2 in Figure 9.4.

$$A_1\left(V_1 - V_2\right) = A_2\left(V_2 - V_w\right) \tag{9.3a}$$

$$\rho A_1\left(\beta V_1 - V_w\right)\beta\left(V_2 - V_1\right) = \gamma\left(A_1\overline{y_1} - A_2\overline{y_2}\right) \tag{9.3b}$$

Equation 9.3b is the momentum equation for a general cross-section that includes the momentum correction coefficient. Small slopes and negligible drag between points 1 and 2

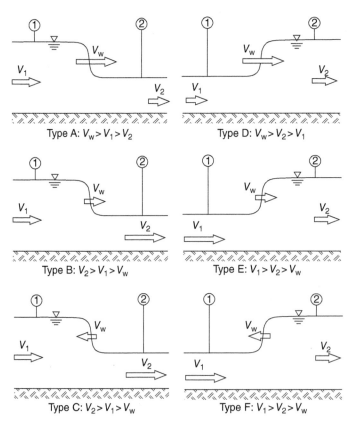

Figure 9.4 Types of surges as categorized by Jain. *Source:* From Jain (2001), used with permission of John Wiley & Sons.

is assumed. The right-hand side of Equation 9.3b is the pressure forces' contribution to the balance. By consulting Table 6.1, one can expand the right-hand side to any of the standard cross-sections. From this point forward, we take the momentum coefficient $\beta = 1$ and consider the rectangular cross-section.

These substitutions result in a momentum expression for the rectangular cross-section as Equation 9.3c.

$$\rho y_1 (V_1 - V_w)(V_1 - V_2) = \frac{1}{2}\gamma\left(y_2^2 - y_1^2\right) \tag{9.3c}$$

From Equation 9.3a, $\beta = 1$, and rectangular cross-section, V_w is given by Equation 9.4.

$$V_w = \frac{V_2 y_2 - V_1 y_1}{y_2 - y_1} \tag{9.4}$$

Substituting the right-hand side for V_w from Equation 9.4 into Equation 9.3c results in Equations 9.5.

$$y_1\left[V_1 - \frac{V_1 y_1 - V_2 y_2}{y_1 - y_2}\right](V_2 - V_1) = \frac{g}{2}\left(y_1^2 - y_2^2\right) \tag{9.5a}$$

Rearranging Equation 9.5a results in Equation 9.5b, and further rearrangements result in Equations 9.5c and 9.5d.

$$\frac{y_1 y_2}{y_1 - y_2}(V_2 - V_1)^2 = \frac{g}{2}\left(y_1^2 - y_2^2\right) \tag{9.5b}$$

$$(V_2 - V_1)^2 = \frac{g}{2}\frac{(y_1 - y_2)^2 (y_1 + y_2)}{y_1 y_2} \tag{9.5c}$$

Solving 9.5c for $V_1 - V_2$ results in Equation 9.5d.

$$V_1 - V_2 = \pm |y_1 - y_2| \sqrt{\frac{g(y_1 + y_2)}{2 y_1 y_2}} \tag{9.5d}$$

Substituting for $(V_1 - V_2)$ from Equation 9.5d into Equation 9.3c recasts the momentum equation (rectangular channel) to appear as Equation 9.5e.

$$y_1(V_1 - V_w)\left[\pm |y_1 - y_2| \sqrt{\frac{g(y_1 + y_2)}{2 y_1 y_2}}\right] = \frac{g}{2}\left(y_1^2 - y_2^2\right) \tag{9.5e}$$

Now, Equation 9.5e can be solved for $(V_1 - V_w)$ to give Equation 9.6a.

$$V_1 - V_w = \pm\sqrt{\frac{g}{2}\frac{y_2}{y_1}(y_1 + y_2)} \tag{9.6a}$$

Adding Equations 9.5d and 9.6a result in Equation 9.6b.

$$V_w - V_2 = \pm \sqrt{\frac{g}{2} \frac{y_1}{y_2} \left(y_1 + y_2 \right)}$$

(9.6b)

Equations 9.6 applies to each of the six cases in Figure 9.4. Table 9.1 summarizes the relations between V_1, V_2, and V_w. Table 9.1 contains notations designed to help one generalize between Jain's general surge approach (2001) and the specific sluice gate analysis of Graf and Altinakar (1998).

Coming back to Equations 9.6, notice that if the disturbance is small ($y_1 \approx y_2 = y$), then $V_1 \approx V_2$ and Equations 9.6 becomes identical to Equation 9.7a for a rectangular channel.

$$V_w - V = \pm \sqrt{gy}$$

(9.7a)

For a general cross-section, Jain (2001) shows that one can substitute the hydraulic depth D for y in Equation 9.7a resulting in Equation 9.7b.

$$V_w - V = \pm \sqrt{gD}$$

(9.7b)

The term under the radical is denoted as the wave celerity.

Table 9.1 Types of surges in open channels.

Jain (2001) Type	Graf and Altinakar (1998) type and (generalizations beyond the simple sluice gate)	Sign of			Relation among V_1, V_2, and V_w	Sign of V_w
		$(V_1 - V_w)$	$(V_2 - V_1)$	$(V_2 - V_w)$		
Col 1	Col 2	Col 3	Col 4	Col 5	Col 6	Col 7
A	c' (initially similar to dam-break downstream)	–	–	–	$V_w > V_1 > V_2$	+
B	b (downstream gate closure)	+	+	+	$V_2 > V_1 > V_w$	+
C	d (initially similar to the dam-break reservoir)	+	+	+	$V_2 > V_1 > V_w$	–
D	b' (release at some point downstream)	–	+	–	$V_w > V_2 > V_1$	+
E	c (Jump moving downstream)	+	–	+	$V_1 > V_2 > V_w$	+
F	a (tidal bore; jump moving upstream)	+	–	+	$V_1 > V_2 > V_w$	–

Notes: The left to right direction is positive. Downstream is point 2. The c' and b' cases in Col 2 indicate results similar to the sluice gate of Graf and Altinakar (1998) but result from controls at a distance upstream or downstream. The dam-break similarities are valid at the time of the break but do not hold after the initial break. The depiction of some velocities may be opposite that shown in cases of Figure 9.4, but the magnitudes in Col 7 holds.

Source: Used with permission from Jain (2001), and Graf and Altinakar (1998).

Example 9.1 (Adapted from Jain 2001).
A river is flowing at a depth of 2.4 m and a velocity of 1 m/s. It meets a tidal bore, which abruptly increases the depth to 3.6 m. Determine the type of surge. Find the speed with which the bore moves upstream. Then, find the magnitude and direction of the water behind the bore. Use the Jain (2001) analysis.

First, using Jain (2001), the surge is moving upstream and $y_1 < y_2$. It is a Jain type F from Figure 9.4. We write Equation 9.6a as follows, realizing that V_w is the smallest of the three velocities from Table 9.1.

$$V_w = V_1 - \sqrt{\frac{g}{2}\frac{y_2}{y_1}(y_1 + y_2)}$$

Substituting values in the above relationship results in $V_w = -5.64$ m/s.
Since $V_2 > V_w$, we wrote Equation 9.6b as follows.

$$V_2 = V_w + \sqrt{\frac{g}{2}\frac{y_1}{y_2}(y_1 + y_2)}$$

Table 9.1 was used to reason the appropriate sign for the $\pm$ terms in Equations 9.6. Substituting values in the above, the velocity below the bore is -1.21 m/s.

Graf and Altinakar (1998) focus on sluice gate operations, and Figure 9.5 shows the effects of opening and closing a gate. Figure 9.6 elaborates on the scenarios of Figure 9.5. Graf and Altinakar (1998) use different notations, but the relationships are derived from continuity and momentum balances. Table 9.1 gives a correspondence between the Jain (2001) types (upper case) and the Graf and Altinakar (1998) types (lower case). Table 9.1 addresses other situations such as moving hydraulic jumps (which, as stated in Chapter 6, should be anchored to a specified location). The positive wave (from upstream or downstream) is a wave that maintains its integrity during propagation. On the other hand, the negative wave diminishes as it propagates upstream or downstream.

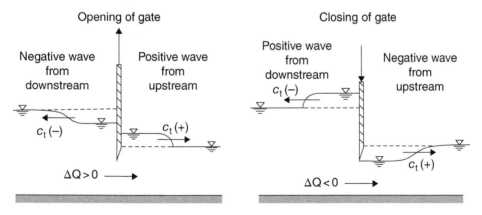

Figure 9.5 Example of for different types of translatory waves due to operation of an in-channel gate. *Source:* From Graf and Altinakar (1998); used with permission of John Wiley & Sons.

The – or + sign of c_t (corresponds with the Jain 2001 V_w) refers to movement upstream (-) or downstream (+).

Example 9.2 Compute Example 9.1 using the Graf and Altinakar (1998) results. This situation is analogous to that of a positive wave moving downstream, case a, in Figure 9.6 (e.g. the wave maintains integrity as it moves upstream).

$$c(-) = U_1 - \sqrt{gh_1}\sqrt{\left(\frac{h_2}{2h_1}\right)\left(1 + \frac{h_2}{h_1}\right)}$$

Substituting $h_1 = 2.4$ m, $h_2 = 3.6$ m, and $U_1 = 1$ m/s into the above relation results in $c(-) = -5.64$ m/s, which agrees with V_w from Example 9.1.

The reader may wonder why Jain (2001) presents six scenarios, while Graf and Altinakar (1998) give only four scenarios. Graf and Altinakar (1998) focus only on the effects above the opening and closing sluice gates (Figure 9.5). Jain (2001) generalizes to include the impact of operations upstream or downstream (Figure 9.4). The sluice gate scenarios of

(a) **Positive wave** from downstream

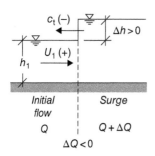

$$c_t(-) = U_1 - \sqrt{gh_1}\sqrt{(h_2/2h_1)(1 + h_2/h_1)}$$

(b) **Negative wave** from upstream

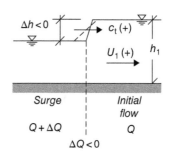

$$c_t(+) = U_1 + \sqrt{gh_1}\sqrt{(h_2/2h_1)(1 + h_2/h_1)}$$

(c) **Positive wave** from upstream

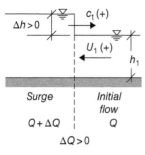

$$c_t(+) = U_1 + \sqrt{gh_1}\sqrt{(h_2/2h_1)(1 + h_2/h_1)}$$

(d) **Negative wave** from downstream

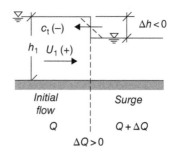

$$c_t(-) = U_1 - \sqrt{gh_1}\sqrt{(h_2/2h_1)(1 + h_2/h_1)}$$

Figure 9.6 Types of surges associated with gate opening/closing. *Source:* From Graf and Altinakar (1998); used with permission of John Wiley & Sons.

Figure 9.5 are useful for visualizing the effects of hydropower generation cycles on downstream flows. HEC-RAS (Brunner 2016) addresses the sluice gate and other in-line structure impacts on channel behavior. The dam-break phenomenon, discussed below, is theoretically like the case of the rapid, complete opening of a sluice gate, except that with the dam-break, the reservoir above the dam is depleting and not constant.

The analysis in Example 9.2 simplified the sluice gate problem to that with a series of pulses. The pulse applies close (upstream and downstream) to the gate. As one moves further away from the gate, the pulse profile degrades into a wave. As shown in Figure 8.9, the diffusion solution shows how a wave degrades from a pulse to a wave.

The Dam-Break Problem[2]

The dam-break has risen to prominence because of several notable catastrophes in the last century. Examples include the Teton dam failure in Idaho in 1976 and the Kelley Barnes dam failure in Georgia in 1978, both of which resulted in numerous casualties. The dam break is an extension of a sluice gate's rapid opening, except that the draining reservoir above the dam introduces time dependency. The S_t. Venant equations did not apply to the sluice gate due to the non-hydrostatic pressure conditions near the gate. The dam-break has been modeled as a plate that suddenly translates downstream, enabling the S_t. Venant equations. This analysis provides a basis for accommodating the reality of the draining reservoir and the advancing downstream wave. For convenience, the St. Venant equations developed in Chapter 8 for the wide rectangular channel are repeated below.

$$\frac{d}{dt}\left(V + 2c\right) = g\left(S_f - S_o\right) \tag{8.43a}$$

$$\frac{d}{dt}\left(V - 2c\right) = g\left(S_f - S_o\right) \tag{8.43b}$$

$$\left(V + c\right) = \frac{dx}{dt} \tag{8.43c}$$

$$\left(V - c\right) = \frac{dx}{dt} \tag{8.43d}$$

In Equations 8.43, c is the celerity $\{(gy)^{1/2}\}$, and V is the velocity. Referring to the discussion of Equation 8.23, Q is replaced by $V-2c$, $V + 2c$, $V + c$, and $V-c$, which are constant. We consider two cases: downstream channel with zero initial depth and downstream channel with finite initial depth.

Dry downstream channel: Consider that the dam is simulated with a plate that is instantaneously accelerated to a speed w. We allow the plate to move at infinite speed but do so in steps to gain an appreciation of how the characteristics can lead to a solution. Figure 9.7a

---

2 This discussion draws heavily from the Jain (2001) discussion of the dam-break problem.

depicts the wall motion simulating the dam failure. The plate's path in the *x-t* plane is represented by the line OBEH in Figure 9.7b. The inverse slope of the line gives the speed of the plate. With an increasing velocity of the plate from *O* to *E*, the inverse slope gradually increases. The inverse slope becomes constant beyond *E* as the plate begins to move at a constant speed w. The velocity of the water in contact with the plate is equal to the plate velocity. The boundary condition is therefore prescribed in terms of velocity along OBEH. Figure 9.7 represents a first step (of three total) toward developing the dry channel dam-break.

Several *C*- characteristics issuing from the line OBEH (slope is d*t*/d*x*) can be drawn, as shown in Figure 9.7b. The slope of the *C*- characteristics may be found using Equation 8.43d. The inverse slope of the first *C*- characteristic OA that divides the undisturbed and disturbed zones is $-C_0$ as $V_0 = 0$ initially. Note that the characteristics issuing from OBEH are perpendicular to OBEH and express d*x*/d*t* at points along OBEH. The undisturbed zone is identified by Zone 1 in Figure 7.6. The slope of the *C*- characteristic BC issuing from point *B* is described by Equation 9.8a.

$$\frac{dx}{dt} = V_B - C_B \tag{9.8a}$$

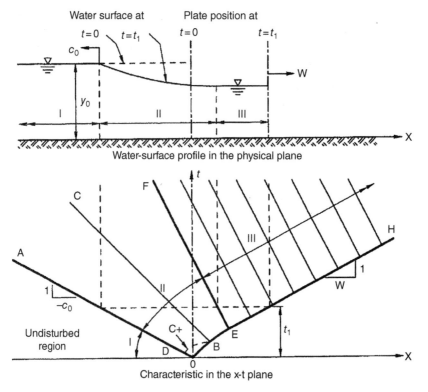

Figure 9.7 Flow development by the movement of a plate. *Source:* From Jain (2001); used with permission of John Wiley & Sons.

A C+ characteristic BD is drawn through point B to the initial C- characteristic OA. From Equation 8.43a, which is valid through point B to the initial C- characteristic OA, we write Equation 9.8b.

$$V_B + 2c_B = V_D + 2c_D = 2c_0 \tag{9.8b}$$

Note that $V_D = V_0 = 0$ and $c_D = c_0$. Also, Equation 9.8b is valid in the entire x-t plane and may be written as Equation 9.8c.

$$V + 2c = 2c_0 \tag{9.8c}$$

From Equations 9.8a and 9.8b, the slope of BC can be expressed in terms of either V_B or c_B. Equation 9.9 shows this rearrangement.

$$\frac{dx}{dt} = \frac{3}{2}V_B - c_0 = 2c_0 - 3c_B \tag{9.9}$$

Because V_B increases with time until $V_B = w$ at point E, the inverse slope given by Equation 9.9 of the C- characteristic issuing from line OBE increases as t increases along line OBE. These characteristics diverge in the zone denoted II in Figure 9.7. Furthermore, the depth in a section in Zone II decreases with time. The characteristics issuing from the straight line EH in the zone denoted III in Figure 9.7 are parallel, as plate velocity is constant along EH. The flow conditions (i.e. V and y) in zone III are therefore constant. The flow conditions are variable in Zone II, which connects Zones I and III, wherein the flow conditions are constant. The constant depth y_E in Zone III is obtained by substituting $V = w$ and $c = \sqrt{gy_E}$ in Equation 9.8c, which results in an expression for y_E shown in Equation 9.10.

$$y_E = \frac{\left(c_0 - \dfrac{w}{2}\right)^2}{g} \tag{9.10}$$

Equation 9.10 suggests that depth decreases as w decreases and becomes zero when $w = 2 c_0$. The inverse slope of the characteristics in Zone III from Equation 9.9 (middle term) for $V_E = w = 2 c_0$ is w. This condition implies that all characteristics in Zone III coalesce into the straight line EH. Zone III disappears, as shown in Figure 9.8. The water surface profile in Figure 9.8a is shown for $t > t_E$, where t_E is the period in which the plate is accelerated from speed zero to $w =$ twice c_0. The depth at the leading edge of the wave is zero. The leading edge that is in contact with the plate is moving with a velocity of 2 c_0. Figure 9.8 shows that if w exceeds twice c_0, the plate loses contact with the water behind it. In other words, if w exceeds twice c_0, the flow field is not affected by the removal of the plate. Note that the C- characteristics of Figure 9.8 are not perpendicular to OBEH because they have a slope of $1/(V - c)$, which is no longer tied to the imaginary wall w. Note that the slope is dt/dx, but the characteristic equation is given in terms of dx/dt. This explains why the inverse slope is used. Figure 9.8 depicts an intermediate step in our development of the dry channel dam-break scenario.

(a)

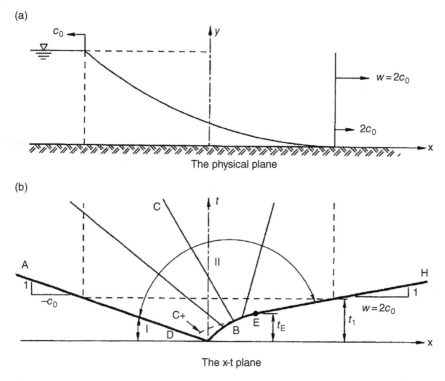

The physical plane

(b)

The x-t plane

Figure 9.8 A limiting case wherein $w = 2c_0$. *Source:* From Jain (2001); used with permission of John Wiley & Sons.

One may simulate the plate's sudden removal by accelerating the plate with an infinite acceleration to a velocity equal to or greater than twice c_0, as depicted in Figure 9.10. It should be noted that the $V = 2\, c_0$ assumes negligible roughness in the dry channel. When roughness is substantial, one should reduce velocity by as much as one half (Jain 2001). In that case, point E in Figures 9.7 and 9.8 coincides with point O at the origin, and the line EH originates from the origin, as shown in Figure 9.9. The slope of the characteristics is 1/$(V–c)$. Equation 9.9 enables the determination of the flow condition at any point $P\,(x, t)$ because the slope dx/dt of the C- characteristic passing through point P is equal to x/t. The velocity V goes to 0 along O-t; thus, the slope is c here. Flow is supercritical to the right of O-t and subcritical to the left of O-t. The water surface profile at any time t is given by Equation 9.9 (right), which one may write as Equation 9.11.

$$\frac{x}{t} = 2\sqrt{gy_0} - 3\sqrt{gy} \qquad (9.11)$$

At any instant, for this rectangular channel and rectangular reservoir, the water surface profile is a parabola. The leading-edge advances at twice c_0, and the trailing edge advances at speed c_0. Again, one of the C- characteristics is colinear with t at $x = 0$ and therefore has an inverse slope of zero. Velocity $V = c$ along the t-axis, corresponding to critical flow along the t-axis. Graf and Altinakar (1998) show that the depth and velocity at $x = 0$ are

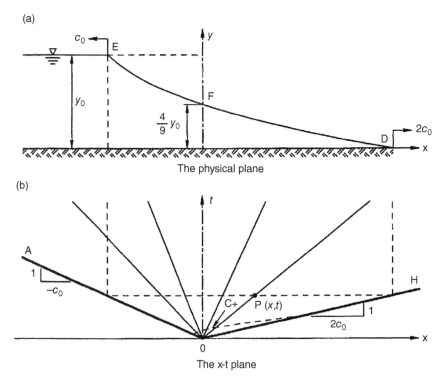

(a)

The physical plane

(b)

The x-t plane

Figure 9.9 Sudden failure of a dam with a dry downstream bed, where the plate moves at an instantaneous speed of $2 c_0$. *Source:* From Jain (2001); used with permission of John Wiley & Sons.

constant and equal to $d = 4 y_0/9$ and $V = 2 c_0/3$ during the reservoir draining phase (when c_0 is constant). The slopes of the C- characteristics in the first quadrant are positive, implying supercritical flow. Going back to our discussion of waves (see Figure 9.4), the wavefront EF is a Type C wavefront as it is propagating upstream. The sloping wavefront FD is moving downstream and therefore is of Type B. Though both wavefronts are propagating at velocity $(V-c)$, V is smaller than c in the former. Velocity V is higher than c in the latter.

Example 9.3 A dam retaining water to a depth of 20 m failed suddenly. The bed downstream was dry and smooth. Find the following: (i) speed of the leading edge; (ii) speed of the receding edge; (iii) depth and velocity in a section 10 km downstream of the dam an hour after the failure.

$$c_0 = \sqrt{g y_0} = \sqrt{9.81 * 20} = 14 \text{ m/s}$$

Speed of leading smooth edge $= 2 c_0 = 28$ m/s
Speed of receding edge $= c_0 = 14$ m/s
Depth at the dam $= (4/9) y_0 = 4 (20)/9 = 8.89$ m
Velocity at the dam $= 2 c_0/3 = 2(14)/3 = 9.33$ m/s

(a)

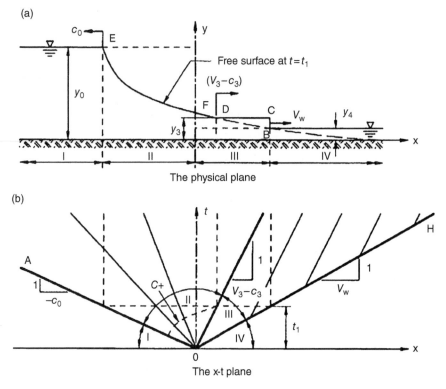

The physical plane

(b)

The x-t plane

Figure 9.10 Sudden failure of a dam with a finite depth in the downstream channel with 0.138 $y_0 > y_4 > 0$. *Source:* From Jain (2001), used with permission of John Wiley & Sons.

The depth y in the section at $x = 10\,km$ and $t = 30$ minutes can be determined from Equation 9.11 as follows.

$$\frac{10*1000}{30*60} = 2*14 - 3\sqrt{9.81*y} \Longrightarrow y = 5.76\,m$$

The velocity at the section may be obtained from Equation 9.8c.

$$V + 2\sqrt{9.81*5.76} = 2*14 \Longrightarrow V = 12.97\,m/s`$$

Finite depth in downstream channel: The case of the finite but limited depth in the downstream channel is slightly different from that of the dry channel described above. Here, we have a surge analyzed by methods discussed above at the leading edge of the profile. We use characteristics as before. Figure 9.10 gives a schematic of the flow situation and the x-t plane with characteristics. There are three distinct depths of interest, ranging from y_0 (initial depth), y_3 (depth of the downstream surge), and y_4 (original downstream depth). Instead of a discontinuity at B (the case with the dry channel), the depth and velocity downstream of the surge in Zone IV are constant. The constant depth and velocity are also true upstream of Zone III's surge.

The continuity and momentum equations across the surge (see Equations 9.3a, 9.3c) for $V_4 \approx 0$, with rearrangements, appear as follows (refer to Equations 9.3–9.6) to give Equations 9.12.

$$y_3\left(V_3 - V_w\right) = -y_4 V_w \tag{9.12a}$$

$$V_w = \sqrt{\frac{g}{2}\frac{y_3}{y_4}\left(y_3 + y_4\right)} \tag{9.12b}$$

The subscripts fit the zones in Figure 9.10. Equation 9.12b may be obtained from Equation 9.8b, where the negative sign before the radical is taken as V_w being positive. From Equation 9.8c, we can write Equation 9.12c.

$$V_3 + 2c_3 = 2c_0 \tag{9.12c}$$

The unknown quantities are y_3, V_3, and V_w. Substitution into Equation 9.12a for V_3 from Equation 9.12c and V_w using Equation 9.12b leads to Equation 9.13, from which y_3 may be determined.

$$\sqrt{\frac{g}{2}\frac{y_3}{y_4}\left(y_3 + y_4\right)} = \frac{y_3\left(2\sqrt{gy_0} - 2\sqrt{gy_3}\right)}{\left(y_3 - y_4\right)} \tag{9.13}$$

V_w and V_3 are determined by substituting y_3 into Equation 9.12b and Equation 9.12c, respectively (note that $c_i = $ sqrt [$g\,y_i$]). A spreadsheet, "DamBrkFinite," is available to solve Equation 9.13 for y_3, V_w, and V_3. On the spreadsheet, for two values of y_0, values of V_3, V_w, and $V_w - V_3$ were examined over a range of y_0/y_4 above and below the 0.138 threshold. When y_0/y_4 exceeded 0.13, $V_w - V_3$ increased. When y_0/y_4 was less than 0.13, $V_w - V_3$ was small and constant. The difference in $V_3 - c_3$ changed from positive to negative as one crossed the 0.138 ratio threshold. A Mathematica program, "DamBrkFinite," is available as well.

The slope of the C- characteristic in Zone III in Figure 9.10 is positive, implying that the flow in Zone III is supercritical, as discussed previously. One of the C- characteristics in Zone II is colinear with the t-axis, and the flow is critical there. Thus, the depth and velocity at section F are $4\,y_0/9$ and $2\,c_0/3$, respectively. As V_3 decreases with an increasing y_4, and V_3 becomes equal to c_3 (critical flow) when y_4 is about $0.138\,y_0$.

It can be shown that the slope of the C- characteristics for Zone III for y_4 larger than about $0.138\,y_0$ is negative, as shown in Figure 9.11. The flow in Zone III is therefore subcritical, and section D translates upstream. The t-axis lies in Zone III, where the flow conditions are steady. The discharge at $x = 0$ continues to be independent of time. It is now a function of depth y_4 and is less than the maximum of $8\,c_0\,y_0/27$. Note that the inverse slope of the C- characteristics in Zone III for y_4 is approximately equal to $0.138\,y_0$. The flow in Zone III is then critical, and section D remains stationary at $x = 0$. If y_4/y_0 is less than 0.138, section D translates downstream. However, it has a range of constant depth and flow, enabling reasonable solutions.

The practical application of the characteristics for both the dry channel and the finite depth channel is limited, beyond providing a rationale for how the solution functions.

(a)

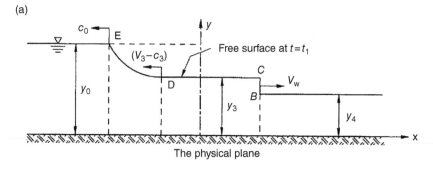

The physical plane

(b)

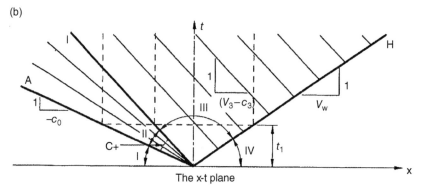

The x-t plane

Figure 9.11 Sudden failure of a dam with a finite depth in the downstream channel with $y_0 > y_4$ >0.138 y_0. *Source:* From Jain (2001), used with permission of John Wiley & Sons.

Software is available for rigorously modeling the dam-break problem. Sturm (2010) provides a concise historical overview of the modeling approach used. DAMBRK (Chow et al. 1988) has been one of the best-known softwares. Computational fluid dynamics (CFD) software such as FLOW-3D CFD software is state-of-the-art. Chaudhry (1993) gives computational details of the CFD modeling approach. HEC-RAS (Brunner, 2016) has a module that incorporates ideas from DAMBRK and other software for simulating a dam failure, and an example is included in the downloads ("BaldCreekDmBrk"). The HEC-RAS example allows for a measured cross-section and models the dam failure process over an hour. HEC-RAS models piping failure and breaches. As with other unsteady flow analyses, the DAMBRK analyses are anything but trivial. Figure 9.12 is a screenshot from "BaldCreekDmBrk," which shows the breach profile. The program is run as an unsteady flow simulation. The profile plot is animated and provides an animated view of the breach. Figure 9.13 provides a screenshot just before and just after the breach.

Oscillatory Waves

An important unsteady flow phenomenon in coastal engineering is that of oscillatory waves. Predictions of wave behavior build from the characteristic equations

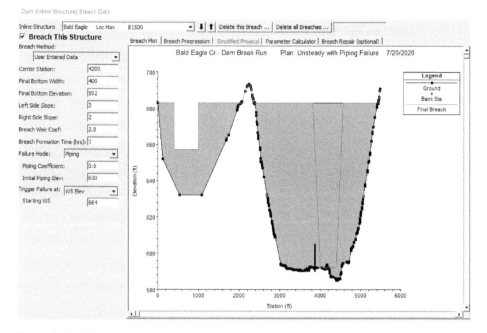

Figure 9.12 Plan of the breach to the Bald Creek Dam HEC-RAS example. The dam is an in-line structure that contains a weir to the left and a breach profile to the right. The piping failure begins near the outflow gate and happens over 1 hour. *Source*: HEC-RAS Computer program. © United States Army Corps of Engineers.

discussed above. The significant variables are wave height, wave period, and depth. Throughout the text thus far, the celerity of a wave has been described as shown in Equation 9.14

$$c = \sqrt{gh}$$ (9.14)

The parameter c is the wave velocity (L/T), g is gravity (L/T^2), and h represents the depth (L). Equation 9.14 is adequate for shallow water flows. Flows discussed in earlier chapters are generally shallow flows. As the period (L = distance between waves) to depth ratio is greater than 20 (Henderson 1966), when the period/depth ratio is on the order of 2, then Equation 9.14 must be written as

$$c = \sqrt{\frac{gL}{2\pi}}$$ (9.15)

If L/y is less than 2, denoting deep water (Henderson 1966), the wave speed is given in Equation 9.16.

$$c = \sqrt{\frac{gL}{2\pi} \tanh\left(\frac{2\pi y}{L}\right)}$$ (9.16)

(a)

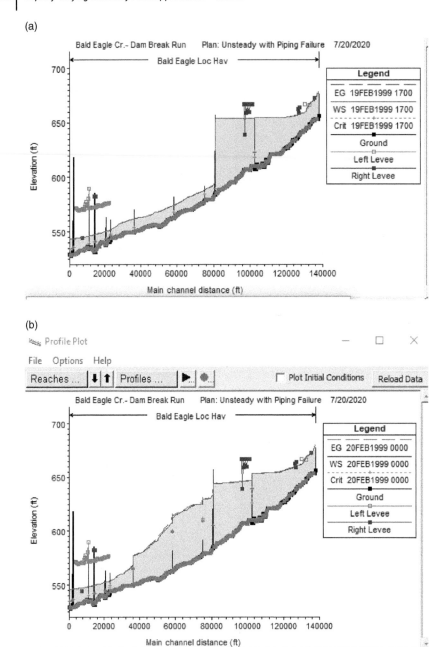

(b)

Figure 9.13 Profile plot of the Bald Creek example (a) just before the breach and (b) just after the breach. *Source:* HEC-RAS Computer program. © United States Army Corps of Engineers.

The most general case of the wave celerity equation includes surface tension effects. Henderson (1966) gives the wave velocity with surface tension (σ, F/L) as follows in Equation 9.17.

$$c = \sqrt{\left(\frac{gL}{2\pi} + \frac{2\pi\sigma}{\rho L}\right) \tanh\left(\frac{2\pi y}{L}\right)} \qquad (9.17)$$

In Equation 9.17, ρ is the mass density (M/L^3) and other symbols are as previously defined.

Computing the wave velocity with equation 9.14 in deep water could lead to an error of 50–80% of the actual wave velocity. Water particles in deep water waves follow a circular or elliptical path. In contrast, shallow-water particles translate back and forth with limited net movement. Henderson (1966) gives a characteristic equation solution to an equation describing particle movement in deep water. He shows that waves break as they encounter shallow water from deep water. Waves break when the wave (H) height from trough to peak relative to the mean depth (y) is 0.78 or greater. In other words, a deep-water wave breaks when the mean depth y decreases such that H/y is 0.78 (Henderson 1966).

Wave energy, a secondary variable, is a substantial consideration for coastal infrastructure engineering. I refer the reader to works, such as that of Henderson (1966), for an accessible treatment of oscillatory wave motion and energy in shallow and deep flows. Because oscillatory waves are infrequent in channels, I leave the detailed treatment of oscillatory waves to texts such as that of Henderson (1966). Cutting to the chase, for most engineering applications, Equation 9.14 suffices for computing wave velocity or celerity.

Summary and Conclusions

Unsteady flow is practically solved using software for solving the continuity and momentum equations. We have only considered limited cases of unsteady flow superficially. The overarching goal was to become conversant in terms and approaches used.

The advance of the irrigation waterfront down the furrow is a type of sluice gate opening, a positive wave from upstream. The advance speed is not the speed predicted by Graf and Altinakar (1998) because of the infiltration effect. Bautista et al. (2012) provide public domain software for modeling surface irrigation surges that include infiltration effects.

We have only examined one wave type of multiple types in one of a variety of situations. Waves originating from sluice gate moves provided the context, which was extended to include other applications such as the tidal bore and moving hydraulic jumps. Channel geometry was assumed rectangular and constant; the cross-section may change with distance. One could refer to the Jain (2001) analysis and develop the hydraulic depth relationship for a nonrectangular channel. Channels may bifurcate and rejoin. Graf and Altinakar (1998), Jain (2001), and Henderson (1966) discuss wave meets, which results in reflections up and downstream. Henderson (1966) provides additional introductory coverage of wave mechanics in rivers, streams, and coastal environments. A similar presentation by

Chaudhry (1993) presents a detailed CFD approach that leads to a 2-D analysis. Sturm (2010) gives a brief overview of 3-D analysis. Time and space preclude more detailed coverage of these topics.

The dam-break is another application of rapidly varying unsteady analysis and is of interest because of looming catastrophe should it occur. Chaudhry's (1993) and Sturm's (2010) presentations mentioned above, also address dam-break issues. HEC-RAS software models dam-break and accommodates measured cross-sectional geometry above and below a dam and allows for non-instantaneous but rapid failure or breaching of a dam. The simple analysis we explored helps elaborate on the capabilities of using the characteristics when they are applicable.

Problems and Questions

1   What is the crucial difference between the velocity of a positive downstream surge and the advance curve of a furrow irrigation scenario?

2   The initial depths upstream and downstream of a sluice gate are 4 and 1 m, respectively. The gate opening is suddenly increased to make the downstream depth equal to 1.5 m. Calculate the velocity of the wave downstream. Assume a rectangular channel.

3   Figure the velocity of the wave above the gate situation of problem one.

4   Take the initial conditions of question two. The gate opening width is suddenly decreased so that the downstream depth is equal to 0.6 m. What is the wave velocity?

5   What is the crucial difference between a "positive wave" and "negative wave?"

6   What is "celerity?"

7   A dam retaining water to a depth of 30 m failed suddenly. The bed downstream was dry and smooth. Find the following: (i) speed of the leading edge; (ii) speed of the receding edge; (iii) depth and velocity in a section 10 km downstream of the dam, an hour after the failure.

8   Repeat Problem 7, assuming the channel downstream was 1.5 m deep.

9   A river is flowing at a depth of 3 m and with a velocity of 2 m/s. It meets a tidal bore, which abruptly increases the depth to 4 m. Determine the type of surge. Find the speed with which the bore moves upstream. Then, find the magnitude and direction of the water behind the bore. Use the Jain (2001) analysis.

10   Solve Problem 9 using the Graf and Altinakar (1998) approach.

11   What are "characteristic" equations, and why are they so powerful?

References

Bautista, E., Schiegel, J.L., and Strelkoff, T.S. (2012). WinSRFR 4.1 Software and User's Manual. USDA-ARS U.S. Arid Land Agricultural Research Center, Maricopa, AZ.

Brunner, G.W. (2016). *HEC-RAS River Analysis System: User's Manual*. Davis, CA: US Army Corps of Engineers, Hydrologic Engineering Center.

Chaudhry, M.H. (1993). *Open-Channel Flow*. Englewood Cliffs, NJ: Prentice-Hall.

Chow, V.T. (1959). *Open-Channel Hydraulics*. New York, NY: McGraw-Hill Publishers.

Chow, V.T., Maidment, D.R., and Mays, L.W. (1988). *Applied Hydrology*. New York, NY: McGraw-Hill Publishers.

Graf, W.H. and Altinakar, M.S. (1998). *Fluvial Hydraulics: Flow and Transport Processes in Channels of Simple Geometry*. Chichester, UK: Wiley.

Hall, W.A. (1956). Estimating irrigation border flow. *Agricultural Engineering* 37 (4): 263–265.

Henderson, F.M. (1966). *Open Channel Flow*. New York, NY: Macmillan Publishing Co.

Jain, S.C. (2001). *Open-Channel Hydraulics*. New York, NY: Wiley.

James, L.G. (1988). *Principles of Farm Irrigation System Design*. New York, NY: Wiley.

Strelkoff, T.S. and Clemmens, A.J. (2007). Hydraulics of surface systems. In: *Design and Operation of Farm Irrigation Systems*, 2e (eds. G.J. Hoffman, R.G. Evans, D.L. Martin and R.L. Elliot), 436–498. MI: American Soc. Agr. & Biol. Engrs., St. Joseph.

Sturm, T.W. (2010). *Open Channel Hydraulics*, 2e. New York, NY: McGraw-Hill Publishers.

Walker, W.R. and Skogerboe, G.V. (1987). *Surface Irrigation: Theory and Practice*. Englewood Cliffs, NJ: Prentice-Hall, Inc.

10

Channel Design Emphasizing Fine Sediments and Survey of Alluvial Channel Sediment Transport

This third treatment of earthen channel design builds on the previous discussion of permissible velocity and tractive force. We also complete the discussion of earthen channel design, particularly earthen channels in fine-textured materials.

We refer to the alluvial channel as a natural stream, active bedforms, and sediment transport. Other authors denote the alluvial channel as a "moveable bottom" channel.

We revisit concepts of energy as a contributor to sediment movement. Momentum effects are involved but are beyond the scope of this text. This chapter explores the alluvial or moveable bottom implications on channel hydraulics, sediment transport, and channel design.

Methods ranging from complex to simple are presented for sediment transport calculations. Unfortunately, space does not allow for detailed explanations for the methods. This chapter, unlike the earlier chapters, provides outlines and sketches rather than detailed discussion. The reader should expect to review the respective method sources and accompanying spreadsheets to become sufficiently familiar with the methods.

Goals

- To define an alluvial channel in the context of natural earthen channels.
- To survey which design approaches are best for designing earthen channels in given situations.
- To analyze the incipient motion of sediments and riprap specification for given conditions; in particular, to refine a design criterion based on shear stress for fine sediments.
- To model the settling velocity of sediments found in natural channels.
- To assess the nature and type of alluvial bedforms.
- To model stage–discharge of alluvial channels.
- To model sediment transport in alluvial channels in equilibrium.
- To evaluate the implications of the location in a drainage network on sustainable channel design.
- To briefly survey techniques for predicting sediment movement through stormwater detention ponds and on infrastructure.

Open Channel Design: Fundamentals and Applications, First Edition. Ernest W. Tollner.
© 2022 John Wiley & Sons Ltd. Published 2022 by John Wiley & Sons Ltd.
Companion website: www.wiley.com/go/tollner/openchanneldesign

- To consider alternative empirical channel design methods based on regional channel characteristics.
- To consider a contemporary approach to design a sustainable channel given a prescribed rate of sediment transport.

Alluvial Channel vs. Earthen Channel and Other Preliminaries

An alluvial channel denotes a low-order natural stream (e.g. river) with a broad network of lower-order tributaries, any one or more of which can have substantial sediment-producing activity ongoing given time (see Knighton 1998). The sediment size distribution is like the riverbed material size and would be in rough equilibrium. Sediment delivery versus river stage is approximately constant with time. In higher-order streams where sediment supply would be more episodic, the analysis applies but is a quasi-equilibrium snapshot. Channel aggradation and degradation is in the near term in higher-order streams.

Recall that the permissible velocity threshold values for various earthen materials were based on practitioner surveys. The tractive force approach attempts to bring a more theoretical basis (refer to Figure 4.4). Figure 4.4 shows that the tractive force for gravels and larger particles have reasonable precision. Tractive force values for sands and smaller particles become problematic based on the spread in the experimental data. Techniques stemming from alluvial techniques increase the precision in the sand and smaller size range.

Sediment is rarely uniform but typically is a distribution of sizes. Table 10.1 provides a sediment-grade scale with nomenclature and respective size ranges. Even with a permitted velocity constraint, some fine sediment usually moves. Variations in flow further cause sediments to move. Sediments move in a variety of patterns known as bedforms. The bedform type depends on the nature of the flow. Figure 10.1 shows an array of bedforms possible in alluvial channels. Vanoni (2006) gives a detailed description of bedform types as part of a comprehensive treatment of advanced sedimentation engineering. The types of bedforms correlate to flow conditions and descriptors that are further defined below. The bedform affects the total roughness of the channel. This roughness adds to the roughness imparted by the grain size. The Manning's n is variable in alluvial channels due to the alluvial dynamics occurring on the channel wetted perimeter.

Early Approaches to Sediment Transport

Early approaches to sediment transport exploited the concept of excess shear (Simons and Senturk 1992 cited DuBoys, 1879 in French), as manifested in a series of sliding layers. Simons and Senturk cited Schoklitsch (1914, in German), who proved the sliding layer model deficient but found his data correct. Shields (1936) proposed a similitude model. The dimensionless parameters are shown on the Shields diagram mentioned previously. The Shields work, discussed below, remains useful for predicting incipient motion and revetment sizing. Meyer-Peter–Muller (MPM 1948) presented an approach for bedload transport. Their bedload transport formula remains in use due to its simplicity compared with other methods. The MPM and other work not cited here laid the foundation for the contributions

Table 10.1 Sediment grade scale using the Am. Geophysical Union scale.

Size category	Size range (mm)
Very large boulders	4096–2048
Large boulders	2048–1024
Medium boulders	1024–512
Small boulders	512–256
Large cobbles	256–128
Small cobbles	128–64
Very coarse gravel	64–32
Coarse gravel	32–16
Medium gravel	16–8
Fine gravel	8–4
Very fine gravel	4–2
Very coarse sand	2.0–1.0
Coarse sand	1.0–0.5
Medium sand	0.5–0.25
Fine sand	0.25–0.125
Very fine sand	0.125–0.062
Coarse silt	0.062–0.031
Medium silt	0.031c0.016
Fine silt	0.016–0.008
Very fine silt	0.008–0.004
Coarse clay	0.004–0.002
Medium clay	0.002–0.001
Fine clay	0.001–0.0005
Very fine clay	0.0005–0.00024

of Einstein and others. Einstein recognized the need to quantify bedform effects and used fall velocity as an important sediment parameter. We present Einstein's work in detail and subsequent approaches for bedload and suspended load sediment transport.

Incipient Motion

Incipient motion is related to permissible velocity and tractive force, as discussed in earlier chapters. The motion is incipient if the velocity is at the permissible threshold or the tractive force is critical. The permissible velocity approach and critical tractive force approach imply no sediment movement, which is technically not valid. Hjulstrom (1935) defines velocity thresholds delineating erosion, transport, and settlement regimes. The sediment was uniform and noncohesive. The nonuniformity of sediment particle sizes limits the practical use of Hjulstrom's

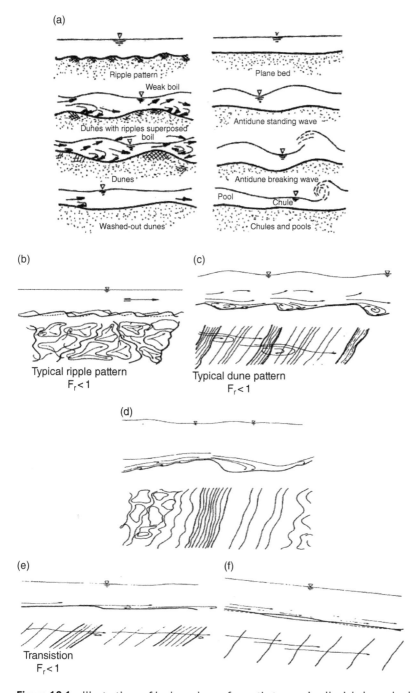

(a)

Ripple pattern

Weak boil

Dunes with ripples superposed

boil

Dunes

Washed-out dunes

Plane bed

Antidune standing wave

Antidune breaking wave

Pool   Chule

Chules and pools

(b)

Typical ripple pattern
$F_r < 1$

(c)

Typical dune pattern
$F_r < 1$

(d)

(e)

Transistion
$F_r < 1$

(f)

Figure 10.1 Illustrations of bed roughness forms that occur in alluvial channels along with accompanying water surface profiles. Case A represents an array of bedforms found in sand channels. Cases B and C represent the lower flow regime bedforms. Case D shows the transition from ripples to dunes. Cases E and F represent a transition from ripples to a plane bed. *Source:* from Simons and Senturk 1992; used with permission of Water Resources Publications.

chart. Yang (1996) summarized several other methods for determining incipient motion. Incipient motion analysis in the following discussion is typically limited to noncohesive particles. Einstein (1950) presented an elegant probabilistic approach, which is discussed below.

Incipient motion analysis typically leads to a discussion of a diagram known as the Shields diagram. Shields (1936) performed a dimensionless analysis to describe the incipient motion of sediment particles. He found the crucial factors to be shear stress, τ; the difference in sediment density and water density, $\rho_s - \rho_f$; particle diameter, d_{50}; kinematic viscosity, ν; and gravitational acceleration, g. These quantities were grouped into two dimensionless parameters shown in Equations 10.1a–d.

$$d\frac{\left(\frac{\tau_c}{\rho_f}\right)^{\frac{1}{2}}}{\nu} = \frac{dU_*}{\nu} \tag{10.1a}$$

$$\tau *_{cr} = \frac{\tau_c}{d\left(\rho_s - \rho_f\right)g} = \frac{\tau_c}{d\gamma_f\left(\frac{\rho_s}{\rho_f} - 1\right)} = \frac{\tau_c}{d\left(\gamma_s - \gamma_f\right)g} \tag{10.1b}$$

In Equations 10.1, τ_c is the critical tractive force at incipient motion and γ is the unit weight of water. Figure 10.2 shows a plot of Equations 10.1, which is known as the Shields diagram. Conditions above the Shields line indicate particle transport, while there is no movement below the line. The scheme describes a variety of sediments well based on the fit of multiple datasets. The right-hand (asymptotic) side of the diagram represents a rough boundary with larger (e.g. gravel) particles. In contrast, the left side describes a laminar boundary with silt-clay particles. The middle transition zone is descriptive of sands.

One difficulty in using the Shields diagram is that the dependent variables (e.g. τ_c or grain size) appear in both the ordinate and abscissa. A third parameter, presented in Vanoni (2006), Equation 10.1c, was developed to determine the grain size, d, and the corresponding shear stress.

$$dsh = \frac{d}{\nu}\sqrt{0.1\left(\frac{\gamma_s}{\gamma_f} - 1\right)gd} \tag{10.1c}$$

Parameter dsh is an auxiliary variable shown on versions (e.g. Figure 10.2) of the Shields diagram. The critical shear stress required to move larger particles understandably increases as one goes to the Shields diagram's right side, and vice versa. The presence of smaller sediment particles can cause effective critical shear stress to be lower than predicted.

Further improvements to the Shields diagram were proposed by Julien (2010). He developed an expression for a nondimensional particle size expressed, as shown in Equation 10.1d.

$$d_s = \left(\frac{\left(\frac{\gamma_s}{\gamma} - 1\right)gd^3}{\nu^2}\right)^{\frac{1}{3}} \tag{10.1d}$$

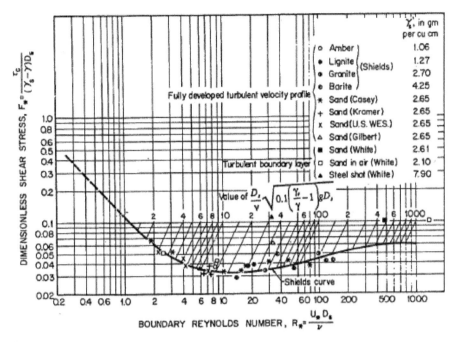

Figure 10.2 The Shields diagram for incipient motion determination. *Source:* from Simons and Senturk 1992; used with permission of Water Resources Publications.

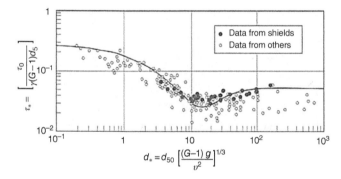

Figure 10.3 The alternate form of the Shields diagram for direct determination of critical shear stress. *Source:* from Julien 2010; used with permission of Cambridge University Press.

Julien (2010) plotted the nondimensional particle size d_s versus the nondimensional critical tractive force, τ_c, resulting in an alternative Shields diagram (Figure 10.3). This modification effectively decouples the grain size and critical tractive forces, making the alternative chart more straightforward to use.

The alternative Shields diagram lends to designing channels where sediment movement must be near zero. A fish raceway in an aquacultural design is an example of the Shields (or modified) diagram used in design work. The modified Shields diagram lends to riprap or revetment design.

Example 10.1 (Adapted from Graf and Altinakar 1998)

A channel is being constructed with a slope imposed by the terrain of $S_f = 0.01$. The channel conveys $30\,\text{m}^3/\text{s}$ at a temperature of 14 C without causing any erosion. A grain size analysis indicates a noncohesive soil with $d_{50} = 50\,\text{mm} = 0.05\,\text{m}$ and a specific gravity of 2.65. The angle of repose of this material is $\phi = 37°$. The manning n is estimated to be 0.025.

For water at 14 C, the density is $999.1\,\text{kg/m}^3 \approx 1000\,\text{kg/m}^3$, and kinematic viscosity is $1.186 \times 10^{-6}\,\text{m}^3/\text{s}$. The density of the sediment is $2650\,\text{kg/m}^3$.

Next, compute the dimensionless particle size using Equation 10.1d.

$$d_s = \left(\frac{\left(\frac{\gamma_s}{\gamma} - 1\right)gd^3}{\nu^2}\right)^{\frac{1}{3}} = \left(\frac{\left(\frac{2650}{1000} - 1\right)9.81 \times 0.05^3}{\left(1.186 \times 10^{-6}\right)^2}\right)^{\frac{1}{3}} = 1121$$

From Figure 10.3, the nondimensional critical tractive force is $\tau_c = 0.055$.

From Equation 10.1b

$$\tau_c = \tau *_{cr} g\left(\rho_x - \rho\right)d_{50}$$

One can write the critical tractive force as $\tau_v = g\rho R_h S_f$. Equating this expression with the previous expression results in the following.

$$R_h = \frac{\tau *_{cr} d_{50}}{S_f}\left(\frac{\rho_s - \rho_f}{\rho_f}\right) = \frac{0.055(.05)}{0.01}\left(\frac{2650 - 1000}{1000}\right) = 0.45$$

One may now use the Manning equation to compute the velocity.

$$\text{A.} \rightarrow V = \frac{1}{n}R_h^{0.67}S^{0.5} = \frac{1}{0.025}0.45^{0.67}0.01^{0.5} = 2.34\,\text{m/s}$$

The required cross-sectional area is $A = Q/V = 30/2.34 = 12.82\,\text{m}^2$. The wetted perimeter is $A/R_h = 12.82/0.45 = 28.49\,\text{m}$. A wide channel approximation would lead to depth $= R_h = 0.45$. Then, wetted perimeter $28.49 = 0.45\,(2) + \text{Width}$, leading to Width $= 27.6\,\text{m}$. This easily meets the wide channel approximation. Graf and Altinakar (1998) also used the permissible velocity method ($V_c = 2.5\,\text{m/s}$) and found the width to be $23.1\,\text{m}$ and depth $= 0.52\,\text{m}$. Add freeboard to these values using methods pertinent to the cross section as was done in earlier chapters.

The sediment transport folder contains a spreadsheet entitled "IncipientMotion WideChannel." One may use this spreadsheet to compute the critical velocity from the normal depth calculated using permissible velocity or tractive force software. "IncipientMotionWideChannel" calculates the critical velocity using a known particle size distribution and normal depth. Suppose the critical velocity is less than the velocity at normal depth. In that case, the channel is not sustainable according to the incipient motion analysis. In this case, rerun the permissible velocity or tractive force analysis using more stringent permissible velocity or tractive force values. Suppose the critical velocity is near equal or greater than the velocity computed from the design methods. In that case, the channel should be sustainable as designed. The "IncipientMotionWideChannel" sheet has a page for wide channel design at the critical velocity for sand or gravel materials.

Let us revisit a fundamental question. Of the three approaches to earthen or alluvial channel design discussed, what method should one consider? If precision in inputs is lacking, as it frequently is, the permissible velocity method is adequate. If one has a defined particle size (e.g. an example of more precision in known inputs) larger than sand, tractive force works. Approaches based on the Shields (or modified Shields) diagram suffice. If one has defined sand, silt, or clay-sized particles, the Shields diagram is the best approach. We further discuss the Shields and related approaches below.

Riprap or Revetment Specification

The modified Shields diagram lends to a straightforward approach to specifying riprap size. One may assume a nondimensional particle size d_s to be in the fully rough condition without computing d_s. The allowable d_{50} is computed from the nondimensional tractive force read from the ordinate of the modified Shields diagram (Figure 10.3). We also compare this value with the tractive force associated with Lane (1953) as computed in Chapter 4 (refer to Figure 4.1) using the critical tractive force vs. particle size in Figure 4.4a. To meet a flow target (per meter of width), we iterate to find the depth required to satisfy the design flow. The Manning n is computed using a simple Strickler-type formula. "RevetmentSizing" in the "SedimentTransport" folder makes these computations. The spreadsheet "BagnoldBrownlieWilcockKenworthy" contains two pages that employ Shields-based design approaches for channel design.

Example 10.2

A wide channel with a side slope of 20° using gravel with an angle of repose of 35°. The specific gravity of the gravel is 2.65. The slope is 0.001. Find the gravel size and depth meeting the Shields criteria that transports $3\,m^3/s$-m.

Enter the above information into the spreadsheet "RevetmentSizing." Use the solver to find that the depth is 1.62 m and the $d_{50} = 20$ mm. The Shields criteria tractive force is slightly below the Lane (1953) approach, which is desired for a conservative solution.

A channel that is not "wide" requires a solution of depth as a function of R and other hydraulic elements, depending on the cross section. Mathematica notebooks, "DepthFromHydraulicRadiusRectangle.nb," "DepthFromHydraulicRadiusTrapezoid.nb," and "DepthFromHydraulicRadiusParabola.nb" compute the depth from a known hydraulic radius for the respective cross sections. Add the freeboard as above. One would modify the "RevetmentSizing" spreadsheet to have a set width, total flow, and modified Manning equation (based on total area) to compute the R. Then, use the above Mathematica notebooks to get the channel depth (see the exercises at the end of the chapter). One may compare the details with the IncipientMotionWideChannel (gravel material).

The Shields approach to specifying particle size for riprap (revetment) is a logical extension of techniques presented in Chapter 4. One should compare designs presented by those methods with designs based upon the Shields-type approaches.

The above analysis does not account for fine particles smaller than d_{50}. Yang (1996) describes a process called armoring. Channel armoring leaves larger particles behind while removing smaller particles. Armoring is a natural process that occurs with alluvial channels with noncohesive sediments. Incipient motion analysis, while implying zero sediment transport, should be viewed as a near-zero transport. Richardson et al. (2001) give an additional discussion of revetment design specific to bridge piers.

Bedform Descriptions and Analysis

Modeled or measured stage–discharge relationships are necessary for computing sediment transport. Strategies for adding the roughness component to the grain component towards a stage–discharge relation involve starting with a grain roughness and adding a form roughness due to bedforms. Partitioning may appear as a hydraulic radius due to grains and the additional part due to bedforms. Table 10.2 summarizes strategies used by a variety of

Table 10.2 Roughness partitions used to compute stage–discharge relationships in alluvial channels.

No.	Grain and bedform partition	The linkage between the grain roughness and bedform roughness[a]	Comments	Investigator
1	$R = R' + R''$		Hydraulic radius	Einstein (1950) and colleagues
2	n = loosely related to $n' + n''$	$n'' = d^{1/6}/\text{constant}$ n depends on bedform type and sediment size.	Manning roughness. Grain roughness as a function of sediment size $d^{1/6}$. N is a function of bedform type and sediment size.	Simons and Richardson (1971)
3	$S = S' + S''$	Graphical relationship between τ_*' sand τ_*, depending on the bedform type.	The slope then applies similarity to hypothesize grain tractive force to total tractive force	Engelund (1966, 1967)
4	$k_s = k_s' + k_s''$	$k_s' = 3d_{90}$ $ks'' = 1.1\Delta\left(1 - e^{-\frac{25\Delta}{\lambda}}\right)$	Equivalent sand grain roughness where Δ/λ is the steepness of the bedforms and Δ is the bedform height.	Van Rijn (1984)
5	$f = f/f_0$	$f_0 = \dfrac{8}{5.75\text{Log}\dfrac{12y_0}{2.5d_{50}}}$ $\dfrac{f}{f_0} = 1.2 + 8.92\dfrac{\Delta}{y_0}$ $\dfrac{\Delta}{y_0} = f(\text{Shields})$	Friction factor approach	Karim and Kennedy (1990)

[a] Note that the single prime represents the grain component, and the double prime represents the bedform contribution.

investigators. We focus on the Einstein (1950) method and the Van Rijn (1984) methods for modeling stage–discharge in alluvial channels below. Bedform modeling has improved between the work of Einstein (1950), Van Rijn (1984), and Vanoni (2006), with further improvements needed.

Sediment transport is approached by partitioning between bedload and suspended load. Einstein (1950) and Van Rijn (1984), discussed below, follow the partitioning approach. Einstein (1950) uses an elegant probabilistic approach (presented below) to describe sediment particles' motion. Others use more mundane (but more straightforward) regression approaches based on variables such as discharge, velocity, energy slope, tractive force, and stream power (product of velocity and tractive force) to describe total sediment load. We revisit regression approaches in more detail below. The Einstein transport parameters may sometimes be adopted to total load with success. Barfield et al. (1977) used the parameters to fit total load transport. Yang (1996) summarizes older historical methods of sediment transport prediction. We discuss Laursen (1958), and Graf (1971), Bagnold (1966 and modifications in 1980), Brownlie (1981), and Wilcock and Kenworthy (2002) as exemplars of the regression approach. The later regression methods are the methods of choice for routine sediment transport estimation.

Sediment Fall Velocity

Fall velocity is a vital descriptor regarding sediment transport. Table 10.1 provides nomenclature for particles of diverse sizes. The fall velocity derives from a balance between the drag force as the particle moves, counterbalanced by the particle's buoyant weight, as shown in Equation 10.2.

$$C_d \frac{\rho A_f w_f^2}{2} = \frac{(\gamma_s - \gamma)(\pi d^3)}{6}$$
(10.2)

The nomenclature of Equation 10.2 is as follows:

C_d = the drag coefficient (see Figure 10.4a),
ρ = the fluid density (M/L^3),
A_f = the projected area (L^2),
w_f = the particle fall velocity (L/T),
γ_s = the sediment unit weight (F/L^3), and
γ = the unit weight of water, and d is the particle diameter (L).

Equation 10.2 can be simplified and solved for fall velocity, as shown in Equation 10.3.

$$w_f = \sqrt{\frac{\frac{4}{3}\left(\frac{\gamma_s}{\gamma} - 1\right)gd}{C_d}}$$
(10.3)

(a)

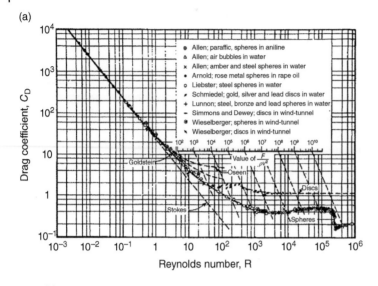

(b)

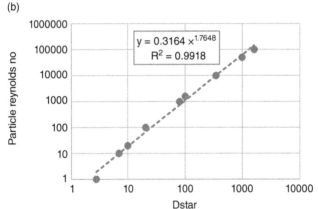

Figure 10.4 Fall velocity relationships: (a) Drag coefficient vs. particle Reynolds number. *Source:* From Simons and Senturk 1992; used with permission of Water Resources Publications. (b) A plot of the dimensionless particle diameter vs. particle Reynolds number at a temperature of 24 C. *Source:* Adapted based on Sturm (2010).

The drag coefficient is predictable in the laminar range. The parameter $C_d = 24/Re$, where Re is the particle Reynolds number ($Re = w_f \, d/v$ with v defined as the kinematic viscosity, L^2/T). Substituting the definition for C_d and Re into Equation 10.2 leads to the Stokes formula for fall velocity shown in Equation 10.4.

$$w_f = \frac{\left(\dfrac{\gamma_f}{\gamma} - 1\right) g d^2}{18v} \tag{10.4}$$

(c)

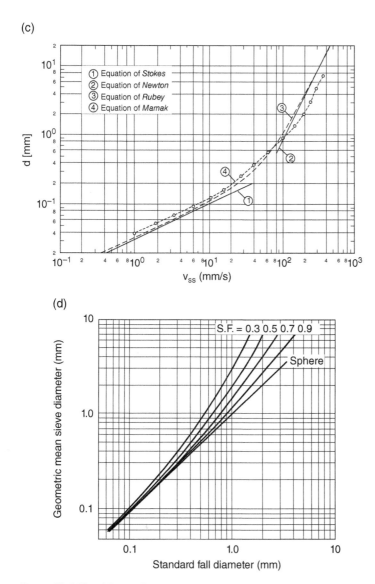

(d)

Figure 10.4 (Cont'd) (c) Selected equations for settling velocity at approximately 20 C. *Source:* From Graf and Altinakar 1998; used with permission of John Wiley & Sons. (d) Correction for particle shape depending on the shape factor. *Source:* From the US Interagency Committee 1957.

The Stokes range includes medium silt and clay particles. Outside the Stokes range, one can solve for fall velocity by iteration using Figure 10.4a. Sands generally lie within the transition range of Figure 10.4a. White (2005) writes a best-fit relationship between drag coefficient and particle Reynolds number shown in Equation 10.5 that lends to a numerical iteration.

$$C_d = \frac{24}{Re} + \frac{6}{1+\sqrt{Re}} + 0.4 \tag{10.5}$$

An alternative to an iterative solution is available. The solution uses a creative dimensionless analysis. Equation 10.6 shows one approach, which effectively eliminates the fall velocity.

$$\frac{3}{4} C_d Re^2 = \frac{\left(\frac{\gamma_s}{\gamma} - 1\right) g d^3}{v^2} \tag{10.6}$$

One may further define a dimensionless particle size d_s, previously given as Equation 10.1d, repeated below.

$$d_s = \left(\frac{\left(\frac{\gamma_s}{\gamma} - 1\right) g d^3}{v^2}\right)^{\frac{1}{3}} \tag{10.1d}$$

Figure 10.4b shows a relationship between d_s and particle Reynolds number. One can compute d_s, read the Re from Figure 10.4b, then back out w_f from the particle Reynolds number.

Other approaches include Engelund and Hansen (1967) for sand (Specific gravity = 2.65), using a simplified expression for C_d shown in Equation 10.7.

$$C_d = \frac{24}{Re} + 1.5 \tag{10.7}$$

Substituting Equation 10.7 into Equation 10.3 leads to Equation 10.8, from which one can solve for w_f.

$$Re = \frac{w_f d_s}{v} = 8\left(\sqrt{1+0.0139 d_*^3} - 1\right) \tag{10.8}$$

Rubey (1933) presented a computation for w_f shown in Equations 10.9a and b, using an expression F for dealing with larger particles' turbulent flow.

$$w_f = F\left[dg\left(\frac{\gamma_s}{\gamma} - 1\right)\right]^{\frac{1}{2}} \tag{10.9a}$$

$F = 0.79$ if $d < 0,001$ m. if $d > 0.001$ m

$$F = \left[\frac{2}{3} + \frac{36v^2}{gd^3\left(\frac{\gamma_s}{\gamma}-1\right)}\right]^{\frac{1}{2}} - \left[\frac{36v^2}{gd^3\left(\frac{\gamma_s}{\gamma}-1\right)}\right]^{\frac{1}{2}} \qquad (10.9b)$$

This formula and other related formulas are plotted in Figure 10.4c.

C_d is independent of Reynolds number for large particles (larger sands, gravels, and beyond). Yang (1996) proposed approximating w_f with a simple function of particle size (specific gravity of 2.65), as shown in Equation 10.10.

$$w_f = 3.32d^{\frac{1}{2}}\left(m/s, d > 0.002m\right) \qquad (10.10)$$

The above relationships assume a spherical particle and temperatures around 15–20 C. The kinematic viscosity of water is $\approx$ constant in this range. Particles are rarely spherical. A shape factor computed using the orthogonal dimensions a, b, and c is suggested, as shown in Equation 10.11.

$$S.F. = \frac{a}{\sqrt{bc}} \qquad (10.11)$$

An interagency committee proposed correcting the mean sieve diameter to a standard fall velocity using Figure 10.4d. These fall diameters are used in the above relationships to arrive at the effective fall velocity w_f.

The spreadsheet "FallVelocity" contains relationships for fall velocity in SI units. Fall velocity computed in imperial units using methods discussed above is used in the spreadsheet, "GrafLaursenEinsteinTotalLoad." Both spreadsheets are contained in the Excel "SedimentTransport" directory. The later spreadsheet is discussed in detail in the sediment transport section below.

Key assumptions contained in the above equations involve specific gravity and temperature. The specific gravity is 2.65, and the temperature is in the range of 10–20 C. One can adjust specific gravity where this term is explicit, and one can adapt temperature via the kinematic viscosity. The spreadsheet utilizes Figures 10.4b–d via nonlinear regression fits for these respective relationships.

A Probabilistic Approach to Sediment Transport

On an individual particle basis, the likelihood of moving is probabilistic. The definition of crucial transport parameters in a probability sense provides an insight into a most elegant theory which supports a practical objective. Einstein (1950) observed that particles randomly lift, then move a distance before coming to rest again. Some particles move a considerable distance before coming to rest, while larger particles barely move at all if they move. Einstein (1950) provides such an elegant approach to describing sediment saltation. The presentation below follows that provided by Graf (1971). The exchange probability p is a function of the ratio of effective particle weight/hydrodynamic lift, as shown in Equation 10.12.

$$1 > \frac{k_2\left(\rho_s - \rho\right)gd^3}{0.5C_L\rho k_1 d^2 u_b^2\left(1+\eta\right)}$$

(10.12)

In Equation 10.12, C_L is the lift coefficient $= 0.178$,

d is the sediment diameter,

u_b is the velocity at the bed,

k_1 and k_2 are scaling constants,

ρ_s is the sediment density,

ρ is the water density. The velocity u_b (velocity near the channel bottom) is given by Equation 10.13.

$$u_b = u_* 5.75 \text{Log} \left[\frac{30.2 * 0.35X}{\Delta} \right]$$

(10.13)

Δ is k_s/x, where k_s is the d_{65} of the sediment. X is read from the log velocity correction distribution (see Figure 10.5a), enabling u_b to be calculated rough, smooth, and transition boundary conditions.

The value of k_s may seem arbitrary. Sturm (2010) summarizes the results of multiple investigators showing that k_s has been taken as d_{65} (sand) to $2d_{65}$ (sand and gravel) to $3.5d_{84}$ (coarse sand and gravel).

u_* is $\sqrt{(gRS)}$ is for computing, $\delta = \frac{11.5v}{u_*}$, which is divided into k_s resulting in k_s/δ, the horizontal axis of the pressure correction curve shown in Figure 10.5b.

Substituting Equation 10.14a into Equation 10.12 results in Equation 10.13.

$$1 > \frac{1}{1+\eta}\left(\frac{\rho_s - \rho}{\rho} \frac{d}{R_h' S} \right)\left[\frac{2k_2}{\left(0.178k_1\right)\left(5.75\right)^2} \right]\left[\frac{1}{\text{Log}^2\left(\frac{10.6X}{\Delta} \right)} \right]$$

(10.14a)

We are introducing abbreviations that simplify Equation 10.14a to Equations 10.14b–e.

$$1 > \frac{1}{1+\eta}\Psi B \beta_x^{-2}$$

(10.14b)

$$\Psi = \left(\frac{\rho_s - \rho}{\rho} \frac{d}{R_h' S} \right)$$

(10.14c)

$$B = \left[\frac{2k_2}{\left(0.178k_1\right)\left(5.75\right)^2} \right]$$

(10.14d)

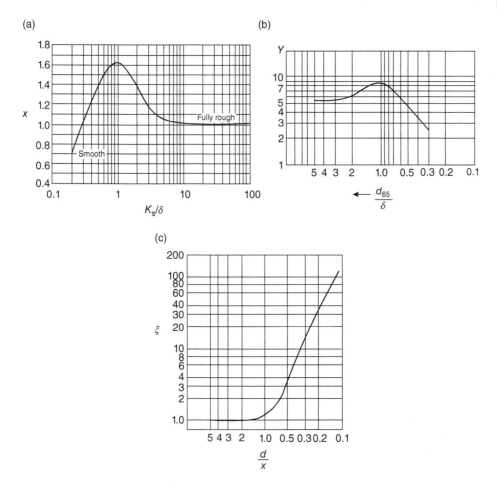

Figure 10.5 Factors used by the Einstein method for computing sediment load: (a) Velocity correction factor, x; (b) Correction factor, y; and (c) Hiding factor. *Source:* After Einstein (1950).

$$\beta_x^{-2} = \left[\frac{1}{\text{Log}^2\left(\dfrac{10.6X}{\Delta}\right)}\right] \tag{10.14e}$$

We introduce a hiding factor for smaller particles partially hidden by larger particles. Likewise, we present a pressure correction for the lift factor that adjusts the lift coefficient for particles in mixtures with different roughness. Figure 10.5 shows these correction factors. Einstein (1950) determined these corrections with the caveat that further improvement was needed. Equation 10.15 is further simplified by introducing the hiding and pressure corrections in Figure 10.5 along with other simplifications as follows in Equation 10.15.

$$\left|1+\eta\right| > \xi Y B' \frac{\beta^2}{\beta_x^2} \Psi \tag{10.15}$$

ξ is the hiding factor read from Figure 10.5c.

Y is the pressure correction read from the pressure correction graph based on d_{65}.

$$B' = B / \beta^2 \text{ where } \beta = \text{Log} \tag{10.16}$$

Note that lift is always positive, but η may be positive or negative, which explains why the absolute value appears on the left-hand side of Equation 10.15. We can write Equation 10.15 as Equation 10.16a after introducing additional simplifications.

$$\left(\frac{1}{\eta_0} + \eta_*\right)^2 > B_*^2 \Psi_*^2 \tag{10.16a}$$

$B_* = B' / \eta_0 = 0.143$ (experimentally determined)

$$\Psi_* = \epsilon Y \left(\frac{\beta^2}{\beta_x^2}\right) \Psi$$

$\eta_0 = 0.5$ based on data analyzed by Einstein (1950)

$\eta_* = \eta / \eta_0$

One may write the limiting case for motion as Equation 10.17.

$$\left(\eta_*\right)_{\text{Lim}} = \pm B_* \Psi_* - \frac{1}{\eta_0} \tag{10.17}$$

Given that η can be described as a random normal variate, we can write an expression for the probability of particle movement as a function of ψ shown in Equation 10.18a.

$$p\psi = 1 - \frac{1}{\sqrt{\pi}} \int_{-B_* \Psi_* - \frac{1}{\eta_0}}^{+B_* \Psi_* - \frac{1}{\eta_0}} e^{-t^2} \tag{10.18a}$$

When integrated, Equation 10.18a appears as Equation 10.18b.

$$p\Psi = 1 - \sqrt{\pi} / 2\left(-\text{Erf}\left(2 - 0.143\Psi\right) + \text{Erf}\left(2 + 0.143\Psi\right)\right) \tag{10.18b}$$

Einstein (1942) developed a bedload equation that appears in Equation 10.19.

$$\frac{g_s i_s}{A_L k_2 \gamma_s d^4} = \frac{i_b p}{k_1 k_3 d^2} \sqrt{\frac{g(\rho_s - \rho)}{d\rho}} \tag{10.19}$$

In Equation 10.19, i_s the fraction of suspended load by size fraction,

I_b is the fraction of suspended load by size fraction, and

g_s is the suspended load rate in weight per unit time.

The probability p is the fraction of time during which, at any one spot, the instantaneous lift exceeds the weight of the particle. If transport is not intensive, the probability of erosion p is small. Einstein (1950) interpreted that p relates to the distance $A_L d$ which a particle travels between consecutive resting places. If p is small, travel distance is virtually constant, amounting to a single step length λ. Einstein found that λ (dimensionless) was about 100. If p is large, $p(1-p)$ particles are deposited after traveling $2*\lambda$ while p^2 particles continue in motion, and so on. One can write this probability relation as an infinite series, as shown in Equation 10.20:

$$A_L d = \sum_{n=0}^{\infty}(1-p)p^n(1+1)\lambda = \frac{\lambda}{1-p} \tag{10.20}$$

Introducing Equation 10.20 into Equation 10.19, and separating p (and denoting it PΦ) on the left and side results in Equation 10.21.

$$\frac{p(\Phi)}{1-p(\Phi)} = \left(\frac{k_1 k_3}{k_2 \lambda_b}\right)\left(\frac{i_s}{i_b}\right)\left(\frac{i_s}{i_b}\right)\left(\frac{g_s}{\gamma_s}\right)\sqrt{\frac{\rho}{\rho_s-\rho}}\sqrt{\frac{1}{gd^3}} \tag{10.21}$$

Writing $\Phi = \left(\dfrac{g_s}{\gamma_s}\right)\sqrt{\dfrac{\rho}{\rho_s-\rho}}\sqrt{\dfrac{1}{gd^3}}$ and A$_*$ $= \left(\dfrac{k_1 k_3}{k_2 \lambda_b}\right)$, Equation 10.21 may be rearranged, as shown in Equation 10.22.

$$\frac{p(\Phi)}{1-p(\Phi)} = A_*\left(\frac{i_s}{i_b}\right)\Phi = A_*\Phi_* \tag{10.22}$$

Equation 10.22 enables a solution to the integral in Equation 10.18 for a value of p as a function of sediment size and flow conditions. Einstein (1950) provided Equation 10.23 for an estimation of the probability of transport.

$$p(\Phi) = \frac{A_*\Phi_*}{1+A_*\Phi_*} \tag{10.23}$$

Using experimental data of several investigators, Einstein estimated that A$_*$ is 43.5. We calculate $p(\Phi)$ and $p(\psi)$ values in several flow conditions and sediment sizes in the spreadsheet "GrafLaursenEinsteinTotalLoad" in the downloads. We further discuss the spreadsheet in the stage–discharge section below, and we refer to the probability discussion while there. Figure 10.6 presents an experimentally determined relationship between Φ_* and ψ_*.

Gessler (1965), in a study of channel armoring, presented data demonstrating that the likelihood of particle movement was related to the ratio of the critical to the actual tractive force acting on the particle. Buffington and Montgomery (1997) present data compiled from eight decades of incipient motion studies. They compared, calculated, and observed critical tractive force values stratified by initial motion definition. Median grain size type (surface, subsurface, or lab mixture), relative roughness, and flow regime were figured into the analysis. They constructed a traditional Shields plot from data representing the bed's initial motion critical tractive force values. The resulting graph revealed uncertainty of

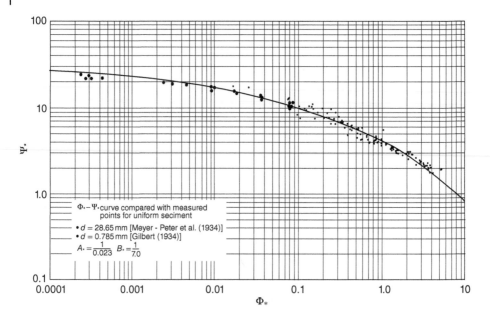

Figure 10.6 Einstein Φ_* vs. Ψ_* plot. *Source:* From Einstein (1950).

what constitutes incipient motion in sediment mixtures. Yang (1996) recast Gessler's data from the likelihood of the probability of movement. The probability function began with a 50% chance of moving at the critical tractive force.

The probability increases as the actual to critical tractive force ratio exceeds one. Yang (1996) found that the particle has a 95% chance of moving when the actual/critical tractive force ratio reached two. There is some chance of movement when the actual to critical tractive force ratio is less than one. Fluid turbulence gives rise to the probability of sediment particle lift and saltation transport. Einstein's pioneering work (1950 and others) undoubtedly influenced subsequent work, such as that presented by Gessler (1965). The fluid turbulence is probabilistic and correlated to the mixing length discussed in Chapter 4. The turbulence explains the probabilistic nature modeled by Einstein (1950) and the experimental findings of Gessler (1965) and Yang (1996).

The Anon (2007) summarized results showing that in coarse beds with a wide range of sizes (especially sand–gravel mixtures). The fines may begin to move at flows much smaller than flows required to move the coarse grains. If the probability of movement was greater than 0.65, the channel was deemed unstable. The Einstein (1950) insight that sediment moves according to probability lives on.

Einstein (1950)–Laursen (1958)–Graf (1971) Stage–Discharge and Other Hydraulic Calculations

The spreadsheet "GrafLaursenEinstein" begins by assigning a hydraulic radius R' based on the grain roughness. The spreadsheet computes values for a range of hydraulic radii.

Calculate u'_* associated with the grain roughness. The thickness of the laminar sublayer (δ) can be calculated with Equation 10.24.

$$\delta = \frac{11.6v}{u'_*} \tag{10.24}$$

The kinematic viscosity (ν) is 10^{-5} ft^2/s. Now, one can calculate the mean velocity near the bottom using Equation 10.13. The sediment transport intensity factor (ψ) may be calculated using Equation 4a. To get to the bedform contribution, Einstein and Barbarossa (1952) related the ratio of velocity over u_*'' (due to bedforms) to the shear transport parameter Ψ in Figure 10.7. Figure 10.7 does not convey knowledge of bedform type, although the right-hand side is said to be antidunes. One can back out u_*'' from this relationship and calculate the R″ due to bedforms. One may then compute the R $=$ R′ + R″ and compute the overall u_* friction velocity. Then, get the overall depth, area, wetted perimeter, and Q. We now have a stage–discharge point. The spreadsheet "GrafLaursenEinstein" makes these computations. Note that the graphical relations are included via curve fits performed using nonlinear regression techniques.

Some additional terms in the hydraulic calculations are included in the "GrafLaursenEinstein" spreadsheet involved in the Einstein bed and suspended load calculations. They include X, a characteristic distance $= 0.77$ ks/x if $k_s\delta/x > 1.8$ and X $= 1.39\delta$ if $k_s\delta/x < 1.8$. Experimental data

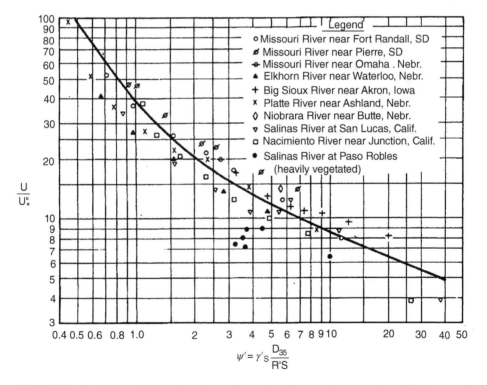

Figure 10.7 Einstein's relationship between the ratio of the velocity/grain-based friction velocity and the sediment transport parameter Ψ'_s. *Source:* Adapted from Einstein and Barbarossa (1952).

support these relationships. The pressure correction term Y (see Figure 10.5b) discussed in the probability section is tabulated. A correction term β_x is computed $= \log (10.6 \, X/(k_s/x))$. The β term (see Equation 10.15) provides an adjustment to the sediment diffusion parameter related to suspended load transport discussed below. Finally, the Einstein transport parameter P_E is computed using Equation 10.25, which is part of the suspended solids determination discussed below.

$$P_E = \frac{1}{0.434} \log \left(\frac{30.2D}{\dfrac{k_s}{x}} \right) \tag{10.25}$$

Einstein (1950) computed bedload based on computations for multiple sediment sizes. The bedload is calculated for each diameter, then totaled over all the particle sizes. One calculates the adjusted shear transport intensity Ψ^*, which is modified from Ψ by the parting hiding factor ξ (see Figure 10.5c), the β ratio, and the pressure correction Y (see Figure 10.5b). These adjustments all relate to the notion that larger particles partially obscure smaller particles. One may then compute the bedload for the size fraction (i_s) from the transport function Φ (see Figure 10.6).

The suspended load transport stems from a balance between the fall velocity versus the turbulent intensity in the flow. A Fickian convection–diffusion model provided the basis for the Einstein (1950) analysis. The gradient of sediment concentration with depth times a diffusion coefficient (ε) is proportional to the turbulent shear stress in the flow, which is counterbalanced by the particle fall velocity (w_f), as shown in Equation 10.26.

$$\overline{w'c'} = -\epsilon_s \frac{dC}{dz} = w_f C \tag{10.26}$$

Compare the turbulence's discussion generated by the mixing length hypothesis (Chapter 5) with the left-hand side of Equation 10.26. Further, the diffusion coefficient is not constant. Still, it is linearly proportional to the shear in the flow, which is linearly proportional to the depth, resulting in Equation 10.27.

$$\tau = \rho \varepsilon \frac{du}{dz} = \frac{u_*}{\kappa z} \tag{10.27}$$

In Equation 10.27, u is velocity, k is the von Karman coefficient (usually 0.4), with other terms being previously defined. The diffusion coefficient may be expressed, as shown in Equation 10.28.

$$\epsilon_s = \beta \kappa u_* \frac{z}{y_o \left(y_o - z \right)} \tag{10.28}$$

Equation 10.28 is a parabolic equation giving maximum sediment concentration at mid-depth.

Substituting Equation 10.28 into the right two terms of Equation 10.26 and then integrating and simplifying results in sediment concentration as a function of depth, as shown in Equation 10.29, is referred to as the Rouse equation.

$$\frac{C}{C_a} = \left[\frac{(y_o - z)}{z} - \frac{a}{(y_o - a)}\right]^{R_o}$$

(10.29)

$$R_o = \frac{w_f}{\beta \kappa u_*}$$

(10.29a)

The Rouse number (R_o) uses a $\beta = 1$ in the Einstein approach. The suspended load for a given size class follows Equation 10.33, where we use Equations 10.29.

$$g_{ss} = \int_a^{y_o} Cu\,dy = \int_a^{y_o} C_a \left[\frac{(y_o - z)}{z} - \frac{a}{(y_o - a)}\right]^{R_o} 5.75 u'_* \log\left(\frac{30.2y}{k_s} \atop x\right) dy$$

(10.30)

Equation 10.30 is a formidable integral. Einstein (1950) substituted $A_E = a/y_o$ and further rearranged Equation 10.33 to parse it into two integrals, as shown in Equation 10.31.

$$g_{ss} = 5.75 C_a u'_* y_o \left(\frac{A_E}{1 - A_E}\right)^{R_o}\left[\log\left(\frac{30.2y_o}{k_s} \atop x\right)\int_{A_E}^1 \left(\frac{1-y}{y}\right)^{R_o} dy + 0.434 \int_{A_E}^1 \left(\frac{1-y}{y}\right)^{R_o} \ln y\,dy\right]$$

(10.31)

$$I_1 \dots I_2$$

Integrals I_1 and I_2 were integrated numerically by Einstein (1950). These integrals were evaluated using a Mathematica program notebook, "EinsteinSedimentIntegrals." A non-linear regression program was employed to fit the integrals' values for a range of R_o and A_E. The reference sediment concentration C_a is defined in Equation 10.32.

$$C_a = \frac{1}{11.6} \frac{i_s g_s}{u'_* a'}$$

(10.32)

Note that the reference concentration is related to the bedload transport rate. The reference depth a' is set to be 2× the sediment size d. As before, constants are experimentally determined.

The spreadsheet then evaluated the suspended sediment load using Equation 10.33.

$$i_{ss} g_{ss} = i_{ss} g_{ss}\left(P_E I_1 + I_2\right)$$

(10.33)

The suspended and bedload components are added together for the total bed material load.

The transport probabilities P (Ψ_*) and P (Φ_*) are computed and shown with the Einstein results section of the spreadsheet. The agreement is remarkable. The Φ vs. ψ diagram (Figure 10.6) can be regarded as the probabilities of sediment movement. We systematically evaluate this spreadsheet in the next section. Figure 10.6, discussed above, shows the relationship between Ψ_* and Φ_*. Variations of this relationship have appeared in several sediment transport models.

Using the "GrafLaursenEinstein" spreadsheet, one can compare sediment transport under conditions where the tabulated critical and actual tractive force (based on $\sqrt{\gamma RS}$) are given. The permissible velocity and actual velocity are aligned near the alignment of the tractive forces. Sediment transport is near but not zero when velocity and tractive forces are near critical.

The Laursen (1958) approach uses the Einstein (1950) method for computing stage–discharge. He then computed total load and bedload for size fractions using an empirical relationship between u_*/w_s and the total load parameter and the bed load parameter shown in Figure 10.8. The respective empirical functions are then converted to a concentration using an equation said to be "more intuitive than rational," as shown in Equation 10.34.

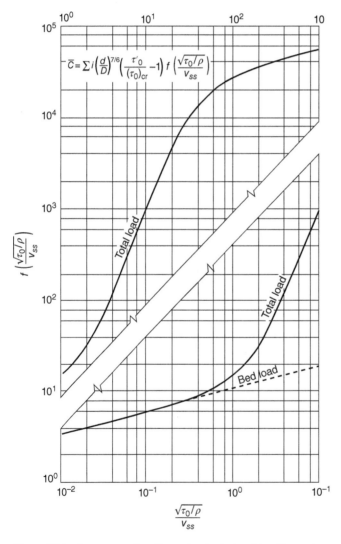

Figure 10.8 Laursen total sediment load relationships. *Source:* Laursen 1958; used with permission of American Society of Civil Engineers (ASCE).

$$\overline{\%C} = \sum i \left(\frac{d}{y_0}\right)^{\frac{7}{6}} \left[\frac{\tau_0'}{(\tau_0)cr} - 1\right] fct\left(\frac{u_*}{w_f}\right) \tag{10.34}$$

In Equation 10.34, $\overline{\%C}$ is the volumetric sediment concentration (%) and the $fct\left(\dfrac{u_*}{w_f}\right)$ applies to both bedload and total load. The sediment transport in terms of weight per unit time may be computed from Equation 10.35.

$$q_s = q\overline{\%C}\,/\,265 \tag{10.35}$$

Equation 10.35 implies that the specific gravity of the sediment is 2.65. The term q is the water flow rate in terms of weight per time. The Laursen (1958) approach seems to provide reasonable results.

Graf (1971) also uses the Einstein (1950) stage–discharge computation. He relates a modified dimensionless transport parameter (see Equation 10.21) to the sheer intensity parameter (see Equation 4a). The Graf approach applies the modified transport parameter to predict a concentration C that reflects the size class's total sediment load. The modified transport parameter is given by Equation 10.36.

$$\Phi_A = \frac{\overline{Cu}R}{\sqrt{\left[\dfrac{\rho_s - \rho}{\rho}\right]gd_{50}^3}} \tag{10.36}$$

Nonlinear regression was used on data from several investigators to produce a relation between the transport parameter Φ_A and shear intensity factor ψ. A fundamental assumption was that the combined bed and suspended load by Equation 10.37.

$$\Phi_A = 10.39\left(\Psi_A\right)^{-2.52} \tag{10.37}$$

The spreadsheet "GrafLaursenEinstein" computes the sediment concentration C from Equations 10.14a, 10.36, and 10.37 for each sediment size.

Van Rijn (1984) Stage–Discharge and Total Load

Sturm (2010) and Van Rijn (1984) provide the primary guidance for the discussion below. Van Rijn (1984) used a single size (a modified d_{50} based on an empirical adjustment involving the d_{15} and d_{84}). Van Rijn partitioned roughness k_s into a grain component (k_s') and a bedform component (k_s''). He set $k_s' = 3d_{90}$ based on his review of available data. Van Rijn computed the grain shear velocity (u_*') using a Keulegan-type equation, as shown in Equation 10.41.

$$u_*' = \frac{V}{5.75\log\dfrac{12R}{3d_{90}}} \tag{10.38}$$

V is the velocity, computed from q/y, with q being the flow per unit width and y being depth. R is the total hydraulic radius, equal to depth in the wide channel.

The final velocity for a given depth is given by Equation 10.39.

$$V = 5.75 u\_* \log 12 R / \left(3d\_90 + k\_s^{\,''} \right) \tag{10.39}$$

In Equation 10.39, $u*$ is the overall friction velocity given by $(gRS)^{0.5}$.

$k_s^{''}$ is the bedform contribution to roughness, provided by Equation 10.40.

$$k\_s^{\,''} = 1.1\Delta \left(1 - e^{\wedge}\left(-25\Delta / \lambda \right) \right) \tag{10.40}$$

Δ is the bedform height, and λ is a bedform spacing parameter, taken to be $7.3\,y_o$. Auxiliary computations are required for computing Δ and are as follows. First, we need a dimensionless transport number T, defined as shown in Equation 10.41a.

$$T = \frac{u_*'^2}{u_{*C}^2} - 1 \tag{10.41a}$$

u_{*C} comes from a modified Shields diagram shown in Figure 10.3 based on a dimensionless grain size d_s, computed using Equation 10.41b.

$$d_s = \left[\frac{(SG - 1) g d_{50}^3}{v^2} \right]^{\frac{1}{3}} \tag{10.41b}$$

v is the kinematic viscosity (L^2/T)
SG is the specific gravity of the sediment
g is gravity (L/T^2)

We can now compute the bedform height using Equation 10.41c.

$$\Delta = 0.11 y_o \left(\frac{d_{50}}{y_o} \right)^{0.3} \left(1 - e^{-0.5T} \right) \left(25 - T \right) \tag{10.41c}$$

Sturm (2010) provides an alternative to Equation 10.44c in the case of large rivers. Note that van Rijn (1984) attempts to estimate the bedform type and bedform dimensions, whereas Einstein (1950) used a generic function to calculate the bedform contribution to roughness[1]. Van Rijn (1984) sheds light on the nature of the bedforms based on the dimensionless particle diameter and the Transport parameter T, which Figure 10.9 shows. Simons and Richardson (1966) had tried to classify bedforms (see Figure 10.10). They used a concept known as stream power (tractive force × flow velocity) vs. the

---

1 As an aside, Simons and Senturk (1992) summarize some 38 approaches for describing the roughness of alluvial channels based on bedforms and grain roughness. We only scratch the surface of this deep topic in this text.

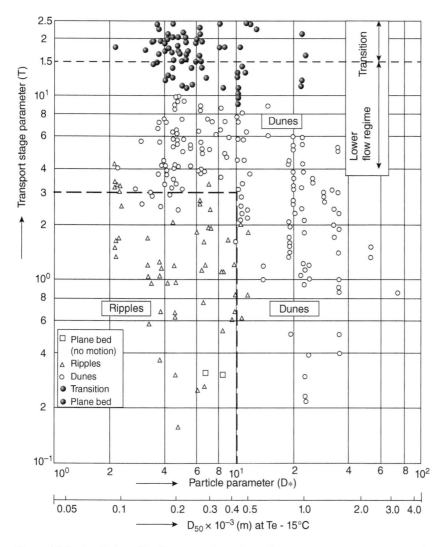

Figure 10.9 Prediction of bedform type from the sediment transport parameter *T* and dimensionless particle diameter d_s where

$$T = \frac{\tau'_*}{\tau_{*c}} - 1 = \frac{u'^2_*}{u^2_{*c}} - 1$$

τ'_* = Shields parameter for grain shear stress, τ_{*c} = critical value of Shields' parameter, u'^2_* = field friction velocity, and u^2_{*c} = critical value from the modified Shields diagram for the corresponding d_*. *Source:* From van Rijn 1984; used with permission of American Society of Civil Engineers (ASCE).

median fall diameter. It is evident from Figures 10.9 and 10.10 that the prediction of bedform types is not settled. However, progress is noticeable when compared to Einstein (1950). The Anon (2007) uses a stream power-based method for channel design that attempts to account for bedforms without directly addressing bedform contributions to flow resistance.

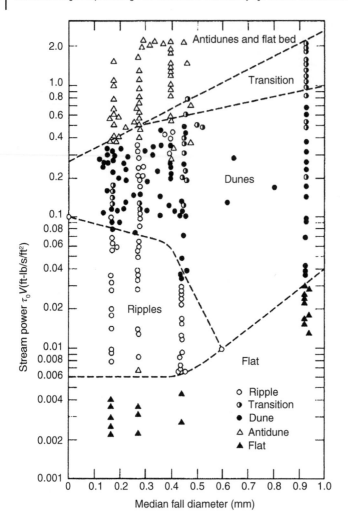

Figure 10.10 Prediction of bedform type from sediment fall diameter and stream power. *Source:* From Simons and Richardson (1966).

The van Rijn (1984) discharge–stage method is implemented on a spreadsheet, "VanRijnSD_TL," in the downloadables. The spreadsheet sets a target discharge and iterates for a depth that accommodates the required bedform roughness for the given flow and sediment characteristics.

The van Rijn bed material load calculation continues by taking a bed load function like the Einstein approach. They relate it to the critical transport and dimensionless size, which appears as shown in Equation 10.42.

$$\Phi_b = \frac{q_b}{\sqrt{(SG-1)gd_{50}^3}} = 0.053\frac{T^{2.1}}{d_*^{0.3}} \tag{10.42}$$

where q_b is the volumetric transport rate per unit width, and other terms are as previously defined.

The suspended load analysis proceeds similarly to the Einstein approach given above. Van Rijn (1984) modifies the sediment diffusion ratio (ϵ_s, see Equation 10.28) and the proportionality coefficient (β, see Equation 22b below) to slightly modify the Einstein (1950) concentration formula (Equation 10.29).

$$\epsilon_s = \beta \kappa u_* \frac{z}{y_o\left(y_o - z\right)} \tag{10.28}$$

Equation 10.31 is a parabolic equation giving maximum sediment concentration at mid-depth. Substituting Equation 10.28 into the right two terms of Equation 10.26 and then integrating and simplifying results in sediment concentration as a function of depth, as shown in Equation 10.29.

$$\frac{C}{C_a} = \left[\frac{\left(y_o - z\right)}{z} - \frac{a}{\left(y_o - a\right)}\right]^{R_o} \tag{10.29}$$

$$R_o = \frac{w_f}{\beta \kappa u_*} \tag{10.29a}$$

$$\beta = 1 + 2\left(\frac{w_f}{u_*}\right)^2 \tag{10.29a}$$

In Equation 10.29, C_a is a reference concentration at a small distance, a (use $0.05 \cdot y_o$ or half the bedform height), above the bottom. R_o is a collection of terms defined above and denoted as the Rouse number. The β term is a constant of proportionality between ϵ_s and ϵ as defined by van Rijn (1984).

Van Rijn modified the Einstein reference concentration C_a, as shown in Equation 10.42.

$$C_a = 0.015 \frac{d_{50}}{a} \frac{T^{1.5}}{d_*^{0.3}} \tag{10.43}$$

Van Rijn (1984) uses $a =$ half the bedform height instead of twice the sediment size.

Einstein followed the line of thought that the von Karman coefficient decreased as sediment load increased. Van Rijn took the tack of increasing R_o. Based on laboratory data with heavily sediment-laden flows, van Rijn developed the following correction shown in Equation 10.47.

$$\Delta R_o = 2.5 \left[\frac{w_f}{u_*}\right]^{0.8} \left[\frac{C_a}{0.65}\right]^{0.4} \tag{10.44}$$

The final R_o is $R_o + \Delta R_o$, denoted R_o'.

Van Rijn then computes an integrating factor, I_f shown in Equation 10.45, which ties in variables from Equations 10.28 to 10.44.

$$I_f = \frac{\left[\dfrac{a}{y_o}\right]^{R_o'} - \left[\dfrac{a}{y_o}\right]^{1.2}}{\left[1 - \dfrac{a}{y_o}\right]^{R_o'} \left(1.2 - R_o'\right)} \tag{10.45}$$

One can then compute the volumetric suspended load transport (q_s), as shown in Equation 10.46.

$$q_s = I_f y_o V C_a \tag{10.46}$$

The total load is given by summing the bedload solved from Equation 10.42 plus the suspended Load in Equation 10.46. The spreadsheet "VanRijnSD_TL" makes these computations. It converts volumetric transport to mass transport as well.

Total Load by Regression Approaches

Regression approaches are more straightforward than the more elaborate methods discussed above. One may say that regression approaches "smooth over" the erratic probability-based moves. Regression approaches are the go-to methods for practitioners charged with estimating sediment transport in earthen channels.

Yang's method: Yang (1972, 1996) developed the unit stream power as an independent variable that determines total sediment discharge. Yang (1972) based his sediment concentration on one modified sediment diameter close to d_{50}. Our discussion follows Sturm (2010). Unit stream power is the power available per unit weight of the fluid to transport sediment. It is equal to the velocity V and energy slope S. Based on dimensional analysis, Yang found the following dimensionless parameters:

$$C, \frac{VS}{w_f}, \frac{w_f d_{50}}{v}, \text{ and } \frac{u_*}{w_f}$$

Yang defined a dimensionless critical velocity (V_c), as shown in Equations 10.47a and b.

$$\frac{V_c}{w_f} = \frac{2.5}{\log\left(\frac{u_* d_{50}}{v}\right) - 0.06} + 0.66 \text{ for } 1.2 < \frac{u_* d_{50}}{v} < 70 \tag{10.47a}$$

$$\frac{V_c}{w_f} = 2.05 \text{ for } \frac{u_* d_{50}}{v} \geq 70 \tag{10.47b}$$

Using 463 datasets, Yang (1972) regressed the above variables to arrive at Equation 10.48 for total sediment load.

$$\log C_t = 5.435 - \frac{0.286 \log\left(w_f d_{50}\right)}{v} - \frac{0.457 \log u_*}{w_f}$$
$$+ \left(1.799 - \frac{0.409 \log\left(w_f d_{50}\right)}{v} - \frac{0.314 \log u_*}{w_f}\right) \log\left(\frac{VS}{w_f} - \frac{V_c S}{w_f}\right) \tag{10.48}$$

The spreadsheet "YangTotalLoad" contains the calculations. The regression approach gives results comparable to the more intensive bedload plus suspended load approaches. The spreadsheet assumes measured depth and flow data; however, one could model stage–discharge to estimate the flow conditions as with the Einstein–Laursen–Graf approach discussed earlier.

Other regression approaches for predicting total sediment transport tend to perform well. The strategy is to set a critical value of a given parameter and use a power function related to the respective parameter's difference and critical value. The critical value is determined experimentally. Velocity (V_c) and tractive force (τ_c) are covered in earlier chapters. The critical stream power ($\tau_c V_c$) is a combination of velocity and tractive force. Critical discharge (Q_c) relates to critical velocity via the continuity equation. Table 10.3 summarizes these relationships. Yang (1996) shows experimental results for the respective approaches that indicate good correlations between indicated variables of Table 10.3 and sediment transport.

Methods given above are not exhaustive but instead are a sampling of available approaches. All the methods for stage–discharge–sediment transport discussed above imply constancy in sediment supply unless the analysis is regarded as a snapshot in time. In lower-order streams, upstream tributaries may be receiving sediment from headwater watersheds. Alternatively, they may be degrading (e.g. bank or channel erosion). In a headwater analysis (where land-disturbing activities are occurring), one is looking at a snapshot with the assumption of a quasi-equilibrium between transport and deposition/degradation processes. With the snapshot concept established, one should note that even in headwaters without substantial land disturbance, low sediment transport rates are possible. It is recognized that even in the best conservation practices, soils may erode at rates ranging from 1 to 5 ton/(acre-year) according to Renard et al. (1997), providing a low but more or less constant sediment supply.

Bagnold's method: As shown above, sediment transport prediction methods can range from simple to complex. One reasonable compromise between simple and complex approaches that remains in use today is the Bagnold (1966) approach. The channel is idealized as a wide rectangular channel. The flow rate, slope, D$_{50}$, depth, and width of the channel are required for the sediment transport computation.

His method is appealing because it has a physical basis. The method is given by Equation 10.49.

Table 10.3 Summary of selected approaches for analyzing total sediment load, excluding probabilistic approaches, summarized by Yang (1996).

Approach	Relationship	Nomenclature
Discharge (L^3/T)	$q_s = A_1 \left(Q - Q_c\right)^{B_1}$	q_s = sediment discharge per unit width of the channel
Velocity (L/T)	$q_s = A_2 \left(V - V_c\right)^{B_2}$	Q = flow rate per unit width
Energy slope (-)	$q_s = A_3 \left(S - S_c\right)^{B_3}$	V = average flow velocity S = energy or water surface slope
Tractive force (F/L^2)	$q_s = A_4 \left(\tau - \tau_c\right)^{B_2}$	τ = shear stress or tractive force
Stream power (F/L-T)	$q_s = A_5 \left(\tau V - \tau_c V_c\right)^{B_5}$	τV = stream power per unit width VS = unit stream power
Unit stream power (L/S)	$q_s = A_6 \left(VS - V_c\right)^{B_6}$	A_1 through B_6 empirical coefficients. Subscript c denotes a critical value derived by observation or curve fitting.

$$q_t = \frac{\gamma_s - \gamma}{\gamma} P \left(\frac{e_b}{\tan \alpha} + 0.01 \frac{V}{\omega} \right) \tag{10.49}$$

The nomenclature of Equation 10.49 is as follows:

$\gamma_s =$ unit weight of sediment (lb/ft³)
$\gamma =$ unit weight of water (lb/ft³)
$q_t =$ total sediment transport (lb/s/ft)
$\omega =$ sediment fall velocity (ft/s)
$e_b =$ bed load efficiency factor (-), read from Figure 10.11a
$\tan(\alpha) =$ ratio of tangential to normal shear force, read from Figure 10.11b
$V =$ mean flow velocity (ft/s)
$\tau =$ tractive force (lb/ft²)
$P =$ stream power $= \tau V = \gamma dSV$
$d =$ depth of flow
$S =$ channel slope.

The success of the Bagnold approach highlights the efficacy of stream power for estimating sediment transport. Bagnold (1980) added the concept of excess stream power to his original concept, which further improved the method. Martin and Church (2000) added additional data, nondimensional analysis, and used curve fitting to arrive at the following model.

$$q_b = \text{const} \left(P - P_o \right)^{3/2} \left(\frac{D_{50}^{1/4}}{d} \right) \left(\frac{1}{\rho_r^{1/2} g^{1/4}} \right) \tag{10.50a}$$

$$P_o = \text{critical stream power} = 5.75 \left[\tau_S^* \left(\frac{\gamma_s - \gamma}{\gamma} \right) D_{50} \right]^{1.5} \gamma g^{0.5} \text{Log}_{10} \left(\frac{12d}{D_{50}} \right) \tag{10.50b}$$

$P =$ unit stream power $= 1000 * d * S * V$
The parameter $\rho_r = \gamma_s - \gamma = 1650$
$g =$ gravity (9.81 m/s²)
const $=$ a fitting constant
$D_{50} =$ median particle diameter (m). Other nomenclature is defined above.

The "BagnoldModified" sheet of the "BagnoldBrownlieWilcockKenworthy" excel sheet makes this computation. Note that this spreadsheet page also makes Yang (1972) discussed previously.

Brownlie's method: Brownlie (1981) method models the channel as a trapezoidal channel. The flow, top width, slope, depth, side slope z, and D_{50} are required. The Manning n is computed via iteration. A Manning n between 0.02 and 0.04 typically represents antidunes, standing waves, or the plane bed (upper flow regime). Manning n less than 0.02 represents dunes or ripples (lower flow regime, see Richardson et al., 2001). Brownlie gives a depth predictor, which is useful in validating the solution. The equations of Brownlie (1981) are based on a nondimensional analysis centered around stream power. See the "BrownlieYang" page of the "BagnoldBrownlieWilcockKenworthy" excel workbook for Brownlie's sediment transport predictions.

(a)

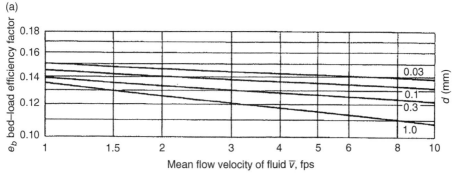

(b)

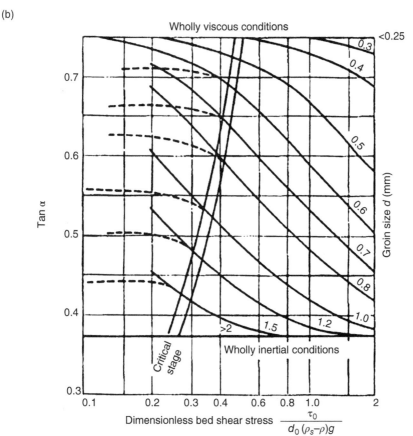

Figure 10.11 Bedload efficiency (e_b, see a) and normal-to-shear force ratio (tan α, see b) functions. *Source:* After Bagnold (1966).

Other simple and straightforward methods for sediment transport computation are also available. The regression-based approaches lead to a relatively straightforward estimation of sediment transport, particularly when the process is based around stream power. Richardson et al. (2001) give additional examples of sediment transport computations applicable to bridge pier and related applications and other diverse applications.

Sediment Measurement

Experimental measurement methods are necessary for validating the sediment transport theories discussed above. Many of the original research works mentioned above discuss equipment for measuring sediment transport rates. Also, Vanoni (2006) gives an exhaustive discussion of approaches for measuring point and depth-integrated suspended sediment loads. He also discusses bedload measurement techniques.

Equipment developed in the 1940s and 1950s remains state-of-the-art. Bedload transport has been quantified by front-open baskets placed on the stream bottom for a finite time. One may measure suspended load at points by taking a volumetric sample over defined time intervals. Velocity measurements are also required. The suspended load assessment requires multiple sample points so that one may integrate over the depth. One would ideally repeat the bedload collection and the depth integration at several points over the stream cross section. Sediment transport measurement is labor-intensive and problematic due to bedform considerations. Suspended load measurement is the most labor-intensive.

The recent development of acoustic Doppler meters promises to be a significant advance in suspended sediment quantification. Richardson et al. (1961) reported an experimental sonic depth sounder for bedform characterization and scour depth near bridge piers. One may estimate depth (bathymetry) with sonar depth sounders. Velocity and, to some extent, sediment concentration using acoustical multibeam sonar surveys may be estimated with acoustic Doppler technology. The acoustic Doppler equipment enables continuous measurements with time, which is not possible with conventional sampling equipment. Assessing variations with time extends a snapshot to include a time dimension. Commercially available equipment is coming available.[2] The worldwide web contains pictures and descriptions of bedload and suspended load measurement equipment.

The sedigraph (sediment concentration vs. time) frequently does not parallel the hydrograph. Rising hydrographs' limbs contain the highest sediment concentrations due to the increased difference between friction slope and bottom slope. On the other hand, the falling hydrograph limb has less friction slope than the bottom slope. The friction slope determines the shear force available to transport sediment. Sampling should, therefore, focus on the time of rising discharge.

Defining the boundary between the zone of bedload transport and suspended load transport remains a challenge. Einstein (1950) somewhat arbitrarily set twice the roughness height as the boundary. Others have put the limit just above the bedforms. Much work remains.

Sediment Routing Through Detention Ponds and Streams

Thus far, we have treated sediment transport as a quasi-equilibrium problem where sediment supplies upstream were equal to sediment transported. Equilibrium is practically never the case. Sediment transported was in balance with sediment on the wetted channel

---

2 A web search will turn up equipment pictures and descriptions. For example, see www.r2sonic.com.

perimeter. Sediment retention ponds are constructed to capture sediment from upstream areas under construction and susceptible to erosion. Eroded sediments typically have many fines ranging from sands to clay. The sediment supply typically exceeds the sediment transport rate of streams and reservoirs. Haan et al. (1994) advocate treating a pond as a reactor model with dead storage and short-circuiting. They also consider the plug flow reactor. Haan et al. (1994) provide equations for continuously stirred reactors and plug-flow reactors with dead storage and short-circuiting. Short-circuiting occurs when flow moves from the inlet to the outlet without contacting part of the reactor volume. Dead storage is the portion of the reactor volume that is short-circuited. One of the models presented is the single continuously stirred reactor, which appears in Equation 10.51.

$$\frac{c_T}{c_o} = 1 - f_1 e^{\frac{f_1 * t}{f_2} \frac{t}{t_d}} \tag{10.51}$$

The terminology for Equation 10.50 is as follows:

C_T = total concentration of short-circuited plus reactor flow
C_o = inflow concentration
t_d = theoretical detention time (= Volume/flow rate)
V = reactor volume
f_1 = fraction of flow going to reactor
f_2 = fraction of reactor that is an active volume.

In addition to the model in Equation 10.50, Haan et al. (1994) presented several model reactors, including the continuously stirred reactor in series, plug flow reactor, plug flow continuously stirred, and a diffusion plug flow reactor. The inflow concentration could be defined over time by a relationship between sediment concentration and time. This relation is called a sedigraph. The various reactors are all chemical or biological systems. Sediment trapping is a physical process. Thus, physically, the basis for the reactor equations is lacking; however, the empirical results seem to agree with experience.

Haan et al. provide worked examples demonstrating various reactors' effects on the ratio of inlet sediment concentration to outlet sediment concentration.

Sediment capture is modeled by hypothesizing that settlement time is the hydraulic residence time of the reactor. Sediment settling faster than the detention time (e.g. the larger sized particles) remains in the pond. Haan et al. (1994) also presented a model for routing a sedigraph in a stream. They assume that one can model sediment as a power function of water discharge as shown in Equation 10.52:

$$c = kq^a \tag{10.52}$$

In Equation 10.51, c is the sediment concentration, q is the water flow rate, and k and a are constants. The parameter, a, typically lies between 0.5 and 1. One can estimate constant from flow and sediment concentration data using regression techniques.

Equation 10.53 gives the mass of sediment is given as

$$q_s = cq \tag{10.53}$$

Parameter q_s is the mass flow of sediment.

The total sediment yield comes from integrating Equation 10.51 to yield Equation 10.54.

$$Y = \int_0^{D_{st}} kq^{a+1} dt \tag{10.54}$$

In Equation 10.55, Y is the total sediment yield for the storm with duration, D_{st}. Since k is a constant, one can remove it from the integral of Equation 10.52 and solve for k, as shown in Equation 10.55.

$$k = \frac{y}{\int_0^{D_{st}} q^{a+1} dt} = \frac{Y}{\sum_{i=1}^n q_i^{a+1} \Delta t_i} \tag{10.55}$$

One may divide the storm into n increments for purposes of evaluating the interval in Equation 10.56. Knowing k and the constant, a, and total sediment yield Y, one can estimate the sedigraph from Equation 10.53.

Haan et al. (1994) cite other related approaches for sediment routing. They also provide extensive discussion on routing sediment from an eroding field to a stream. The SedCad (Warner et al. 1998) software discussed below implement the above procedures for sediment routing.

Software Support for Estimating Sediment Transport

Software support is available for estimating sediment transport. HEC-RAS (Brunner 2016) has a sediment design tool, shown in Figure 10.11. One inserts the sediment characteristics, assumes quasi-equilibrium conditions, and selects one or more sediment transport approaches. HEC-RAS sediment design tool predicts total bed material load using several available methods. Cross-section data in this and the following examples were added using a GIS-HEC-RAS software interface (ESRI 2020).

The effect of the extreme event on alluvial channels renders the quasi-equilibrium analysis discussed above to be nonapplicable. In situations where an accumulation (aggradation) of sediment may occur, or degradation occurs, the channel cross section is no longer constant. Flow conditions change as the cross section changes. Hydraulic analysis of alluvial channels with unsteady flows becomes an iterative analysis that can take substantial computer time. HEC-RAS provides some examples of sediment aggradation/degradation on a regional scale, and Figure 10.12 shows one such case showing what happened over a year along a reach of the HEC-RAS Euclid River example (Figure 10.13).

Software is also available for estimating sediment transport on the field scale. SedCad[3] is a field-scale program that models erosion processes, sediment transport through various best management practices, including detention ponds. SedCad estimates the time

---

3 Version 5 is available from Civil Software Design, P.O. Box 706 Ames Iowa. Other contact information is as follows: email – pschwab@mysedcad.com; Phone/Fax – (515) 292-4115.

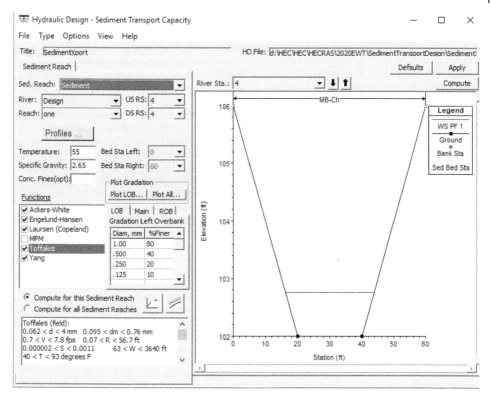

Figure 10.12 Example output from the HEC-RAS sediment design tool. *Source:* HEC-RAS Computer program. © United States Army Corps of Engineers.

required for detention ponds to accumulate the required amount of sediment that mandates pond clean-up. Figure 10.14 shows outputs from a SedCad pond design problem.

Implications of Sediment Transport on Infrastructure

Sediment transport has significant societal implications. Sediment transport over time results in filled reservoirs. Figure 10.15 shows how Lake Meade filled with sediment from 1937 to 1949. Data for the present was not available. In Athens, Georgia, the Barnett Shoals hydroelectric plant has been rendered nearly obsolete due to sediment deposition. Sediment presence impacts civil infrastructure (bridge piers, abutments, and the like) in multiple ways. Sediment presence impacts the expense of treating water for domestic uses. Partitioning between bedload and suspended load can influence the design of water intake structures. Not to be forgotten is that sediment represents soil loss from construction sites and soil/nutrient loss from productive farmlands (Vanoni 2006).

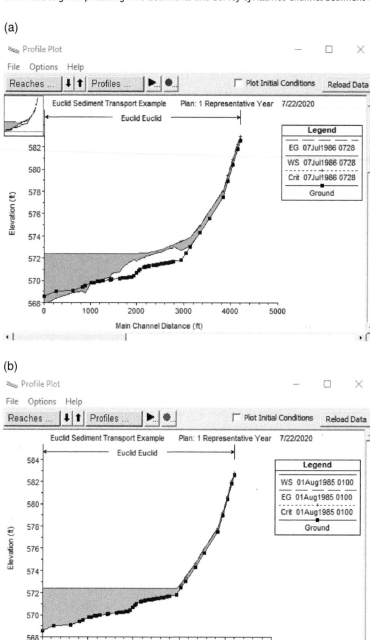

Figure 10.13 Channel profile from HEC-RAS (a) initially and (b) one year later for the Euclid River sediment transport example. *Source:* HEC-RAS Computer program. © United States Army Corps of Engineers.

(a)

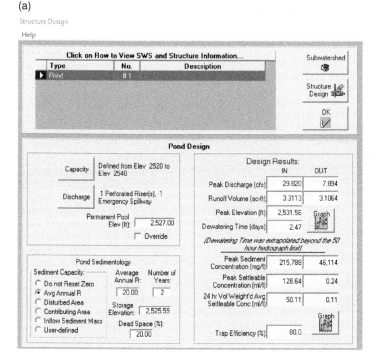

(b)

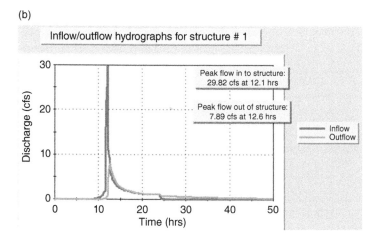

Figure 10.14 The flow of eroded sediment from a small watershed into and out of a small pond equipped with a perforated riser and emergency spillway, showing (a) design details; (b) inflow and outflow hydrographs; and (c) inflow and outflow sedigraphs. *Source*: HEC-RAS Computer program. © United States Army Corps of Engineers.

(c)

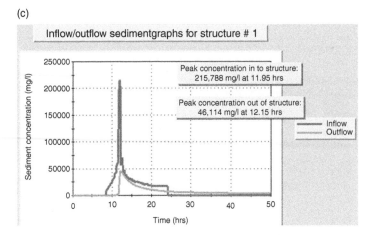

Figure 10.14 (Continued)

Empirical Channel Design Approaches Leading to Sustainable Channels

The past analysis showed that the parabolic cross section is the theoretically expected cross section in earthen or alluvial channels. The parabolic cross section presumed that the critical tractive force was constant in the cross section and, by extension, constant along the reach for a distance. The design is theoretically optimum for one flow rate with a constant tractive force over a given distance. Many natural channels look "somewhat parabolic." However, natural channels tend to have more complex cross sections than a simple parabola. Evidence to be presented below shows that other cross sections tend to evolve into more complex shapes over time. Are there other alternatives to channel design, given the wide day-to-day variation in flows conveyed by materials that might not have a constant tractive force? This section briefly reviews practical channel design techniques and how these techniques are oriented toward sustainable channels. The sustainable channel maintains its cross section over time while providing ecological services and requiring less maintenance.

Classical regime theory: Lacey (1930) published what was called the 'regime theory' for irrigation canal design in the Punjab region of India. The regime theory is not a theory in the strict sense of the term because it does not incorporate physical explanations for the findings (Henderson, 1966). Blench (1957) and Simons and Albertson (1960) published more general versions of the regime theory highlighted below. The regime theory approaches were derived from observations of canals or other channels' performance using various regression techniques. Table 10.4 summarizes the Blench method. The Simons–Albertson (1960) methods are outlined in Tables 10.5a, b. Examples demonstrate these methods.

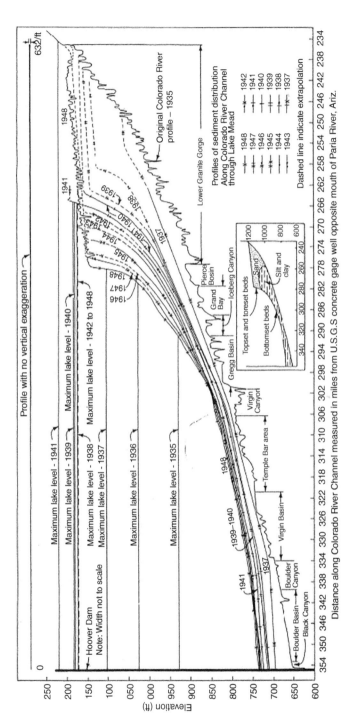

Figure 10.15 Results of a comprehensive survey of Lake Mead: 1948–49 showing how sediment accumulation has occurred since Lake Mead's founding in 1935. *Source:* From Smith (1960).

Table 10.4 The Blench design equation summary for alluvial canal design.

Eq. no.	Design equation	Notation
1	$y = \sqrt[3]{\dfrac{f_s Q}{f_b^2}}$	y is the mean depth (ft)
2	$b = \sqrt{\dfrac{f_b Q}{f_s}}$	Q is flow rate (ft³/s) f_s, side factor, $= 1$ (slight cohesiveness), 0.2 (medium cohesiveness), and 0.3 (high cohesiveness); b is the mean width (ft)
3	$S = \dfrac{f_b^{5/6} f_s^{1/12} y^{1/4}}{3.63 g Q^{\frac{1}{6}}\left(1 + \dfrac{c}{2330}\right)}$	S is the slope (-). This is a derived equation for the equilibrium slope.
4	$f_b = 9.6\sqrt{d}\left(1 + 0.012c\right)$	f_b is the bed factor; c is sediment concentration, (ppm); d is the median sediment diameter (in)
5	$f_b = \dfrac{v^2}{y}, f_s = \dfrac{v^3}{b}$	Relations between bed factor f_b and side factor f_s and velocity/depth.

Source: Adapted from Blench (1957) and Henderson (1966). Chaudhry (1993) provides a table for an earlier method, the Lacey (1930) method, which is a forerunner to the Blench and the Simons and Albertson methods.

Table 10.5a The Simons and Albertson (1960) design equation summary for alluvial canals.

Eq. no	Design equations	Nomenclature	Comments
1	$P = K_1 Q^{1/2}$	P is the wetted perimeter (ft)	Solve for B knowing b and P
2	$b = 0.9P$	R is the hydraulic radius (ft)	
3	$b = 0.92B - 2.0$	S is the channel slope (-)	Solve for depth depending
4	$R = K_2 Q^{0.36}$	Q is the flow rate (ft³/s)	on R
5	$y = 1.21R$ for $R \leq 7\,ft$	b is the mean width (ft)	
6	$y = 2 + 0.93R$ for $R \geq 7\,ft$	B is the surface width (ft)	
7	$A =$ by or PR (use the mean)	v is the kinematic viscosity	Take the mean of these
8	$v = Q/A$	(0.00001 ft²/s)	area approximations
9	$v = K_3 (R^2 S)^m$	A is area (ft²)	Solve for velocity
10	$\dfrac{C^2}{g} = \dfrac{v^2}{gyS} = K_4\left(\dfrac{vb}{v}\right)^{0.37}$	C is the Chezy coefficient y is the mean depth (ft)	Solve for slope Alternate solution for slope

Source: Adapted from Simons and Albertson (1960) and Henderson (1966).
Table 10.5b provides coefficients for the above equations.

Example 10.3 (Adapted from Henderson 1966)

For a channel with a flow rate of 146 ft³/s, determine the cross section and slope by the Blench and Simons–Albertson methods. Use a kinematic viscosity of 1×10^{-5} ft²/s. The bank material is highly cohesive, and the bed is sandy. The medium bed material size is 0.0125 inches. The bed material concentration is 227 ppm.

For the Blench method, $f_s = 0.3$. From Equation 4 in Table 10.4,

$$f_b = 9.6\sqrt{.0125}\left(1 + 0.012(227)\right) = 4$$

From Equations 2, 1, and 3,

$$b = \sqrt{\frac{4(146)}{0.3}} = 44 \ ft$$

$$y = \sqrt[3]{\frac{0.3(146)}{4^2}} = 1.4 \ ft$$

$$S = \frac{4^{\frac{5}{6}}0.3^{\frac{1}{12}}10^{-\frac{5}{4}}}{3.63(32.2)146^{\frac{1}{6}}\left(1 + \frac{227}{2330}\right)} = 0.00055$$

From Equations 5, we can estimate the velocity.

$$v = \sqrt{4(1.4)} = 2.4\frac{ft}{s}$$

$$v = \sqrt[3]{0.3(44)} = 2.34 \ ft/s$$

The mean of the velocity values is $v = 2.37$ ft/s.

Now, we use the Simons–Albertson method summarized in Table 10.5a. The channel has a sandy bed and cohesive banks, which defines the applicable coefficients as column 2 in Table 10.5b.

$$P = 2.6(146)^{\frac{1}{2}} = 31.4 \ ft$$

$$b = 0.9(31.4) = 28.3 \ ft$$

$$28.3 = 0.92B - 2.0 \xrightarrow{yields} B = 33 \ ft$$

$$R = 0.44(146)^{0.36} = 2.65 \ ft$$

$$y = 1.21(2.65) = 3.21 \ ft$$

$$A = \text{mean}(28.3 \times 3.21, 31.4 \times 2.65) = 87 \ ft^2$$

$$v = 146/87 = 1.68 \ ft/s$$

Table 10.5b The Simons and Albertson (1960) design equation coefficient summary.

Design coefficients	Sandbed and banks	Sandbed and cohesive banks	Cohesive bed and banks	Coarse, noncohesive material	Sandbed and cohesive banks with 2000–8000 ppm sediment load
Channel type→	1	2	3	4	5
K_1	3.5	2.6	2.2	1.75	1.7
K_2	0.52	0.44	0.37	0.23	0.34
K_3	13.9	16.0	—	17.9	16.0
K_4	0.33	0.54	0.87	—	—
M	0.33	0.33	—	0.29	0.29

Source: Adapted from Simons and Albertson (1960) and Henderson (1966).

Two methods for computing slope.

$$1.68 = 16\left(2.65^2 S\right)^{0.33} \rightarrow S = 0.000165$$

$$\frac{1.68^2}{32.2(3.21)S} = 0.54\left(\frac{1.68(28.3)}{10^{-5}}\right)^{0.37} - \rightarrow S = 0.000171$$

Taking the mean of the slopes leads to $S = 0.000168$.

The Simons–Albertson channel is narrower and deeper than the Blench channel. The slope of the Simons–Albertson channel is less steep than the Blench channel.

Both methods start with a flow rate, soil condition, and seek to find width and depth to accommodate the flow. These channels tend to be somewhat parabolic in cross section. The slope is the test for channel stability. If the computed slope is steeper than the field slope, the channel aggrades. Likewise, if the field slope is steeper than the calculated slope, the channel experiences degradation. A design on a steep slope requires drop structures so that the reaches between the structures do not exceed the equilibrium slope. Changing the bottom width to attempt an accommodation of slope would take the channel out of the regime; thus, it is unlikely to be successful.

Henderson (1966) summarizes three criticisms of the regime method, and one additional criticism is appended. They are as follows:

*That it applies to a limited range of conditions occurring in India and is inapplicable elsewhere.
*That the silt factor f is poorly defined in that it must be determined by other factors besides the size d, notably the sediment concentration.
*It is not clear to what extent the channel slopes in the Indian data were self-adjusted.
*Tendency to meander is unaddressed.

The first criticism is partially addressed by the Simons–Albertson method. Data coming from somewhere other than India were used. The second criticism was also discussed by subsequent work and by Simons–Albertson. The slope criticism is less easily answered. The question of using steeper slopes than the minimum slope is not easily answered. Furthermore, the dynamics of sediment loading upstream can confound the equilibrium slope determination downstream.

Given materials can occur on a wide variety of slopes. Steep slopes may result in deeply incised channels, and very shallow slopes may result in very wide width/depth ratios. With an incised channel, the material may change to larger aggregates or bedrock. Rosgin (2005) has attempted to develop a classification of streams based on conditions shown in Figure 10.15. Figure 10.15 presents a variable not addressed by the regime theory, channel sinuosity, or tendency to meander. Modern techniques of sustainable channel design add meandering to the suite of design variables. Before looking at how meandering and other indicators in Figure 10.16 are addressed, we look at selected forces at work to explain meandering.

The regime approaches implicitly assume a steady source of sediment. Thus, one would expect low-order streams in a hierarchical stream network to be more suitable for regime-type design approaches than for streams in headwater areas. The sustainable watershed erosion rate may be a suitable criterion for steady sediment transport in headwater areas or further downstream.

Contemporary design methods: Wilcock and Kenworthy (2002) presented a design approach suitable for low levels of sediment transport that incorporates regime theory elements. Given a flow rate and low sediment transport rate (e.g. 0.05 kg/s), one can develop preliminary channel designs to accommodate sediment transport. The Manning n (e.g. one may assume a bedform and use Richardson et al. (2001) to estimate n) and D_{50} are required. Wilcock and Kenworthy (2002) assume the trapezoidal channel has a top width that is an empirical function of the flow rate (e.g. top width $= a*Q^b)^4$ and an assumed side slope z. The constants a and b in the top width formula may also be assumed and estimated from gaged streams in a region. The historic regime methods may also shed light on the top width–flow relationship. Wilcock and Kenworthy (2002) develop a two-fraction transport function (sand, gravel) based on the Shields transport function. They compute the tractive force from the assumed known flow rate. The sediment transport is computed from an empirical function of a tractive force to the critical tractive force ratio. Other details are given in Wilcock and Kenworthy (2002) and on the corresponding page of the "BagnoldBrownlieWilcockKenworthy" spreadsheet[5]. The slope and depth are determined iteratively by constraining the predicted sediment transport to meet the known target value and the computed flow rate to equal the assumed known flow rate. Other hydraulic elements of the trapezoidal channel are then read.

Reflecting on the parabolic method and the regression approaches discussed earlier, a design approach building on the parabolic analysis could be constructed by designing the channel at a tractive force above the critical tractive force to predict sediment transport

---

4 From a study varying flow rate and keeping permissible velocity constant, the topwidth increased by ≈ $a*q^{0.5}$. The same experiment with a trapezoidal cross section resulted in a linear relationship between flow rate and topwidth. Thus, the channel is dubbed a "hybrid" channel.

5 This spreadsheet is based on one prepared by Will Mattison, graduate student who is completing an MS under the direction of Dr. Brian Bledsoe, College of Engineering, University of Georgia, Athens, GA.

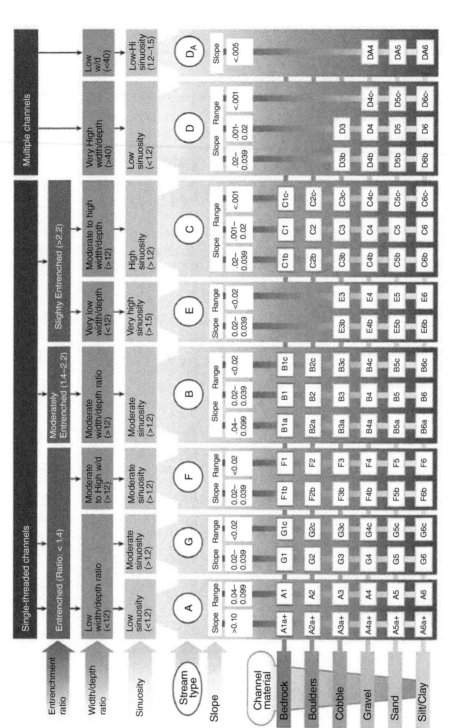

Figure 10.16 Stream classification system schematic. *Source:* Adapted from Rosgen and Silvey (2005).

rate. The desired transport rate (equal to the sustainable rate based on sustainable erosion in the watershed) would increase the tractive force needed to transport the sustainable sediment rate. The parabolic channel would then be designed at the adjusted tractive force level. More experimentation is required to further develop the parabolic design approach with targeted sediment transport.

Forces Impacting Channel Cross Sections – Stream Restoration

Three categories of forces and events affect channel cross section. A cross section where the tractive force is not uniformly applied tends to morph to a cross-section uniformity of tractive force. Figure 10.17 shows how a trapezoidal channel in clay soil changes over time. The channel morphs into a trapezoidal shape, then develops a low-flow channel with a flood

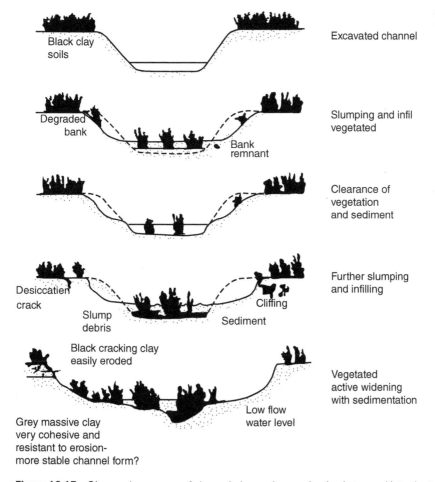

Figure 10.17 Observed sequence of channel change in an urbanized stream. Note the transition from a trapezoidal channel to a parabolic channel with an emerging flood plane and primary channel. *Source:* From Whitlow and Gregory 1989; used with permission of John Wiley & Sons.

plain within the larger channel. Chapter 4 provides a high-level explanation of the parabolic shape. We have no theoretical method beyond tractive force analyses for describing the complex cross section that finally evolves.

A second phenomenon occurs when land-use changes upstream impact sediment loadings. Figures 10.18 shows what happens when uniformly sized sediment-laden flows react in a simulated grass filter. Sediment deposits at the top of the filter in triangular wedges whose hypotenuses have near-constant uniform slopes. Figure 10.19 is a plot of Einstein's sediment transport and shear intensity parameters for the data, with excellent agreement. Similar behavior would be expected in channels with high sediment loading due to development upstream. Historically, during continuous cotton production in the southern United States, streams were clogged with soils. Much of the cotton land transitioned to forests. Sediments in downstream locations then began to mobilize, causing degradation of streams. Pollutants bound with the sediments then began to move toward Lake Lanier, the predominant water supply for the Atlanta, Georgia area.

A third phenomenon impacting cross section is that of natural or imposed channel bends. A flow going around a bend is subjected to Coriolis forces and others that cause secondary currents (refer to Figure 2.13). Secondary currents exert tractive forces. One part of a stream or river tends to be deeper (called the thalweg, derived from German meaning "valley way") due partially to secondary forces. Ippen and Drinker (1962) did detailed flow measurements in a curved trapezoidal channel to document velocity as a function of position in a 60-degree bend. Figure 10.20 shows the results. The imparted tractive force in the curve was 2.8 times the normally expected tractive force.

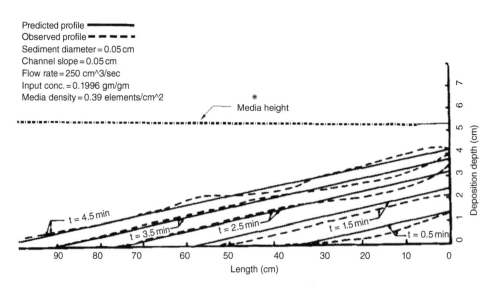

Figure 10.18 Sediment deposition pattern in simulated, rigid vegetation during initial filling. The sediment reaches the media height and then the pattern propagates down stream. *Source:* From Tollner et al. 1977; used with permission of Am. Soc. Agr. & Biol. Engrs.).

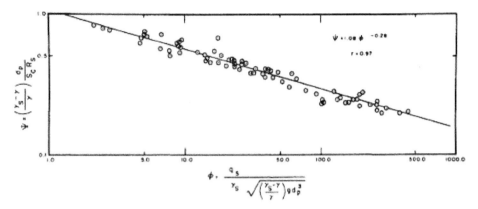

Figure 10.19 Einstein's total transport function as modified for total sediment load in simulated rigid vegetation. *Source:* From Barfield et al. 1977; used with permission of American Society of Agricultural and Biological Engineers (ASABE).

Thus, nonuniform tractive forces, changes in flow and sediment loadings, and natural or imposed bends impact the channel cross section. These phenomena all occur ubiquitously in nature. A design methodology that accommodates these naturally occurring dynamics is, by nature, more sustainable. Considering the nature of the alluvial channel as discussed at the outset, these techniques implicitly assume a degree of sediment supply stability. The above factors impact headwater streams the most. The empirical approaches, therefore, would seem to be more applicable in lower-order streams.

Stream restoration strategies: A significant feature of natural channels is the channel within a channel (see Figure 10.21). Natural channels tend to meander when stable and sustainable. The channel within the flood channel carries a 1- or 2-year flood, while the flood channel may carry a 50 to a 100-year flood. The small channel typically has a sinuous pattern within the flood plain width. The small channel may have a different slope than the valley slope of the flood channel. Also, channels tend to have pools in the meander curves. Between the curves, many channels have riffles (see Figure 10.22). Natural channels are typically specified based on relations derived from observations and regression analysis. Rosgen and Silvey (2005) summarize 41 relationships that work well in the Midwest US for natural channel features. Figure 10.23 summarizes Rosgen relationships.

Referring to Figure 10.16, Rosgen and Silvey (1996, 2005) have classified streams based on entrenchment ratio, width–depth ratio, slope, and channel material in addition to sinuosity or meandering. Channels may be straight and incised, or sinuous, or multiple-braided in morphology. This classification system seems to perform well in the Midwest but less so in the Southeast US. Figure 10.24 presents Richardson and Davis's (2001) results for delineating slope and flow boundaries between meandering, transitional, and braided streams. Not shown is the incised category, which would appear below the meandering boundary.

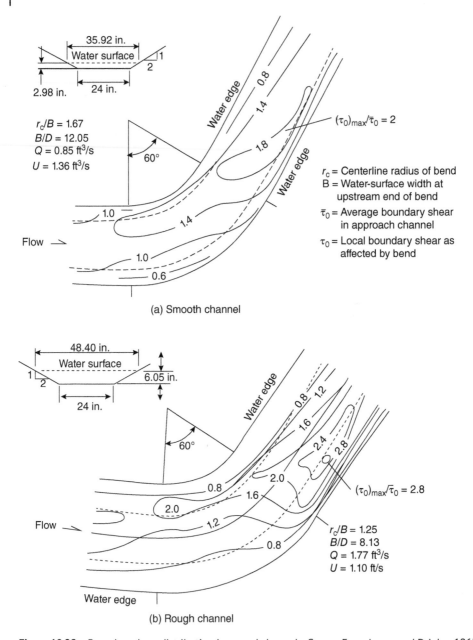

Figure 10.20 Boundary shear distribution in curved channels. *Source:* From Ippen and Drinker 1962.

Different parameters are employed in the southeast US. Figure 2.4, reproduced here as Figure 10.25, shows design relationships. Doll (2002) presented hydraulic relationships found useful in sustainable channel design in the southeast US. Differences in soil

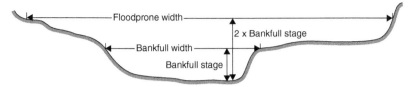

Figure 10.21 Key descriptors of the natural stream showing bank full width and depth. *Source:* From Huffman et al. 2013; used with permission of American Society of Agricultural and Biological Engineers (ASABE).

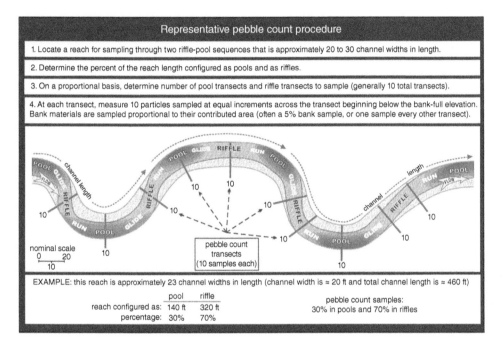

Figure 10.22 Natural channel features with procedures for sampling aggregates in pools and riffles. *Source:* Rosgen and Silvey 2005; used with permission of Wildland Hydrology, Inc.

formation processes and resulting mineralogy would explain why a one-size-fits-all set of parameters does not exist. This regional variation is consistent with the experience of the different regime theories discussed previously.

Designers have creatively designed structures to augment a natural channel's evolution after placing a hypothesized natural channel's footprint on the ground. Figure 10.26 summarizes several structures that facilitate the development of ecological services. FISRWG (1998) and Schiechtl (1980) discuss various bioengineering techniques to protect streambanks and optimize ecological services from restored streams. The long-term stability of these designs warrants further observations, especially in low-order streams.

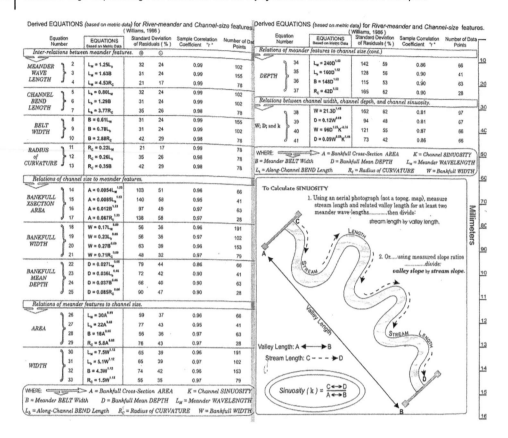

Figure 10.23 Summary of 41 design equations used by Wildland Hydrology in the design of natural channels. *Source:* Rosgen and Silvey 2005; used with permission of Wildland Hydrology, Inc.

Summary and Future Directions

This chapter makes three significant contributions to our discussion of channel hydraulics. We complete the earthen channel design discussion by adding the Shields (or modified Shields) diagram for describing the design of earthen channels in fine-textured soils. The second primary consideration is the analysis of flow and sediment transport in natural alluvial channels. Thirdly, we consider channel design with sediment transport.

Alluvial (or moveable bottom) channel sediment transport with specified channel dimensions and channel design is an extensive body of knowledge that occupies written volumes. The text surveys early methods, complex methods, and regression methods for sediment transport prediction. The spreadsheets provide an avenue to examine and perform the computations. They were our approach to "get around" space limitations in addition to being computational tools. One could take the material herein, plus cited works, and have a good repertoire of sediment transport analysis tools.

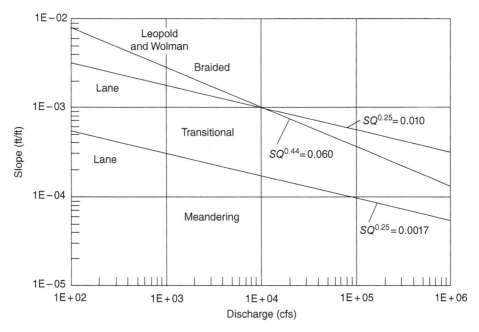

Figure 10.24 Changes in patterns of streams having a stream slope at a given characteristic discharge. *Source:* From Richardson and Davis (2001). © United States Department of Transportation.

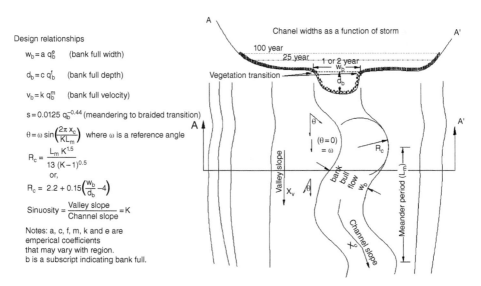

Figure 10.25 Hypothetical natural channel cross section and a plan view with nomenclature and design relationship as summarized. *Source:* Used with permission from Huffman et al. (2013). © American Society of Agricultural and Biological Engineers (ASABE).

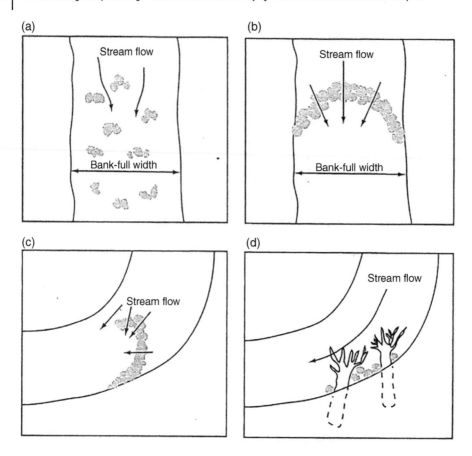

Figure 10.26 Typical structures used for channel stability and habitat improvement during stream restoration, including the following: (a) boulder placement, (b) weirs or vanes, (c) partial vanes or J-hooks, and (d) tree revetment. A commonality among the above structures is that they direct the flow from the bank toward the middle of the stream. *Source:* From Huffman et al. 2013; used with permission of American Society of Agricultural and Biological Engineers (ASABE).

The quasi-equilibrium alluvial channel is a higher-order stream with a steady sediment supply resulting from soil loss from the watershed due to land-disturbing activities or sustainable soil erosion. After defining the context, we assessed the impact of alluvial bedforms on hydraulics and sediment transport. Alluvial sediment mechanics has provided bases for earthen channel design, assuming the sediment supply is consistent with a low sustainable erosion rate in the watershed. For a rapid estimate of sediment total load transport, one might consider the regression methods.

Natural sustainable channel design has defied purely theoretical analysis due to the complexity and magnitude of data required to describe all the sediment transport processes. Techniques building on the regime theory that predicts channel elements and sinuosity are becoming available. The empirical techniques all involve coefficients that depend on the regional location. Our knowledge regarding sustainable channel design and restoration is still deficient but progressing. The notion of defining a critical tractive force, flow rate, stream power, or similar variables, then assessing sediment transport when the actual variable exceeds the corresponding critical variable is becoming popular.

Unsteady sediment-laden flows represent the most general case of open channel hydraulics. Generalizing problems to include two and three dimensions in channels with natural cross sections presents a continuing challenge to researchers and practitioners alike. Society demands that channel engineering solutions must consider human, infrastructural, and ecological impacts. These complex demands require the best in social and technical expertise that society can muster. Topics presented in this text introduce us to the technical challenges toward this end.

Practitioners solve virtually all the problems introduced in this and the past nine chapters using proprietary and public domain software such as those mentioned previously. The spreadsheets serve as valuable back-of-the-envelope methods for fast estimates. Software capability continues to increase. For example, HEC-RAS now has limited 2-D modeling capability. Firms are adding user-friendly interfaces and gathering additional capabilities into single packages, enabling more comprehensive problem solutions. As an example of the latter, one can now license a package (CivilGEO 2020)[6] that brings ESRI GIS (2020),[7] AutoDesk (2020),[8] and HEC-RAS (Brunner 2016) under one umbrella. The comprehensive bundling enables the solution of comprehensive flood plain models, including 2-d flow regions over floodplains. Increasing applications of data analytics and cloud computing are on the horizon. Software and computing developments continue to unfold.

Problems and Questions

1 Is it possible to have zero sediment transport in an alluvial channel? Why or why not?

2 Given a noncohesive sediment d_{50} of 0.1 mm and a slope of 0.001, design a channel to convey 10 m^3/s with near-zero erosion.

3 Given noncohesive sediment with $d_{15} = 0.05$ mm, $d_{35} = 0.1$ mm, $d_{50} = 0.15$ mm, $d_{65} = .5$ mm, and $d_{84} = 1.5$ mm. For a wide rectangular channel with $Q = 10$ m^3/s-m and slope $= 0.001$, compute a stage discharge curve and figure sediment transported at the maximum stage conveying the flow. Use EinsteinLaursenGraf.

4 Do problem 3 using the Yang and van Rijn methods.

5 Compute the fall velocity of the sediment specified in problem 3 using methods available on the FallVelocity spreadsheet. Assume the shape factor is 0.7.

---

6 Do an internet search for "CivilGeoHecRas" and readily find a link to this vendor.
7 Do an internet search for "ESRI ArcGis" and readily find a link to this vendor.
8 Do an internet search for "AutoDesk" and quickly find a link to the vendor.

6   Referring to the spreadsheets in the "tractiveearthen" folder under Excel, evaluate the d_{50} sediment size required to solve Example 4.2, assuming the channel is 10 meters wide.

7   Modify the spreadsheet "RevetmentSizing" to accommodate a total flow for a specified width instead of a flow per unit width.

8   Use the sediment design tool of HEC-RAS to analyze conditions of problem 3.

9   Why would one expect a hysteretic loop in total sediment discharge vs. flow discharge and vs. water surface slope?

10 Consider the channel analyzed in the "GrafLaursenEinstein" spreadsheet. Given the sediment load at maximum flow rate, compute the Blench and Simons–Albertson channel expected to be in regime.

11 Using the Yang spreadsheet for sediment load prediction, couple a discharge stage page onto the spreadsheet to give a comprehensive stage–discharge and sediment load computation.

References

Anon. 2007. Threshold channel design. NRCS National Engineering Handbook, Part 654, Chapter 8, pages 8.1 – 8.102 Stream Restoration Design USDA-NRCS, Washington, DC.

AutoDesk (2020). *111 McInnis Parkway*. San Rafael, CA: AutoDesk, Inc.

Bagnold, R.A. 1966. An approach to the sediment transport problem from general physics. US Geological Survey Professional Paper 422-I.

Bagnold, R.A. (1980). An empirical correlation of bedload transport ratios in flumes and rivers. *Royal Society of London Proceedings* A372: 453–473.

Barfield, B.J., Tollner, E.W., and Hayes, J.C. (1977). Filtration of sediment by simulated vegetation: I. Steady-state flow with homogeneous sediment. *Transactions of ASAE* 21: 540–548.

Blench, T. (1957). *Regime Behavior of Canals and Rivers*. London, UK: Butterworths Scientific Publications.

Brownlie, W.R. (1981). Prediction of flow depth and sediment discharge in open-channels. Report No. KH-R-43A, W.A. Keck Lab, California Institute of Technology, Pasadena, CA.

Brunner, G.W. (2016). *HEC-RAS River Analysis System: User's Manual*. Davis, CA: US Army Corps of Engineers, Hydrologic Engineering Center.

Buffington, J.M. and Montgomery, D.R. (1997). A systematic analysis of eight decades of incipient motion studies, with special reference to gravel-bedded rivers. *Water Resources Research* 33 (8): 1993–2029.

Chaudhry, M.H. (1993). *Open-Channel Flow*. Englewood Cliffs, NJ: Prentice-Hall.

Doll, B.A., Wise-Fredrick, D.E., Brucker, C.M. et al. (2002). Hydraulic geometry relationships for urban streams throughout the piedmont of North Carolina. *Journal of the American Water Resources Association* 38 (3): 641–651.

Doll, B.A., Grabow, G.L., Hall, K.R. et al. (2003). *Stream Restoration: A Natural Channel Design Handbook*. Raleigh, NC: North Carolina State University.

Einstein, H.A. (1942). Formulas for the transportation of bedload. *Transactions of the American Society of Civil Engineers* 107: 561–573.

Einstein, H.A. (1950). The Bedload function for sediment transportation in open channel flows. US Dept. Agric. Soil Conservation Service Tech. Bul; 1026, Washington, DC.

Einstein, H.A. and Barbarossa, N.L. (1952). River channel roughness. *Transactions of the American Society of Civil Engineers* 177: 1121.

Engelund, F. (1966). Hydraulic resistance of alluvial streams. *Journal of Hydraulics Division, ASCE 92* (HY2): 315–326.

Engelund, F. (1967). Closure to "Hydraulic resistance of alluvial streams". *Journal of Hydraulics Division, ASCE 93* (HY7): 287–296.

Engelund, F. and Hansen, E. (1967). *A Monograph on Sediment Transport to Alluvial Streams*. Copenhagen, Denmark: Teknik Vorlag.

ESRI (2020). *ArcGIS Mapping and Analytics Platform*. Redlands, CA: Esri Headquarters.

FISRWG (1998). Stream Corridor Restoration: Principles, processes, and practices. By the Federal Interagency Stream Restoration Working Group (FISRWG)(15 Federal agencies of the US gov't) GPO Item No. 0120-A: SuDocs No. A 57.6/2:EN 3/PT.653. ISBN-0-934213-59-3.

Gessler, J. (1965). The beginning of bedload movement of mixtures investigated as natural armoring in channels, Tech Translation T-5 by E.A. Prych with permission of the Author and the Swill Federal Institute of Technology, W.M. Keck Laboratory of Hydraulics and Water Resources, California Inst. of Tech., Pasadena, CA.

Graf, W.H. (1971). *Hydraulics of Sediment Transport*. New York, NY: McGraw-Hill Publishers.

Graf, W.H. and Altinakar, M.S. (1998). *Fluvial Hydraulics: Flow and Transport Processes in Channels of Simple Geometry*. Chichester, UK: Wiley.

Haan, C.T., Barfield, B.J., and Hayes, J.C. (1994). *Design Hydrology and Sedimentology for Small Catchments*. New York, NY: Academic Press.

Henderson, F.M. (1966). *Open Channel Flow*. New York, NY: Macmillan Publishing Co.

Hjulstrom, F. (1935). The morphological activity of rivers as illustrated by River Fyris. Bulletin of the Geological Institute, Uppsala 25, Chapter 3.

Huffman, R.L., Fangmeier, D.D., Elliot, W.J., and Workman, S.R. (2013). *Soil and Water Conservation*, 7e. St. Joseph, MI: Am. Soc. Agr. & Biol. Engrs.

Ippen, A.T. and Drinker, P.A. (1962). Boundary shear stress in curved trapezoidal channel. *Journal of Hydraulics Division, ASCE 88* (HY5): 143–180.

Julien, P.Y. (2010). *Erosion and Sedimentation*, 2e. New York, NY: Cambridge University Press.

Karim, M.F. and Kennedy, J.F. (1990). Menu of coupled velocity and sediment discharge relations for rivers. *Journal of Hydrologic Engineering, ASCE* 116 (8): 978–996.

Knighton, D. (1998). *Fluvial Forms and Processes: A New Perspective*. London, UK: Oxford University Press.

Lacey, G. (1930). Stable channels in alluvium. *Proc. Inst. Civil Engrs,* 229, Paper 4736.

Lane, E.W. (1953). Progress report on studies on the design of stable channels of the Burear of Reclamation. *Proceedings of the ASCE* 79.

Lane, E.W and Carlson, E.J. (1953). Some factors affecting the stability of canals constructed in coarse granular materials. In *Proceedings, Minnesota International Hydraulics Convention, Joint meeting of IAHR and ASCE Hydraulics Division*, ASCE, NY.

Laursen, E.M. (1958). The total sediment load of streams. *Proceedings of the American Society of Civil Engineers 84* (HY1): 1530-1–1530-36.

Martin, Y. and Church, M. (2000). Re-examination of Bagnold's empirical bedload formulae. *Earth Surface Processes and Landforms* 25: 1011–1024.

Meyer-Peter, E. and Muller, R. (1948). Formula for bedload transport. International Association for Hydraulic Structure, Second Mtg., Stockholm.

Renard, K.G., Foster, G.R., Weesies, G.A. et al. 1997. *Predicting soil erosion by water: A guide to conservation planning with the Revised Universal Soil Loss Equation (RUSLE)*. USDA-ARS Agriculture Handbook No. 703, U.S. Dept. of Agriculture, Washington, DC.

Richardson, E.V. and Davis, S.R. (2001). *Evaluating scour at bridges*. Report No. HEC-18. US. Dept. of Transportation, Federal Highway Administration, Washington, DC.

Richardson, E.V., Simons, D.B., and Posakony, G.J. (1961). *Sonic Depth Sounder for Laboratory and Field Use*, Geological Survey Circular 450. Washington, DC: US Geological Survey.

Richardson, E.V., Simons, D.B., and Lagasse, P.F. (2001). *River engineering for highway encroachments: Highways in the river environment*. Report No. FHWA NHI 01-004 HDS 6., US Dept. of Transportation, Federal Highway Administration, Washington, DC.

Rosgen, D.L. and Silvey, H.L. (1996). *Applied River Morphology*, 2e. Ft. Collins, CO: Wildland Hydrology, Inc.

Rosgen, D.L. and Silvey, H.L. (2005). *The Reference Reach Field Book*, 2e. Ft. Collins, CO: Wildland Hydrology, Inc.

Rubey, W.W. (1933). Settling velocities of gravel, sand, and silt particles. *American Journal of Science*, 5th series 25 (148): 325–338.

Schiechtl, H. (1980). *Bioengineering for Land Reclamation and Conservation*. Edmonton, Alberta, CA: University of Alberta Press.

Shields, I.A. (1936). Application of similarity principles and turbulence research to bedload movement. A translation from German by W.P. Ott and J.C. van Vchelin, U.S. Soil Conservation Service Cooperative Lab., California Inst. of Tech., Pasadena, CA.

Simons, D.B. and Albertson, M.L. (1960). Uniform water conveyance channels in alluvial materials. *Proceedings of the American Society of Civil Engineers* 86 (HY5): 33.

Simons, D.B. and Richardson, E.V. 1966. *Resistance to flow in alluvial channels*. Prof. Paper 422-J. US Geological Survey, Washington, DC

Simons, D.B. and Richardson, E.V. (1971). Flow in alluvial channels. In: *River Mechanics*, vol. 1, chapter 9 (ed. H.W. Shen). Ft. Collins, CO: Water Resources Publications.

Simons, D.B. and Senturk, F. (1992). *Sediment Transport Technology: Water and Sediment Dynamics*. Littleton, CO.: Water Resources Publications.

Smith, W.O. (1960). A comprehensive survey of Lake Mead: 1948-49. US Geological Survey Professional Paper 295 DOI 10.3133/pp295.

Sturm, T.W. (2010). *Open Channel Hydraulics*, 2e. New York, NY: McGraw-Hill Publishers.

Toffaletti, F.B. (1968). A procedure for computation of the total river sand discharge and detailed distribution, bed to surface. Committee on Channel Stabilization, US Army Corps of Engineers Waterways Experiment Station, Technical Report No. 5, Vicksburg, MS.

Tollner, E.W., Barfield, B.J., Vachirakornwatana, C., and Haan, C.T. (1977). Sediment deposition patterns in simulated grass filters. *Transactions of ASAE* 20: 940–944.

US Interagency Committee on Water Resources. (1957). *Some fundamentals of particle size analysis, a study of methods used in measurement and analysis of sediment loads in streams.* Report No. 12. Subcommittee on sedimentation, Water Resources Council, Govt. Printing Office, Washington, DC.

Van Rijn, L.C. (1984). Sediment transport III: Bedforms and alluvial roughness. *Journal of Hydrologic Engineering, ASCE* 110 (12): 1733–1754.

Vanoni, V.A. (ed.) (2006). *Sedimentation engineering*, 2e. Reston, VA: ASCE-EWRI Task Committee for the preparation of the Manual on Sedimentation of the Sedimentation Committee of the Hydraulics Division, ASCE.

Warner, R.C., Schwab, P.J., and Marshall, D.J. (1998). *SedCad 4 Design Manual and User's Guide*. Ames, IA: Civil Software Design.

White, F.M. (2005). *Viscous Fluid Flow*. New York, NY: McGraw-Hill Publishers.

Whitlow, J.R. and Gregory, K.J. (1989). Changes in urban stream channels in Zimbabwe. *Regulated Rivers* 4: 27–42.

Wilcock, R.R. and Kenworthy, S.T. (2002). A two fracton model for the transport of sand-gravel mixtures. *Water Resources Research* 38 (10): 1194.

Yang, C.T. (1972). Unit stream power and sediment transport. *Journal of Hydraulics Division, ASCE* 108 (HY6): 1805–1826.

Yang, C.T. (1996). *Sediment Transport: Theory and Practice*. New York, NY: McGraw-Hill Publishers.

Appendix A

Software and Selected Solutions

Excel®

The downloadables website provides one or more Excel (Anon. 2014) spreadsheets for most chapters. Table A.1 summarized the programs. The most sophisticated tool is the solver or goal seek functions, required for many applications. The user should save original copies and develop custom protection to keep from inadvertently corrupting the workbooks. Packages mentioned below are mostly available on Windows, MAC, or other operating system platforms. For users not familiar with Excel, we recommend YouTube videos readily available on the web.

If you do not see the Excel Solver utility, go to (Excel 10 or 13, Windows) the file tab, find "options" in the left column. Click on "manage add-ins" and move the solver from the inactive to active status. The solver should then be visible under the Data tab; similar steps also apply to the Apple® version of Excel.

Required inputs and required guess solutions are color coded on the spreadsheets. Many of these spreadsheets have multiple pages. One usually provides required data inputs on the first page. Moreover, they are automatically included on the second page when appropriate. The solver for each page (where applicable) does not require any additional configuration.

The user should go through the spreadsheet line by line to fully appreciate the analyses and check the results, which are provided "as is." The provided references provide additional relevant information for the later chapters, as the text may not include all equations.

Mathematica®

The downloads contain selected Mathematica (Wolfram 2020) files to benefit those who have access to the software. Table A.2 lists notebooks for many of the channel analysis problem types. This software provides symbolic solutions as far as algebraically possible. The computer algebra approach sometimes yields insights into solution difficulties one may experience with Excel. Solutions are straightforward. Those desiring to view the Mathematica files and who do have these software licenses may obtain (free) players from www.wolfram.com/cdf-player and www.wolfram.com/cdf-player.

Open Channel Design: Fundamentals and Applications, First Edition. Ernest W. Tollner.
© 2022 John Wiley & Sons Ltd. Published 2022 by John Wiley & Sons Ltd.
Companion website: www.wiley.com/go/tollner/openchanneldesign

Mathematica has a rather steep learning curve; however, resources are available. The website https://www.wolfram.com/language/fast-introduction-for-math-students/en/ gives a quick tutorial. A search reveals several YouTube video tutorials. A Schaum's outline on Mathematica (Don 2019) is available, which I recommend to those desiring to move up the Mathematica learning curve.

HydroCAD

The HydroCAD software (Anon 2011) is an example of software systems enabling surface runoff generation from various surface conditions (using the NRCS Curve Number method). One can route the runoff through streams, culverts, and reservoirs. Reservoirs may be customary surface reservoirs or underground storage. The storages may have complex outlet systems that are commonly used in sediment control and water quality management schemes.

HY-8 culverts

A public domain culvert program, HY-8 (Anon. 2012), is used to solve culvert hydraulics problems. It is available without charge for Windows from the US Federal Highway Administration website at http://www.fhwa.dot.gov/engineering/hydraulics/software/hy8/. Excellent documentation is available. Flow in culverts is readily analyzed using HY-8. HY-8 version 7 contains aquatic organism passage and plunge pool analysis. Consult YouTube videos for a quick introduction to HY-8. HY-8 is relatively intuitive and easy to learn. Table A.3 contains several HY-8 example applications.

HEC-RAS

Finally, we provide an HEC-RAS (Brunner 2016) project file sets for problems ranging from a simple channel to an unsteady flow dam-break analysis. HEC-RAS is the choice when analyzing a natural channel with a variety of flow situations. HEC-RAS is available to the public at the following website: www.hec.usace.army.mil. HEC-RAS has a design tool and fundamental flow analyses that can be handy for "what-if" solutions. Uniform flow, stable channel design, bridge pier analysis are available templates. One should keep all the HEC-RAS files for a project in a single location for the best results. Table A.4 contains files for several HEC-RAS projects.

To start HEC-RAS, open the program Under "File," open the project of interest. Go to "Geometric Data" and inspect the cross sections. Then under "Run," open the "Steady flow" tab. One may open the "Design Tools" tab for using the design tools, which the DesignTools project demonstrates. HEC-RAS is an advanced channel design program. We give a brief introduction showing how to use some channel design functions, insert bridges and culverts, and perform gradually varied flow analysis. Examples of unsteady flow and sediment transport analysis are included for study on your own. YouTube videos are available.

HEC-RAS requires that cross sections be entered by hand or entered via various GIS platforms. One can model a prismatic cross-section by adding one cross section, copying for a given station, and adjusting the elevation with an included tool. Then, add cross sections in the middle via an interpolation tool. HEC-RAS comes with a rich set of example solutions and copious documentation. A complete demonstration is not possible in this text. Unsteady flow simulations are particularly complex, data-intensive, and time-consuming. The examples provided herein are drawn from the extensive set of models that come with the program download. The intent is not to master every facet of HEC-RAS but to give an overview of HEC-RAS capabilities. Graduate students should gain the most benefit from the HEC-RAS examples, but motivated undergraduates can as well.

HEC-RAS generates sets of files, with the project file being the information storage map. Experience has shown that each project should be stored in a separate folder. The main menu indicates some but not all of the files comprising a project. Storage in a folder facilitates moving a project from one platform to another.

All software is provided "as-is" with no real or implied warranty. If used in design work, the user is solely responsible for results obtained with the workbooks and other software contained herein. That said, the author would greatly appreciate notification of any errors or discrepancies.

Software Summary Tables

Table A.1   Excel file summary.

Excel file	File function
SolverTemplate	File for learning how to solve iterative equations
Trapezoidalchanneldesign	Trapezoidal lined channel design
TriangularCurbdesign	Design of a triangular channel with one side $z = 0$ and one side $z =$ user choice
Circular channel design	Design of channel of circular cross-section
Parabolicchanneldesign	Parabolic cross-section design
CircularchanneldesignGeneralManningPowell	Cross-section design of cross-sections using the Powell option for handling laminar and transition zone flows
ParabolicchanneldesignManningPowell	
TrapezoidalchanneldesignManningPowell	
TriangularCurbchanneldesignManningPowell	
CircularChannelNormalCritical	Normal and critical flow analysis
ParabolicChannelNormalCritical	
TrapezoidalChannelNormalCritical	

Table A.1 (Continued)

Excel file	File function
OptimumchannelsComparisons	Optimum channel cross-sections
OptimumCircularchanneldesignDover2	
OptimumRectangularchanneldesign	
OptimumTrianglechanneldesign	
OptimumTrapezoidchanneldesign	
SubOptimumTrapezidchanneldesignSetZ	
AdvancedLiningCosts	
ParabolicWaterwayDesignEarthenPermVelStabCap	Permissible velocity channel design
TrapezoidalWaterwayDesignEarthenPermVelStabCap	
TriangularWaterwayDesignEarthenPermVelStabCap	
FHA_NRCSRipRap	Riprap design with roughness based on riprap size
RIPRAPParabolicWaterwayDesignEarthenPermVel	
RIPRAPTrapezoidalWaterwayDesignEarthenPermVel	
RIPRAPTrianguarwaterwaydesignEarthenPermVel	
SimpleResRouting	Reservoir and stream routing (ResReachRoute)
MuskingumReach	
ParabolicGrassWaterwayDesignPermVel	Grassed waterways – Permissible velocity
TrapezoidalGrassWaterDesignPermVel	
TriangularGrassWaterwayDesignPermVel	
TractiveParabolicGrassWaterwayDesign	Grassed waterway – Tractive force
TractiveTrapezoidalGrassWaterwayDesign	
TractiveTriangularGrassWaterwayDesign	
TractiveErosionMats	Earthen channels – Tractive force
TractiveGranularClay	
CompoundChannelSpecEnergy	Specific energy in a compound channel
GateStaticsHorzVert	Statics analysis of submerged rectangular gate
GateStaticHorzVertCircularHatchComp	Statics analysis with circular hatch
EnergyTransitionOpTrapezoidKnownQ	Design of reservoir transitions to an optimum cross-section
Ene1rgyTransitionOpTrapezoidKnownY	
EnergyTransitionOptRectangleKnownQ	
EnergyTransitionOptRectangleKnownQ	
EnergyTransitionOptSemicircleKnownQ	
EnergyTransitionOptSemicircleKnownY1	
NatTempLiningAutomaticErodibleSoils	Temporary linings design
NatTempLiningAutomaticNon_ErodibleSoils	

(Continued)

Table A.1 (Continued)

Excel file	File function
EnergyTestGeneralCircu;arSectionWithForces	Forces on sluice gates in channels of an indicated cross-section
EnergyTestGeneralParabolicWithForces	
EnergyTestGeneralTrapWithForces	
GeneralCircleDragCylinders	**Flows in indicated cross-sections with cylinders normal to the flow
GeneralParabolicDragCylinders	
GeneralTrapDragCylinders	
BasicSequentDepthTrapezoid	Sequent depths
MathematicaComparisonBasicSequentDepthCircle	
MathematicaComparisonBasicSequentDepthParabola	
EnergyCircularChoke	Channel transition/choke analysis
EnergyParabolicChoke	
EnergyTrapezoidChoke	
WeirsSimple	Basic weir analysis
DirectStepCircular	Gradually varied flows via direct step and standard step methods
DirectStepParabolic	
DirectStepTrapezoid	
StandardStepPrismatic	
StandardStepNonprismatic	
SpatiallyvariedExample	Spatially varied flow example
GrafLaursenEinsteinTotalLoad	Sediment transport in alluvial channels
FallVelocity	
IncipientMotionWideChannel	
RevetmentSizing	
StageDischargeVanRijn	
VanRijnTotalLoad	
YangTotalLoad	
BagnoldBrownlieWilcockKenworthy	
ViscosityDensity	Viscosity worksheet
trapezoidchannelJunction90Modn	90° trapezoidal channel intersection
ReservoirRouting	Simple Pond Routing
CulverDesign_Check	NRCS Wisconsin Culvert software
DmBrkFinite	Solution for dam-break with finite depth downstream

Table A.2 Mathematica notebooks.

Mathematica Notebook file	File function
CircleD0 CircleY PartiallySubMCircHelements	Circular Channel known diameter, known depth, and cross-section partially covered with sediment.
GeneralSluiceGate	Energy
EarthenParabolicPV EarthenTrapezoidalPV EarthenTriangularPV	Earthen channels – Permissible velocity constraint
DamBreakFinite	A solution to the equations for dam-break in a finite stream
GeneralCircleNumericalHYDExponents GeneralParabolicNumericalHYDExponents GeneralTrapNumerical GeneralTrapNumericalHYDExponents	Gradually varied flow calculations
SequentDepthsCircular SequentDepthsParabolic SequentDepthTrapezoidal	Hydraulic jump sequent depth calculations
Optimumcircle optimumParabolicchannel2 OptimumRectangularchannel OptimumTrapezoid OptimumTriangularchannel OptimumTriangularcurbchannel GeneralTrapCost	Optimum channel parameters with various constraints
ParabolaWettedPerim ParaboicAnalysisDesign ParabolicSY ParabolicTY ParabolicY	Parabolic analysis and solutions for slope and depth with limiting velocity, and solutions for top width and depth with limited velocity

(Continued)

Table A.2 (Continued)

Mathematica Notebook file	File function
MuskingumParametersParabola MuskingumParametersTrapezoid	Muskingum routing parameters
Diffusion	DiffusionEquation (diffusion routing)
EinsteinSedimentIntegrals EinsteinProbabilitySedimentIntegral DepthFromHydraulicRadiusParabola DepthFromHydraulicRadiusParabolaNarrow DepthFromHydraulicRadiusRectangle DepthFromHydraulicRadiusTrapezoid	Sediment transport solutions Depth from hydraulic radius for indicated cross-sections
TrapBY TrapBY2z TrapezoidGrassPV TrapSY TrapSYexact TrapY TrapYZ1Z2 TrapSY2z	Trapezoidal channel solutions with limiting velocity or general case. The 2z problems indicate a trapezoidal cross-section with different side slopes.
TriangpickZ TriangSY TroamgSYz1z2 TriangZY	Triangular channel solutions
ParabolicGrassPV TrapezoidGrassPV TriangularGrassPV	Grassed waterway solutions using Permissible Velocity
Veldistlinearalphabeta Veldistlog10alphabeta Veldistlog10alphabeta2 Veldistlogalphbeta Veldistparabolicalfabeta Veldistpoweralphbeta	Alpha and Beta coefficients arising from linear, log10, log, parabolic, and power function velocity profiles

Table A.3   HY-8 culvert files.

HY-8 files	File function
ExampleCulvert	Culvert example
ExampleCulvert2	Culvert example two
ComplexFlowProfile	Culvert with complex flow inside the pipe

Table A.4   HEC-RAS sample channel projects.

HecRas files	Notes
BasicChannel.prj	Design of a basic channel. These files comprise one model, and all must be contained in the same folder. Load the project file; others load as needed.
DesignFeature.prj	Project set up for the application of some of the design tools. Load the DesignFeatures.prj file. We consider uniform flow and then rerun for a stable channel
Tollner1	Alternative file for the design features as Tollner1.
BridgeCulv1	Project to evaluate bridge and culvert editors
WeirBridgeCuvChan.prj	Project set up to include a bridge and culvert to demo the bridge and culvert editors
GradVariedFlow.prj	Project set up for a gradually varied flow analysis
HydraulicJmp.prj	Simple hydraulic jump
BeaverCr.prj	Unsteady flow example based on the HEC-RAS Beaver Creek example
EuclidExample.prj	Sediment transport example based on the HEC-RAS Euclid Example
SedimentXport.prj	Sediment transport Design Functions
BaldCreekDmBrk.prj	Dam break example

Projects are arranged in separate folders; all files associated with the projects must remain in the same folder for best results.

Selected Symbolic Solutions

One may design channels with known flow rates not to exceed set velocities by constraining the hydraulic radius or slope. Hydraulic radius is the most often used constraint. Knowing the velocity and flow rate, one can solve for the area from the continuity equation. One can solve for the required hydraulic radius from Manning's equation, knowing the slope. Selected solutions using Mathematica®12 follow.

A. Determining bottom width and depth in a trapezoidal channel with known slope, side slope, and permissible velocity

Soln1stexact = Solve[{Rexact == Rvl, A == Avl}, {b, d}]

$$R_{exact} = \frac{bd + d^2 z}{b + 2d\sqrt{1 + z^2}} \text{ and } A = bd + d^2 z$$

$$\left\{ b \rightarrow \frac{1}{Rvl} \left(Avl - \frac{2\,Avl}{4 + 3z^2} - \frac{2\,Avl\,z^2}{4 + 3z^2} - \frac{Avl\,z\sqrt{1 + z^2}}{4 + 3z^2} + \frac{1}{4 + 3z^2}\sqrt{1 + z^2} \right. \right.$$

$$\left. \sqrt{4\,AvlRvl^2 \left(4 + 3z^2 \right)\left(-z - 2\sqrt{1 + z^2} \right) + Avl^2 \left(-z - 2\sqrt{1 + z^2} \right)^2} \right),$$

$$d \rightarrow \frac{1}{2Rvl \left(4 + 3z^2 \right)} \left(-Avl \left(-z - 2\sqrt{1 + z^2} \right) - \sqrt{4\,AvlRvl^2 \left(4 + 3z^2 \right)\left(-z - 2\sqrt{1 + z^2} \right)} \right.$$

$$\left. \left. + Avl^2 \left(-z - 2\sqrt{1 + z^2} \right)^2 \right) \right\}$$

$Avl = \dfrac{q_{set}}{v_{pset}}$ = limiting crossectional area for not exceeding the permissible velocity

at a design flowrate

$Rvl = \dfrac{nv_{pset}}{\left(\varphi\sqrt{Slope} \right)^{1.5}}$ = limiting hydraulic radius for not exceeding the permissible velocity

moreover, z is the trapezoid side slope ($H{:}V$); b is the bottom width (L); d is the depth (L); Slope is the channel slope (L/L); and, φ is the dimensional constant (SI, 1, or English, 1.49).

The solution above is one of four possible solutions that are not complex or extraneous *unless one attempts to move water at a velocity higher than physically possible.*

B. Determining depth and side slope in a triangular Channel with known slope and permissible velocity

If $b = 0$ (e.g., a triangular channel), we have the following exact and approximate solutions:

$$\text{Solnlstexact} = \text{Solve}\Big[\big\{\text{Rexact} == \text{Rvl}, \ A == \text{Avl}\big\}, \{d, z\}\Big]$$

$$\left\{\left\{\begin{array}{l} d \to \dfrac{\dfrac{1}{2}\text{Avl}\sqrt{1 + \dfrac{\left(\text{Avl} - \sqrt{\text{Avl}^2 - 64\,\text{Rvl}^4}\right)^2}{64\,\text{Rvl}}} + \dfrac{1}{2}\sqrt{\text{Avl}^2 - 64\,\text{Rvl}^4}\,\sqrt{1 + \dfrac{\left(\text{Avl} - \sqrt{\text{Avl}^2 - 64\,\text{Rvl}^4}\right)^2}{64\,\text{Rvl}^4}}}{2\,\text{Rvl}}, \\[6mm] Z \to \dfrac{\text{Avl} - \sqrt{\text{Avl}^2 - 64\,\text{Rvl}^4}}{8\,\text{Rvl}^2} \end{array}\right\},\right.$$

$$\left.\left\{\begin{array}{l} d \to \dfrac{\dfrac{1}{2}\text{Avl}\sqrt{1 + \dfrac{\left(\text{Avl} + \sqrt{\text{Avl}^2 - 64\,\text{Rvl}^4}\right)^2}{64\,\text{Rvl}}} - \dfrac{1}{2}\sqrt{\text{Avl}^2 - 64\,\text{Rvl}^4}\,\sqrt{1 + \dfrac{\left(\text{Avl} + \sqrt{\text{Avl}^2 - 64\,\text{Rvl}^4}\right)^2}{64\,\text{Rvl}^4}}}{2\,\text{Rvl}}, \\[6mm] Z \to \dfrac{\text{Avl} + \sqrt{\text{Avl}^2 - 64\,\text{Rvl}^4}}{8\,\text{Rvl}^2} \end{array}\right\}\right\}$$

$$\text{SolnApprox} = \text{Solve}\Big[\big\{R == \text{Rvl}, \ A = \text{Avl}\big\}, \{d, z\}\Big]$$

$$\left\{\left\{ Z \to \dfrac{0.25\,\text{Avl}}{\text{Rvl}^2}, \ d \to 2.\text{Rvl} \right\}\right\}$$

Note: The approximate solution fails when Z is less than 2 or 3.

The second of the two exact solutions is usually the appropriate solution. The approximate solution comes about when assuming $R = d/2$.

C. Determining slope and depth in a triangular channel with known flow rate, permissible velocity, and side slope (Z)

Using an approximation for the hydraulic radius $(=d/2)$, a reasonably simple symbolic solution for channel slope and depth, useful in applications such as terrace channel design, is available. Six solutions are possible, and the one having nonnegative and noncomplex is as follows:

$$\left\{\text{Slope} \to \dfrac{2.519\ 84\,n^2\,V_{pset}^{8/3}\,Z^{2/3}}{Q_{set}^{2/3}\,\phi^2}, \ d \to \dfrac{\sqrt{Q_{set}}}{\sqrt{V_{pset}}\ \sqrt{Z}}\right\}$$

Solutions in terms of slope and other variables are more involved or are not readily obtainable.

References

Anon. 2011. *HydroCAD-10 Owner's Manual*. HydroCAD Solutions LLC. Chocorua, NH. www. hydrocad.net (accessed 25 April 2021).

Anon. (2012). *HY-8 User's Manual*. Washington, DC: Federal Highway Administration. http:// www.fhwa.dot.gov/engineering/hydraulics/software/hy8/index.cfm (accessed 25 April 2021).

Anon. (2014). Microsoft Excel. Redmond, WA.

Brunner, G.W. (2016). *HEC-RAS River Analysis System User's Manual*. Davis, CA: US Army Corps of Engineers, Institute for Water Resources Hydrologic Engineering Center.

Don, E. (2019). *Mathematica and the Wolfram Language*, Schaum's Outlines. New York, NY: McGraw-Hill Publishers.

Wolfram, S. (2020). *Mathematica, V. 12.1*. Champaign, IL: Wolfram Scientific. http://www.wolfram.com/mathematica/new-in-12/ (accessed 25 April 2021).

Appendix B

Solution Charts for Vegetated Waterways Using the Permissible Velocity Method

Note: Solution charts in SI units appear below. One may convert to imperial units using 3.28 ft/m.

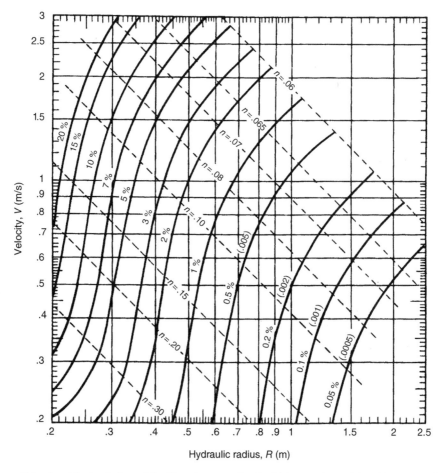

Figure B.1 Graphical solution of the Manning equation for high retardance (class A) vegetated waterways. *Source:* Courtesy of NRCS (1966).

Open Channel Design: Fundamentals and Applications, First Edition. Ernest W. Tollner.
© 2022 John Wiley & Sons Ltd. Published 2022 by John Wiley & Sons Ltd.
Companion website: www.wiley.com/go/tollner/openchanneldesign

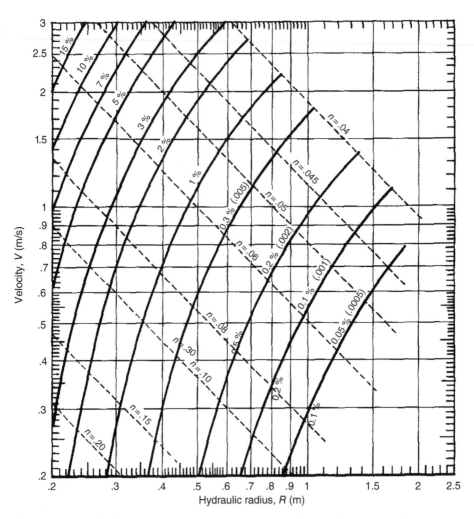

Figure B.2 Graphical solution of the Manning equation for vegetated waterways, high retardance, class B. *Source:* Courtesy of NRCS (1966).

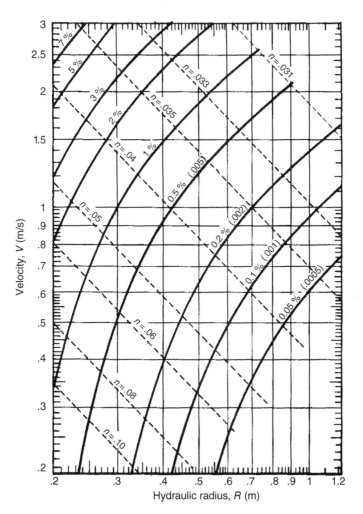

Figure B.3 Graphical solution of the Manning equation for vegetated waterways, moderate retardance, class C. *Source:* Courtesy of NRCS (1966).

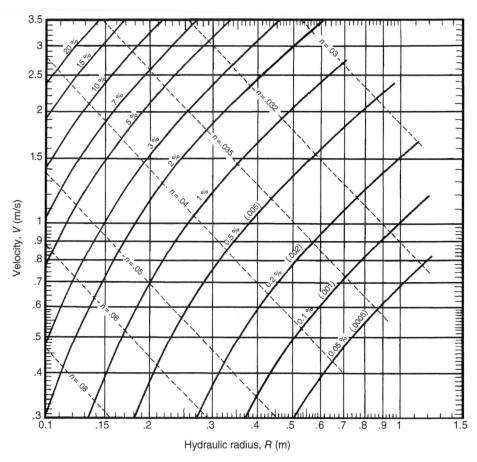

Figure B.4 Graphical solution of the Manning equation for vegetated waterways, low retardance, class D. *Source:* Courtesy of NRCS (1966).

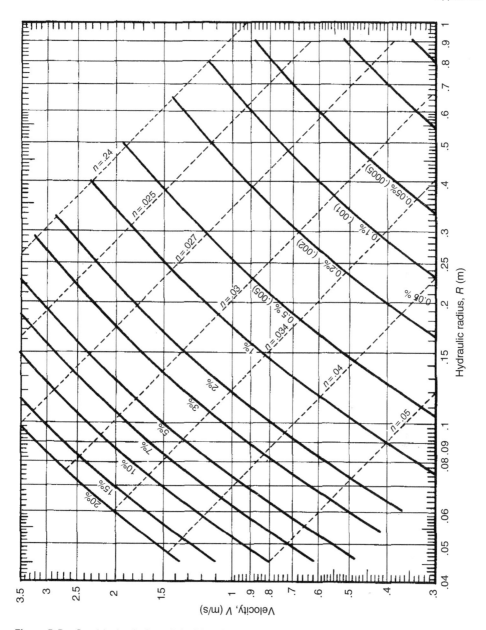

Figure B.5 Graphical solution of the Manning equation for vegetated waterways, low retardance, class D. *Source:* Courtesy of NRCS (1966).

Reference

NRCS (1966) Handbook of Channel Design for Soil and Water Conservation. SCS-TP-1961.

Appendix C

Selected Cost Data for Channel Excavation and Lining Materials

Data in the tables below were assembled over the years by requesting local authorities' data for the cost of excavation and channel lining materials. The data represent 2012–2013. Data do not account for site or job-specific factors such as equipment mobilization, excavation difficulties such as rocks or trees, or labor considerations. Beware of the units. One should consult local expertise such as county public works departments, Natural Resources Conservation Service engineers, or experienced consulting engineers. For vegetation selection, include ecologists, extension specialists, or other experts.

Costing table

Item	Item cost	Unit	Application rate	Source
Vegitation				
Seed	$0.79	LB	600 lb/acre	ACC
Agriculture lime	$0.11	LB	13 tons/acre	ACC
Fertilizer	$0.47	LB	9 tons/acre	ACC
Straw Bales	$3.64	EA	600 bales/acre	ACC
Rock Lining				
Type 1 Rip Rap	$18.55	TN	1 ton/yard^2	ACC
Concrete				
3000 PSI Concrete Mix	$85.94	CY	Varies by channel	ACC
Portland Cement 92.6LB Bag	$10.46	EA	1 bag/CY concrete	ACC
#4 Rebar 20' Stick	$7.60	EA	Varies by channel	ACC
Earthen/Excavation				
Fill Dirt	$14.67	CY	Varies by channel	ACC
Excavation	$5.70	CY	Varies by charnel	RS Means '09
Temporary Linings				
Jute Mesh	$1.79	SY	Varies by channel	CalDOT
Excelsior	$2.90	SY	Varies by channel	CalDOT

Note: All costs adjusted for inflation.
CY = cubic yards; SY = square yards; EA = Each.

Open Channel Design: Fundamentals and Applications, First Edition. Ernest W. Tollner.
© 2022 John Wiley & Sons Ltd. Published 2022 by John Wiley & Sons Ltd.
Companion website: www.wiley.com/go/tollner/openchanneldesign

Cost table

Item	Item cost	Unit	Application rate	Category	Source
Seed	$0.74	LB	600 lb/acre	Vegetation	ACC
Agriculture lime	$0.10	LB	13 tons/acre		ACC
Fertilizer	$0.44	LB	9 tons/acre		ACC
Straw Bales	$3.48	EA	600 bales/acre		ACC
Type 1 Rip Rap	$17.70	TN	1 ton/square yard	Rock lining	ACC
3000 PSI Concrete Mix	$82.00	CY	Varies by channel	Concrete	ACC
Portland Cement 92.6LB Bag	$9.99	EA	1 bag/CY conrete		ACC
#4 Rebar 20' Stick	$7.25	EA	Varies by channel		ACC
Fill Dirt (use existing)	$14.00	CY	Varies by channel	Earthen/ excavation	ACC
Excavation	$5.25	CY	Varies by channel		RS MEANS
Jute Mating	$6500.00	AC	Varies by channel	Temporary lining	CalDOT
Excelsior	$10 500.00	AC	Varies by channel		CalDOT

Source: Based on Athens-Clarke County Georgia public works department; California Department of Transportation, and RS Means (2009), as indicated above, 2009.
AC = acre; LB = pound, CY = cubic yards; SY = square yards; EA = Each.

Item	Unit cost	Assumption
Excavation	$2.50/CY	
Excavation, including disposal	$3.50/CY	
Backfill	$3.50/CY	
Rock-revetted side slopes	$50.00/CY	Riprap thickness = 3'
Rock lining	$70.00/CY	
Concrete lining	$225.00/CY	Concrete thickness = 8"

Source: Data from Yucca Valley Engineering Department. www.yucca-vallley.org/pdf/engineering/Chapter_IV.pdf.
See above for abbreviations.

Lining	Cost	Source
Grass	$100/acre	Athens-Clarke County
Straw	$2088/acre	Athens-Clarke County
Jute Mesh	$6500/acre	California DOT
Excelsior	$10 000/acre	California DOT
Concrete	$9.11/yard^2	Athens-Clarke County
Type 1 Rip-Rap	$1.77/ft^2	Athens-Clarke County
Excavation	$5.25/yard^3	RS Means

Source: Adapted from Athens-Clarke County Georgia public works department; California Department of Transportation, and RS Means (2009), as indicated above, 2009.

Department of Transportation and Public Works, Engineering Division 2020.

Item	Item	Quantity	Item cost	Total cost
1	Seed Fescue	19 LB	$0.74 LB	$14.06
2	Rye	580 LB	$0.49 LB	$284.20
3	Agriculture time	1170 LB	$0.10 LB	$117.00
4	Fertilizer	580 LB	$0.44 LB	$255.20
5	Straw Sales	19 EA	$3.48 EA	$66.12
6	Type A Silt Fence	95 LF	$0.29 LF	$27.55
7	C-POP 36′ × 50′	120 LF	$0.98 LF	$117.60
8	Graded Aggregate Base	12 TN	$13.75 TN	$165.00
9	Recycled Asph Cone 9.5 mm	3 TN	$110.00 TN	$330.00
10	Fill Dirt (use existing)	−CY	$14.00 CY	$0.00
11	Type 1 Rip Rap	27 TN	$17.70 TN	$477.90
12	36′ RCP	32 LF	$34.06 LF	$1089.92
13	24′ CMP	112 LF	$12.47 LF	$1396.64
14	GDOT 1120 18′	1 EA	$328.13 EA	$328.13
15	GDOT 1120 36′	1 EA	$815.00 EA	$815.00
16	3000 PSI Concrete Mix	64 CY	$82.00 CY	$5248.00
17	Portland Cement 92.6LB Bag	7 EA	$9.99 EA	$69.95
18	#4 Rebar 20′ Stick	32 EA	$7.25 EA	$232.00
19	Jumbo Brick	800 EA	$0.39 EA	$312.00
20	Mortar Mix 75 LB Bag	5 EA	$8.25 EA	$41.25
			Total	$11 387.50
		20% Contingency		$2277.50
		Total estimated cost		$13 665.00

Lining	Application rate
Fertilizer	9 tons/acre
Lime Ca(OH)$_2$	13 tone/acre
Seed	600 lbt/acre
Straw sales	600 lbs/acre

Lining	Cost	Source
Fascue	$0.74/lb	ACC DOT
Straw	$3.48/bale	ACC DOT
Rye	$0.49/lb	ACC DOT
Type 1 Rip-Rap	$1.77/ft^2	ACC DOT
Jumbo Brick	$0.39/EA	ACC DOT
3000 PSI Concrete Mix	$82/CY	ACC DOT
Mortar Mix 75 lb bag	$8.25/EA	ACC DOT
Jute Mesh	$6500/acre	California DOT
Excelsior	$10 000/acre	California DOT

Source: Data from Athens-Clarke County, Georgia public works department (2009).

Excavation, Bulk, Dozer	Cost (par Bulk C.Y.)
Common Earth	$3.30
Clay	$5.25
For 300′ Haut	
Sand and gravel	$5.60
Sandy day and loam	$5.75
Common Earth	$6.40
Clay	$10.50

Source: Data from Athens-Clarke County department of public works, California DOT, as indicated, 2009. See notes above for abbreviations.

Item	Unit	Unit cost
Excavation (First 2000 CY)	Cubic Yard	$25.00
Excavation (Over 2000 CY)	Cubic Yard	$20.00
Rip Rap	Cubic Yard	$65.00
Solid Sodding	Square Yard	$7.00
Gravel Bed	Cubic Yard	$75.00
Reinforced Concrete	Cubic Yard	$800.00
Top Soil	Cubic Yard	$50.00
3* Concrete Lining Coarse	Square Yard	$17.00
Erosion Matting	Square Yard	$2.50
Much – Fine Shredded Hardwood	Cubic Yard	$60.00
Wetland Seed Mix	Pound	$200.00
Brush Layering	Square Yard	$150.00
16* Natural Fiber Roll	Linear Feet	$20.00

Source: Data from Standard Prices for Cost Estimating, December 2010. City of Rockville, Department of Public Works. Retrived from: http://www.rocvillemd.gov/e-gov/pw/Standard_Prices_Cost_Est_Permit.pdf.

Appendix D

Design Strategy Summary for Uniform Flow Channels

Appendix D.1   Design summary overview for steady flow lined and unlined channels.

Design strategy	Pros	Cons	Chapter reference
Standard lined channel – Manning and Chezy	Primary design method for general purpose design Chezy can be modified for nonwater liquid flows	Critical and supercritical flows may occur	Chapter 2
Optimum cross-section design	Potentially the most economical	Not for unlined channels with slopes greater than around approximately 0.2% with many linings.	Chapter 2
Suboptimum cross-section	It can accommodate soil conditions in unlined channels.	See optimum channel cons.	Chapter 2
NRCS riprap	Suitable for short drainage ways that may have a very steep slope.	Empirical	Chapter 2
FHA riprap	Suitable for short drainage ways down relatively steep drainage ways.	Empirical	Chapter 2
Permissible velocity – earthen and grassed waterways	Straightforward Suitable for small sediments such as fine sand or smaller Good for mixed low maintenance vegetation when vegetation is involved. Stability-Capacity designs are available.	Not precise – based on survey data Not conservative	Chapters 2 and 3

(Continued)

Open Channel Design: Fundamentals and Applications, First Edition. Ernest W. Tollner.
© 2022 John Wiley & Sons Ltd. Published 2022 by John Wiley & Sons Ltd.
Companion website: www.wiley.com/go/tollner/openchanneldesign

Appendix D.1 (Continued)

Design strategy	Pros	Cons	Chapter reference
Temporary liners	Suitable for protecting earthen channels and grassed waterways during construction	Frequently highlights the risk associated with experiencing the design storm within the construction and vegetation establishment window	Chapter 3
Tractive force – earthen and grassed waterways	Reasonably precise for particles larger than fine gravel. Suitable for highly maintained waterways such as bioswales Stability–capacity design available	Requires precise data for vegetation It can result in low velocities. Not recommended for vegetated waterways carrying high-sediment waters. Soils are not differentiated.	Chapter 4
Regime theory approaches	Suitable for checking lower order streams in the natural design context	Old approach – in higher order streams, the method would at best provide a snapshot	Chapter 10
Incipient movement – Shields or modified Shields approaches	Suitable for testing movement of sediment when sediment size is well controlled. Can design wide channels with this technique	Requires very consistent sediment size.	Chapter 10
Alluvial design approaches (Einstein, Laursen, Graf), van Rijn, Yang among others	Accounts for bedforms. Accounts for sediment transport. Depths are a byproduct of the analysis. See Chapter 10 for discussion of the selected methods.	Bedforms are hard to predict. Better for design checking rather than design. Sediment transport remains highly empirical	Chapter 10

This table is designed to help narrow down available design methods to two or three in a given situation. Depending on client preferences, one should also try various linings and consider multiple routes.

Index

Open Channel Design: Fundamentals and Applications, First Edition. Ernest W. Tollner.
© 2022 John Wiley & Sons Ltd. Published 2022 by John Wiley & Sons Ltd.
Companion website: www.wiley.com/go/tollner/openchanneldesign